Food Biology Series

Mycotoxins in Food and Beverages

Innovations and Advances

Part II

Books Published in *Food Biology* series

1. **Microorganisms and Fermentation of Traditional Foods**
   Ramesh C. Ray, Montet Didier (eds.), 2014
2. **Bread and Its Fortification: Nutrition and Health Benefits**
   Cristina M. Rosell, Joanna Bajerska, Aly F. El Sheikha (eds.), 2015
3. **Authenticity of Foods of Animal Origin**
   Ioannis Sotirios Arvanitoyannis (ed.), 2015
4. **Fermented Foods, Part I: Biochemistry and Biotechnology**
   Didier Montet, Ramesh C. Ray (eds.), 2015
5. **Foodborne Pathogens and Food Safety**
   Md. Latiful Bari, Dike O. Ukuku (eds.), 2015
6. **Fermented Meat Products: Health Aspects**
   Nevijo Zdolec (ed.), 2016
7. **Fermented Foods of Latin America: From Traditional Knowledge to Innovative Applications**
   Ana Lucia Barretto Penna, Luis A. Nero, Svetoslav D. Todorov (eds.), 2016
8. **Lactic Acid Fermentation of Fruits and Vegetables**
   Spiros Paramithiotis (ed.), 2016
9. **Microbial Enzyme Technology in Food Applications**
   Ramesh C. Ray, Cristina M. Rosell (eds.), 2016
10. **Acetic Acid Bacteria: Fundamentals and Food Applications**
    Ilkin Yucel Sengun (ed.), 2017
11. **Fermented Foods, Part II: Technological Interventions**
    Ramesh C. Ray, Montet Didier (eds.), 2017
12. **Food Traceability and Authenticity: Analytical Techniques**
    Didier Montet, Ramesh C. Ray (eds.), 2018
13. **Seafood Safety and Quality**
    Md. Latiful Bari, Kozi Yamazaki (eds.), 2018
14. **Food Molecular Microbiology**
    Spiros Paramithiotis (ed.), 2019
15. **Mycotoxins in Food and Beverages: Innovations and Advances, Part I**
    Didier Montet, Catherine Brabet, Sabine Schorr-Galindo, Ramesh C. Ray (eds.), 2021

Mycotoxins in Food and Beverages
Innovations and Advances
Part II

Editors

Didier Montet
Researcher and Expert in Food Safety
UMR Qualisud, CIRAD, Montpellier, France

Catherine Brabet
Senior Researcher, UMR Qualisud, CIRAD,
Montpellier, France

Sabine Schorr-Galindo
Professor, Senior Researcher, UMR Qualisud,
Univ. Montpellier, Montpellier, France

Ramesh C. Ray
Retired Senior Principal Scientist (Microbiology)
ICAR - Central Tuber Crops Research Institute,
Regional Centre, Bhubaneswar, India

CRC Press
Taylor & Francis Group
Boca Raton London New York

CRC Press is an imprint of the
Taylor & Francis Group, an **Informa** business
A SCIENCE PUBLISHERS BOOK

First edition published 2021
by CRC Press
6000 Broken Sound Parkway NW, Suite 300, Boca Raton, FL 33487-2742

and by CRC Press
2 Park Square, Milton Park, Abingdon, Oxon, OX14 4RN

CRC Press is an imprint of Taylor & Francis Group, LLC

ISBN: 978-1-032-00837-0 (hbk)
ISBN: 978-1-032-00839-4 (pbk)
ISBN: 978-1-003-17604-6 (ebk)

Typeset in Palatino Roman
by Innovative Processors

Foreword

Reducing mycotoxins contamination of the worldwide food and feed chains is a major challenge towards improving human and animal health. Mycotoxins are responsible for a variety of toxic effects including the induction of cancer, and digestive, blood, kidney and nerve defects. One quarter of the world's food crops, including many basic foods, are affected by mycotoxin producing fungi. The mycotoxin problem is particularly important for human health in tropical areas, such as Sub-Saharan Africa, where crops are susceptible to contamination with the carcinogenic aflatoxins and fumonisins.

Globalization of trade has complicated the way we deal with mycotoxins in that regulatory standards often become bargaining chips in world trade negotiations. While developed countries have numerous mycotoxin regulations and a well-developed infrastructure for enforcing food quality standards, people in developing countries are not protected by food quality monitoring and enforcement of safe standards within their countries. Food commodities being exported are expected to comply with *CODEX Alimentarius* standards, and may indirectly result in higher risk of mycotoxin exposure in developing countries because the best quality foods leave the country and the contaminated food/feed is used for domestic consumption (Logrieco et al. 2018*). In developed countries, these problems are generally invisible but their management is an economic burden for producers, processors and consumers. These costs include the obvious ones of sampling and analysis to meet regulatory and contract requirements but also include larger, hidden costs of destroyed or returned shipments and the time and expense of sourcing and purchasing replacement items.

Multidisciplinary integration of know-how and technology present in this book is requested to better address the broad requirements for reducing mycotoxins in the agro-food chain.

Differences in environmental conditions in European as well as in other continents significantly influence the distribution of specific toxigenic fungi and related mycotoxicological risks. Emerging problems due to climate

---

* Logrieco, A.F., Miller, J.D., Eskola, M., Krska, R., Ayalew, A., Bandyopadhyay, R., Battilani, P., Bhatnager, D., Chulze, S., De Saeger, S., Li, P., Perrone, G., Poapolathep, A., Rahayu, E., Shepard, G., Stepman, F., Zhang, H. and Leslie, J.F. 2018. The Mycotox Charter: Increasing awareness of, and concerted action for minimizing mycotoxin exposure worldwide. *Toxins* 10(149): 1–17.

change and new mycotoxin/commodity combinations add further concern. In addition, trans-global transposition and trade exchanges of plant products significantly contribute to the spreading of toxigenic fungi population structure and relative mycotoxin risk worldwide.

For this reason there is a great need for studying mycotoxin management at global level along chains and to develop sustainable and inexpensive prevention methods, as bicontrol agents, for reducing toxigenic fungi contamination and related mycotoxins. In addition, new molecular approaches which help us to understand and control host/pathogen or environment/fungus interactions will be implemented to prevent mycotoxin production at its biological origin. The use of mycotoxin-detoxifying agents, as adsorbent materials for mycotoxin removal from feedstocks, may play an important way to improve animal welfare.

This book represents a comprehensive and broadly interdisciplinary research that is needed to better minimize mycotoxin contamination along chains by the prevention and decontamination as well as to understand the mycotoxin risk by furthering and updating information about toxicology.

Antonio F. Logrieco

Preface to the Series

Food is the essential source of nutrients (such as carbohydrates, proteins, fats, vitamins, and minerals) for all living organisms to sustain life. A large part of daily human efforts is concentrated on food production, processing, packaging and marketing, product development, preservation, storage, and ensuring food safety and quality. It is obvious therefore, our food supply chain can contain microorganisms that interact with the food, thereby interfering in the ecology of food substrates. The microbe-food interaction can be mostly beneficial (as in the case of many fermented foods such as cheese, butter, sausage, etc.) or in some cases, it is detrimental (spoilage of food, mycotoxin, etc.). The *Food Biology* series aims at bringing all these aspects of microbe-food interactions in form of topical volumes, covering food microbiology, food mycology, biochemistry, microbial ecology, food biotechnology and bio-processing, new food product developments with microbial interventions, food nutrification with nutraceuticals, food authenticity, food origin traceability, and food science and technology. Special emphasis is laid on new molecular techniques relevant to food biology research or to monitoring and assessing food safety and quality, multiple hurdle food preservation techniques, as well as new interventions in biotechnological applications in food processing and development.

The series is broadly broken up into food fermentation, food safety and hygiene, food authenticity and traceability, microbial interventions in food bio-processing and food additive development, sensory science, molecular diagnostic methods in detecting food borne pathogens and food policy, etc. Leading international authorities with background in academia, research, industry and government have been drawn into the series either as authors or as editors. The series will be a useful reference resource base in food microbiology, biochemistry, biotechnology, food science and technology for researchers, teachers, students and food science and technology practitioners.

Ramesh C Ray
Series Editor

Preface

This second volume is the continuation of Volume I of the book **"Mycotoxins in Food and Beverages: Innovations and Advances".**

The coordinators of this second volume want to thank Dr. Antonio Logrieco from the Institute of Sciences of Food Production (ISPA), National Research Council (CNR) located at Bari (Italy) for writing the preface. His opinions on the history of mycotoxins as well as his knowledge of the analysis of the problem of mycotoxins by world experts are of the greatest importance.

Strategies used to reduce and control food contamination by toxigenic fungi and mycotoxins are primarily based on pre- and post-harvest preventions. However, when the presence of mycotoxins cannot be avoided, detoxification methods have been developed and applied. These methods should not harm the food quality and safety, and must comply with current regulatory requirements. This second volume includes a series of chapters that discuss different approaches for mycotoxin control in food, in particular biocontrol strategies.

Professor Ijabadeniyi et al. from Durban University (South Africa) made an inventory of the food processing and decontamination approaches that permit us to control mycotoxins. They reviewed the physical decontamination methods (sieve-cleaning, sorting, milling, steeping, extrusion, etc.), the use of microorganisms and chemical compounds such as ammonia, hydrogen peroxide, sodium bisulfite, ozone and organic acids, and aqueous plant extracts as mycotoxin decontamination agents. They also presented some more modern strategies such as cold plasma treatment, pulsed electric field technology, high pressure and nanotechnologies.

Dr. Isaura Caceres et al. from INP/ENSAT in Toulouse (France) described preventive and curative methods to reduce aflatoxin B1 and ochratoxin A contamination. According to them, good agricultural practices together with the HACCP system are the first strategies to prevent fungal and mycotoxin contamination at different crop stages. They proposed two types of curative approaches that are decontamination methods to remove or degrade the targeted mycotoxins and detoxification methods for mycotoxin transformation leading to a reduced toxicity, through the application of physical, chemical or biological processes.

Dr. Giuseppina Avantaggiato et al. from CNR-ISPA in Bari (Italy) summarized the advances on mycotoxin-detoxifying agents and carried out a modern critique of these agents focusing on adsorbent materials for mycotoxin removal from feedstocks. Based on their extensive experience of EU projects, they explained the EU and EFSA approaches for the evaluation and authorization of mycotoxin-detoxifying agents. They evaluated the non-nutritive mycotoxin-adsorbents, raw clay minerals, modified clay minerals, nanoparticles as activated carbons and carbon nanostructures or chitosan polymeric nanoparticles, nanoclays and some other binders.

Dr. Abderahim Ahmadou et al. from the Institute Polytechnique Rural de Formation et Recherche Appliquée (IPR/IFRA) de Katibougou in Bamako (Mali) and CIRAD in Montpellier (France) clarified the interesting effects that adsorbents can have when they are used with raw materials contaminated by mycotoxins. Most of the adsorbents described in this chapter have shown good results for mycotoxin removal but with certain limits. The authors explained that the most suitable area of application for these binders is animal feed.

Dr. Rayane Hamrouni et al. from Aix Marseille University (France) described *Trichoderma* species as a study model for the production of secondary metabolites active on plants and fungal pathogens, including mycotoxigenic fungi. *Trichoderma* strains are thus considered as an important biopesticide.

Dr. Caroline Strub et al. from Montpellier University (France) discussed the potential of organic amendments and amended soils for biocontrol of phytopathogenic and mycotoxigenic fungi. They described the results of the meta-barcoding analysis of three organic amendments and corresponding amended soils with the aim to use them in large scale on cereal crops such as maize, wheat or barley. They emphasized on the several microbial families of interest as biocontrol agents that could act in the protection of soil and plants against the phytopathogenic and/or mycotoxigenic fungi.

Dr. Asma Chelaghema et al. from Montpellier University (France) described the main mycotoxigenic filamentous fungi and their mycotoxins and discussed the use of plant extracts as biocontrol agents for mycotoxin decontamination. They explained some of the mechanisms of action involved in the inhibition of fungal growth and/or mycotoxin contamination by plant extracts.

Even if now modern analytical devices allow for analyzing more than 400 mycotoxins, it remains difficult to evaluate their toxicities and their impact on human health, in particular combined toxicity of multiple mycotoxins which reflect feed and food contamination. Much data exist on individual toxicology of major mycotoxins but not enough toxicological data exist about their combined effect.

Dr. Nolwenn Hymery from the University of Brest (France) and Dr. Isabelle Oswald from Toxalim team in Toulouse (France) reviewed the toxicity *in vitro* and *in vivo* of regulated mycotoxins and emerging mycotoxins, as well as the *in vitro* effects of fusariotoxin mixtures. They also evaluated dietary exposure to mycotoxins through the estimation of daily intake, and health risks. They concluded that mycotoxin toxicological combined effects are unpredictable based on their individual effects, and relatively scarce toxicological concern was associated to mycotoxins exposure.

Professor Anil Kumar et al. from the Asian Institute of Technology in Bangkok (Thailand) analyzed the effect of mycotoxins on the intestinal microbiome and propose their possible roles in combating mycotoxins. They considered the gut as a bioreactor that could have a great influence on the fate of mycotoxins. In case of severe action, they proposed to restore microbiome by probiotic administration, especially those which possess mycotoxin reducing ability.

The new challenge of the effect of climate change on toxigenic fungi and their production of mycotoxins is analyzed by Dr. Yasmine Hamdouche et al. from the Abdelhamid Ibn Badis University in Mostaganem (Algeria). They summarized the principal climate change factors, and how these parameters will influence growth of mycotoxigenic fungi and mycotoxin production. They also showed us the difficulties that scientists will have to face to create analytical methods and to evaluate the toxinogenicity in the near and far future.

Dr. Didier Montet and Dr. Joël Guillemain (France) reviewed the resistance of genetically modified plants to fungi and highlight the advantages to using these plants against mycotoxin-producing fungi. They particularly discussed the question: Does Bt corn reduce or not insect damage that could in turn reduce mycotoxins?

We wish all of our readers an excellent time reading this book.

Didier Montet
Catherine Brabet
Sabine Schorr-Galindo
Ramesh C. Ray

Contents

CHAPTER

1

Trichoderma Species: Novel Metabolites Active for Industry and Biocontrol of Mycotoxigenic Fungi

Rayhane Hamrouni[1,2*], Nathalie Dupuy[1], Josiane Molinet[1] and Sevastianos Roussos[1*]

[1] Aix Marseille Univ, Avignon Université, CNRS, IRD, IMBE, Marseille, France

[2] Univ. Manouba, ISBST, BVBGR-LR11ES31, Biotechpole Sidi Thabet, 2020, Ariana, Tunisia

1. Introduction

During the last several years, substantial advancements in green chemistry principles have developed for the valorization of crop residues and cleaner production such as biopesticides (Saravanakumar et al. 2016). The market of biopesticides is growing quickly in comparison to conventional chemical pesticides that cause significant losses in agriculture, and their repeated use promotes the development of chemically resistant pathogen strains (Shah and Pell 2003, Keswani et al. 2016, Hamrouni et al. 2019a). Strong growth in the use of the biopesticides market is observed annually at a rate of 44% in North America, 20% in Europe and Oceania, 10% in Latin and South American countries and 6% in Asia (Tranier et al. 2014, De la Cruz Quiroz et al. 2015). Biopesticides are offering a more ecological manner for the management of pests, but they also face several challenges, such as increased barriers for the use of biological products, a lack of awareness towards the use of agricultural biological products, and most importantly a lack of global availability (Olson 2015, Zachow et al. 2016). To enhance the feasibility of using biopesticides, it is necessary to increase the metabolites with an antibiotic effect and biomass of species of interest (Vinale et al. 2014, Roussos et al. 2020).

Biopesticide production mainly uses biological control agents (BCA), because they are used as natural enemies of phytopathogens as mycotoxigenic fungi (Vinale et al. 2014). Furthermore, the use of microorganisms for pest management in agriculture is one of the most effective strategies of biological

*Corresponding authors: sevastianos.roussos@imbe.fr;
Rayhan.hamrouni@gmail.com

control (Loera-Corral et al. 2016). The outcomes of using beneficial microbes are strain dependent and the advantages for the associated plants include:

- Suppression of pathogens,
- Growth promotion,
- Establishment of an antagonistic microbial community in the rhizosphere, and
- Enhanced host resistance to both biotic and abiotic stresses (Howell et al. 2003).

Most of the microorganisms used in biocontrol are filamentous fungi because they are ubiquitous colonizers of their habitats as well as for their secretion capacity for antibiotic metabolites and enzymes, but also some bacteria such as *Bacillus thuringiensis* and *Bacillus subtillis* with biocontrol ability have been used (Harman 2000, Reino et al. 2008).

Fungi belonging to the *Trichoderma* genus are well known producers of secondary metabolites (SMs) with a direct activity against phytopathogens and compounds that substantially affect the metabolism of the plant, they are contributing as much as 50% of BCA's fungi (Reithner et al. 2005, Rubio et al. 2009). Although not essential for their primary metabolic processes, *Trichoderma* strains produce various SMs, including compounds of industrial and economic relevance. The production of SMs has been often correlated to specific stages of morphological differentiation and is associated with the phase of active growth (Thines et al. 2004, Vinale et al. 2006).

SMs show several biological activities possibly related to survival functions of the organism, such as competition against other micro- and macro-organisms, symbiosis, and metal transport. Furthermore, they play an important role in regulating interactions between organisms. Some example of SMs are" **mycotoxins** (SMs produced by fungi that colonize crops capable of causing disease and death in humans and other animals), **phytotoxins** (SMs produced by fungal pathogens that attack plants), **pigments** (colored compounds also with antioxidant activity) and **antibiotics** (natural products capable of inhibiting or killing microbial competitors (Renshaw et al. 2002, Lehner et al. 2013).

In addition, there are other ways to apply biocontrol compounds such as fungal spores, which are the most virulent form, have a long shelf life (Brand 2006, Hamrouni et al. 2019a) and they are cell structures more adapted to grow in a field and resist environmental conditions *in situ* (Shah and Pell 2003). Fungal spores and SMs can be produced by solid state fermentation (SSF). Use of agro-industrial wastes and SSF technology offers an alternative to bio-pesticide production (Hamrouni et al. 2019c, De la Cruz Quiroz et al. 2015, De la Cruz Quiroz et al. 2017a).

SSF can be defined as a fermentation process involving solid material in the absence (or near absence) of free water, but with enough moisture to support the growth and metabolism of the microorganisms (Pandey 2003, De la Cruz Quiroz et al. 2015). It uses agro-industrial residues as substrates for the production of bio-active products of commercial interest; this process falls under waste management technology (Roussos et al. 1997). This technology is receiving renewed attention because of the growing need for new metabolites in various industries such as agriculture (biopesticides), health, food and cosmetics (Thomas et al. 2013) obtaining in this way an inexpensive biotechnological option for modern agriculture in developing countries.

In this paper we summarize the most important metabolites types produced by *Trichoderma* species, emphasizing their biological activities, especially the role that these metabolites play in biological control mechanisms. Some aspects relating to the biosynthesis of these metabolites and related compounds are also discussed. It must be stressed that some of the groups of products mentioned here are the most important fungal metabolite families known.

2. Phylogeny, Biodiversity and Biotechnology of *Trichoderma*

The first description of *Trichoderma* fungi dates back to 1794 (Persoon 1794). In 1865, a link to the sexual state of a *Hypocrea* species was suggested. Hence, in 1969 the development of a concept for its detailed identification was initiated (Rifai 1969). *Trichoderma* are classified as follows:

- **Class:** Euascomycetes (Fungi that tend to form lichen with other organisms)
- **Phylum:** Ascomycota (Fungi that are characterized by their ascus (structure for reproduction)
- **Order:** Hypocreales
- **Family:** Hypocreaceae
- **Genus:** *Trichoderma*

Characterization of the genus *Trichoderma* was based firstly on morphological character such as conidial form, size, color, branching pattern with short side branches, short inflated phialides and the formation of sterile or fertile hyphal elongations from conidiophores. Generally, they produce a broad array of pigments from bright greenish-yellow to reddish in color, although some are also colorless. Similarly, conidial pigmentation varies from colorless to various green shades and sometimes also gray or brown (Hamrouni et al. 2019d). Mycelial form varies from Floccose to arachnoid with white color (Fig. 1). In addition, *Trichoderma* species are free-living fungi which are highly interactive in soil, root, and foliar environments.

In fact, the taxonomic confirmation of the genus *Trichoderma*, based only on morphological markers, can be considered to be limited and of low accuracy, due to the similarity of morphological characters and increasing numbers of morphologically cryptic species (Yedidia et al. 2003). Thereafter, many new species of *Trichoderma* strains were discovered, and by 2006, the genus already comprised more than 100 phylogenetically defined species (Druzhinina et al. 2006). Sometimes, especially in earlier publications, misidentifications of certain species occurred, for example the name *Trichoderma harzianum* has been used for many different species.

In recent years, with the advent of molecular biology methods, it is now possible to identify every *Trichoderma* isolate and determine it as a putative new species (18S rDNA sequence analysis) (Kullnig et al. 2001).

At present, the current diversity of the *Hypocrea/Trichoderma* is reflected in approximately 160 species including; *T. viride, T. virens, T. harzianum, T. asperellum, T. longibrachiatum, T. yunnanense, T. parareesei, T. hamatum, T. atroviride, T. gamsii, T. orientale* and *T. spirale*.

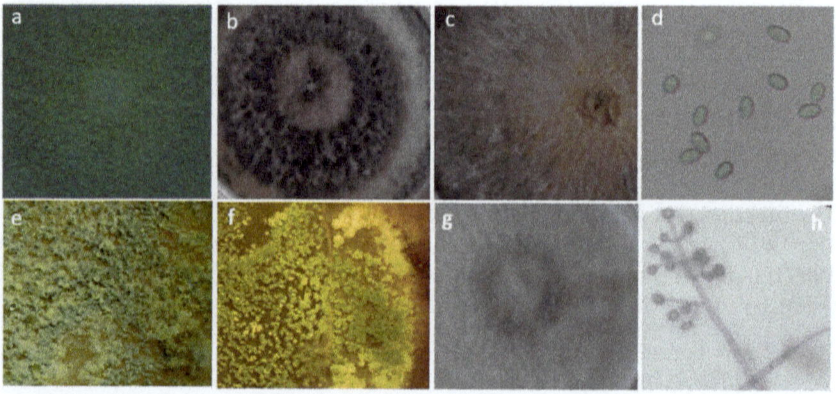

Figure 1. Characteristic features of *Trichoderma* strains. (a) *T. viride*, (b) *T. asperellum* DWG3, (c) *T. harzianum*, (d) conidia of *T. asperellum*, (e) *T. longibrachiatum*, (f) *T. longibrachiatum* during confrontation with *Fusarium* strain, (g) *T. asperellum* Tv 104, (h) *T. harzianum* germination and growing on PDA medium.

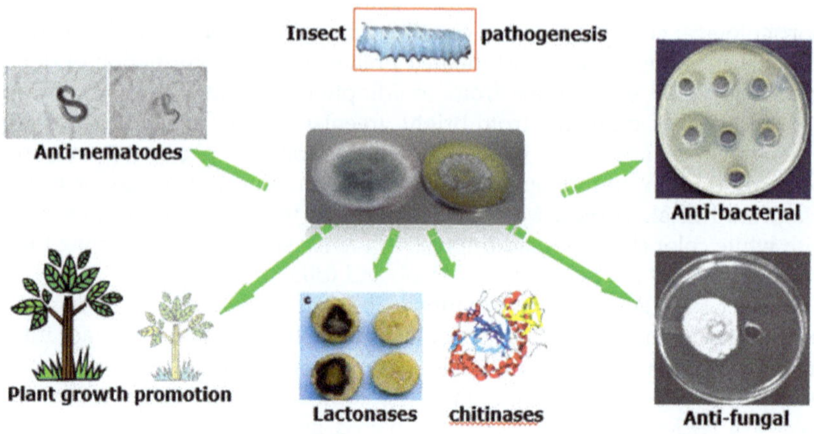

Figure 2. Potentially useful biocontrol activities of *Trichoderma* strains.

Trichoderma are the subject of various industrial applications in agriculture, food, pharmacy and biorefinery. Several species have economic importance as sources of enzymes, antibiotics, plant growth promoters, decomposers of lignocellulosic substrates, and as commercial biopesticides. Actually, strains of *Trichoderma* are commercially available to control plant disease in environmentally friendly agriculture (Harman et al. 2004).

3. *Trichoderma* Strains Used as Biopesticides in Biocontrol

Filamentous fungi have generated special importance due to the fact that they have a higher spectrum of disease control and greater biomass yield. ***Trichoderma* strains** are an important fungus as a biopesticide; these fungi could be a good model for biocontrol application.

After publication of *Trichoderma viride* acting as a parasite on other fungi in 1932 (Weindling 1932), research on antagonistic properties of *Trichoderma* strains progressed rapidly. Nowadays, the most important species in this field are *T. harzianum* (in earlier reports sometimes misidentified as *T. atroviride*), *T. virens*, *T. viride* and *T. asperellum*. In their defensive actions, *Trichoderma* strains apply lytic enzymes (Hamrouni et al. 2019a), proteolytic enzymes (Chen et al. 2009), ABC transporter membrane pumps (Ruocco et al. 2009), volatile metabolites and other secondary metabolites as active measures against their hosts (Fig. 2) or they succeed by their method of impairing growth conditions of pathogens (Reino et al. 2008).

There are several reports that demonstrated BCAs effectiveness against phytopathogens. Based on Table 1 which presents the *Trichoderma* strains most

Table 1. Example of *Trichoderma* strains used for the biological control of fungal phytopathogens (Reino et al. 2008, De la Cruz Quiroz et al. 2015)

Trichoderma strains	*Phytopathogen*	*Diseases*
T. asperellum	*Fusarium oxysporum*	Tomato
T. harzianum and *T. viride*	*Aspergillus flavus* *Fusarium. verticillioides* *Sclerotium ceviporum* *Postia placenta* *Aspergillus niger* *Aspergillus tamari*	Seeds, Onions, Spruce, Pine, Cowpea Banana, Mango and Potato
T. virens	*Rhizopus oryzae* *Rhizoctonia solani*	Cotton seeds, Cucumber, Potato
T. lignorum, *T. hamatum*, *T. pseudokoningii*	*Rhizoctonia solani*	Bean plant
T. harzianum	*Botrytis cinerea* *Mucor piriformis* *Penicillium expansum* *Fusarium oxysporum* *Alternaria alternata*	Strawberry, Apple, Vegetables and ornamental, Potato

used in biological control, *Trichoderma* genus can inhibit phytopathogenic fungi such as the species of *Fusarium oxysporum, Botrytis cinerea, Crinipellis perniciosa, Rhizoctonia solani,* etc. (Verma et al., 2007, Mukherjee et al. 2010). Some of them are mycotogenic fungi especially *Aspergillus flavus (Aflatoxins), Aspergillus niger* (Ochratoxin A), *Penicillium expansum* (Patulin) and *Fusarium oxysporum* (Trichothecenes) and *F. verticillioides* (Fumonisins).

Although, there are other fungi used as biological control, for example *Metarhizium anisopliae* (Van Breukelen et al. 2011), *Beauveria bassiana* (Dalla-Santa et al. 2004), *Epicoccum nigrum* (De la Cruz Quiroz et al .2015) and *Coniothyrium minitans* (Yang et al. 2010). They represent the most used entomopathogenic fungus for marketing because they are non-pathogen and saprophytic fungus. They are mycoparasitic microorganisms considered as a highly effective antagonist against several phytopathogens such as *Leucostoma cincta, Sclerotinia sclerotium, Colletotrichum gloeosporioides, Botrytis cinerea* and *Monilinia laxa* (Viccini et al. 2007, De la Cruz Quiroz et al. 2015).

4. Opportunities of BCA Production Using Solid State Fermentation Technology

Solid State Fermentation technology (SSF) has been useful for biopesticides production, using several kinds of substrate. It is a very simple process to carry out. Bioreactors are glass columns, petri plates or Erlenmeyer flasks. Bioreactors in an industrial scale can be static or stirred, some of them with low costs of production. Agro-industrial residues are generally considered as the best substrates for SSF, mainly due to both the nutritional quality and low cost. Indeed, several processes and products reported the use of sugarcane bagasse as a raw material. It consists of approximately 50% cellulose and 25% each of hemicellulose and lignin. It is an inexpensive material and could be used as a carbon source. Today in Africa, many agro-industrial wastes are disposed of without any use. These wastes can be valorized and used as substrates in the SSF and could generate a positive impact on the world's ecology. These wastes include sugarcane bagasse, coffee husk, vine shoot, Jatropha cake, olive pomace and fruit seeds, and are constituted by cellulose, which represent a potential source of sugars and energy (Pandey et al. 2008, Hamrouni et al. 2019c).

They offer several factors such as carbon and nitrogen sources, mineral salts and high porosity level. Recently, Hamrouni et al. (2019c) reported the possibility to produce spores, lytic enzymes and antifungal metabolites using SSF technology and *T. asperellum* TF1 cultivated on a mixture of vine shoots, Jatropha cake, olive pomace and olive oil as substrate. Likewise, De la Cruz-Quiroz et al. (2017a) highlighted the production of lytic enzymes as amylases, pectinases, chitinases, cellulases, lipases and virulent *Trichoderma* spores using a mixture of sugarcane bagasse, wheat bran, chitin, potato flour olive oil as a substrate by six *Trichoderma* stains with different conditions of culture. In comparison with submerged fermentations several advantages of SSF have been reported. (i) Absence of a liquid phase and substrate of low humidity allows reduced volumes of fermentation reactors, to prevent bacterial contamination, avoid effluents and economize reagents. (ii) Culture media is easier to prepare. (iii) In SSF, agitation is unnecessary and aeration is easier (De la Cruz-Quiroz et al. 2015). Regarding forced aeration effect, it provides that application of controlled forced aeration to remove both CO_2 and the metabolic heat generated during the fermentation is an important aspect on fermentation process production of metabolites (Holker and Lenz 2005). Forced aeration has a high influence on the stimulation of fungal sporulation. First, it promotes induction of early onset of sporulation and second, evaporates excess water, which contributes to an important increase in the sporulation index.

The effect of forced aeration on the production of antifungal metabolites, enzymes and spores by *T. asperellum* strains was also evaluated in a previous work (Hamrouni et al. 2019a) and those results suggested that forced aeration on SSF systems leads to production of high yields.

4.1 *Trichoderma* spore production under SSF

Biopesticides are applied mainly as fungal spores, which are the most virulent form and have a long shelf life (Hamrouni et al. 2019b). There are several reports with the aim of *Trichoderma* strains spore production by SSF to be used in biological control. Generally, high sporulation yields are related to a desiccation process, possibly owed to stress by the culture conditions, causing, thus, an increase in the concentration of conidia (Rodriguez-Fernandez et al. 2012, Carboué et al. 2017). This effect has a direct impact on the technical feasibility of spore production because under such culture conditions, it is possible to use reactors like flasks, trays or bags resulting in energy saving, as far as air control and the laborious handling on the preparation of the fermentation process is concerned.

In this context, Roussos (1985) worked with sugarcane bagasse as support/substrate in SSF. *T. harzianum* production was conducted in Raimbault's columns (Raimbault and Alazard 1980) and a static fermenter (Zymotis), author obtained 3.61×10^{10} and 3.25×10^{11} spores/g substrate, respectively.

Additionally, production of spore from *Trichoderma asperellum* TF1 was evaluated during SSF by Hamrouni et al. (2019b). A sporulation index of 8.5×10^9 spores/g using vine shoots, Jatropha cake, olive pomace and olive oil as substrate and plastic single use bioreactor was recorded. A production of 1.4×10^9 spores/g was also reported during cultivation of *Trichoderma asperellum* T2-10 in polyethylene bioreactors using corn cob as substrate (De la Cruz-Quiroz et al. 2017b).

Motta and Santana (2014) studied spore production from *T. reseei* under SSF conditions in a column bioreactor and obtained a production of 4.41×10^9 spores/g using empty fruit bunches. Similarly, Durand (2003) reported important conidia production by *T. viride* on scale sterile reactor (5.5×10^9 conidia/g) using a sugar beet pulp medium on SSF. On the other hand, Jin and Custis (2011) used straw and rice grain as substrate, SSF was carried out in plastic bags, obtaining a yield of 1.2×10^{11} spores/g.

4.2 *Trichoderma* Biopesticide Formulation

During *Trichoderma* strains growth, product formulation is a critical area for commercial purposes. It is essential to note that only the fungal agent is used neither as a pesticide nor as the sole active agent (Ruffner et al. 2015).

Development of microbial pesticide formulation closely paralleled that of chemical pesticides. But the differences do exist because microbial pesticides do not directly depend on the effect of a poisonous chemical but exploit the activity of living entities. During formulation development, it is essential that the formulated product possesses a long shelf life, which may offer consistent biological controls under critical conditions (Jackson et al. 2010). Furthermore, it is necessary to consider the surface chemistry of pathogen propagation and fungal, as well as, the ecological and environmental factors in order to maximize biocontrol effectiveness (Jackson et al. 2010). Generally, field application of biopesticides products is performed by using a suspension containing a high concentration of antifungal compounds, lytic enzymes and spores. However, there are cases, where the ground fermented matter can be used. This information has been the basis to commence research at the worldwide level by a group led by Dr. Sevastianos Roussos (IRD, IMBE, Aix Marseille University, France). They study the development of new formulations of biopesticides using a new single use bioreactor, SSF process, and fermented matter content secondary metabolites of *Trichoderma* strains (6-PP; peptaibols, etc.), and lytic enzymes (cocktail of chitinases, glucanases, lipases, cellulases, amylases, pectinases).

Commercialized biopesticide products "ESCALATOR®" are made of spores using the mixture of two strains: *T. asperellum* ICCO12 and *T. asperellum* ICC080. It mentioned that optimum doses for a spore formulated product by ESCALATOR are near to $3{\times}10^7$ CFU/g (Colony-Forming Unit). These biopesticides were developed to protect viticulture sector from fungi attacks.

Therefore, in Spain, a formulation marketed under the name TUSAL®, made from *T. harzianum* and *T. viride* cultures, has been prepared by the phytopathology research group of the University of Salamanca and Newbiotechnic S.A. Corporation. It is used to prevent the growth of pathogen soil borne fungi responsible for leaf-falling disease in several crops (Grondona et al. 2004).

5. Lytic Enzymes Produced by *Trichoderma*

The mycoparasitic fungi, *Trichoderma* genus, are shown to be very efficient producers of lytic enzymes responsible for the direct attack against the pathogen of plant diseases (Nakazawa et al. 2009). The process of parasitism (Fig. 3) includes several stages:

- Host-directed growth of *Trichoderma*
- Recognition of the host and attachment
- Synthesis and secretion of toxins metabolites and degradative enzymes hydrolyzing cell wall components
- Penetration of hyphae and lysis of the host.

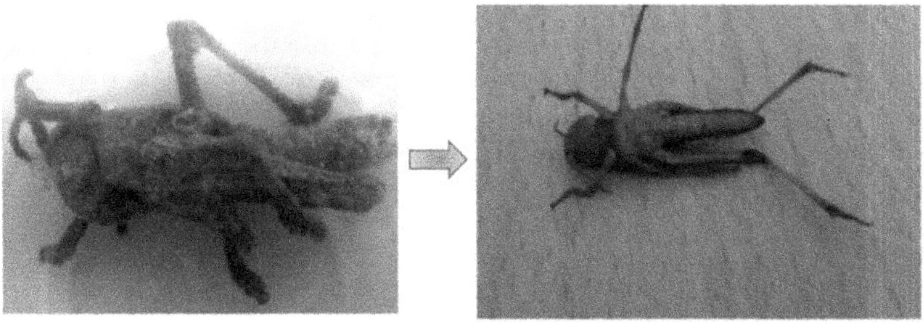

Figure 3. Treatment of *Locusta migratoria* by *Trichoderma harzianum* spores.

Lytic enzymes cocktails that are capable of suppressing a number of fungal phytopathogens that originate in air or soil, are chitinases, glucanases, proteases, amylases, lipases and cellulases. These enzymes are usually extracellular, of low molecular weight and highly stable. They have been reported mainly in isolates of *T. harzianum*. Most of the work has been carried out on strains of *T. viride, T. reesei* and *T. asperellum* (Marra et al. 2006).

5.1 Chitinases

Chitinases of *Trichoderma* strains are likely involved in their antagonistic activity against phytopathogens and in the biocontrol. They are an effective tool for complete degradation of mycelia or conidial walls of phytopathogenic fungi (Solanki et al. 2010).

In 1992, De La Cruz et al. were the first to isolate, purify, and characterize chitinases of *T. harzianum*. After that, several reports explained the chitinolytic system of *Trichoderma* genus. They mentioned that *T. harzianum* may contain seven individual chitinase enzymes. In the well-characterized strain *T. harzianum*, this system included two β(1,4)-N acetylglucosaminidases (102 and 73 kDa), four endochitinases (52, 42, 33, and 31 kDa, randomly cleave b-1,4-glycosidic bonds of chitin) and one exochitinase (40 kDa) (Tanaka et al. 2001).

On the other hand, Kumar et al. (2012) highlighted an increase in chitinase activity through SSF and *T. asperellum* UTP-16 after an optimization work. Furthermore, efficiency of chitinase enzyme was evaluated against the phytopathogenic "*Fusarium* strains" and a mycelial growth inhibition was achieved.

5.2 Glucanases

Glucanases are another hydrolytic enzyme that has many roles in a wide range of different biological systems (Zeilinger and Omann 2007).

β-glucan degrading enzymes are classified according to the type of β-glucosidic linkages: exo-1,3-β-glucanases and exo-1,3-β-glucanases, (β-1,3-glucan: glucanohydrolase). Because 1,3-β-glucan is a structural component of fungal cell walls, the production of extracellular 1,3-β-glucanases has been reported as an important enzymatic activity in biocontrol microorganisms.

In fact, cell walls of phytopathogenic fungi such as *Sclerotium rolfsii, Rhizoctonia solani* and *Pythium* species are composed mainly of β-1.3-glucans and chitin, including also cellulose. Glucanases can degrade the cuticle of insects (Sharma and Pandey 2009). Glucans are also present in the cuticle of nematode eggs, like the phytoparasitic nematodes of the *Meloidogyne* genus whose viability decreases severely when treated with *Trichoderma* species (Muiño et al. 2006). Despite this, De la Cruz Quiroz et al. (2017a) reported an improvement in endo and exo glucanase activities applying SSF process using six *Trichoderma (T. longibranchiatum, T. harzianum, T. yunnanense, and T. asperellum* (T2-10 and T2-31) strains. They indicated that forced aeration during the fermentation process increased the endoglucanase activity, and the best value was observed with *T. yunnanense* (38.32 U/g). This same approach was evaluated by Kalogeris et al. (2003) in SSF process using wheat bran as substrate. The authors presented good results in the enzyme activities for endoglucanase, and exoglucanase from *Thermoascus aurantiacus*. Although *T. asperellum* is one of the less studied species of the genus. A number of β-1.3-endo glucanases and β-1.3-exo glucanases have been purified from this strain (Marcello et al. 2010).

5.3 Proteases

Proteases are extracellular enzymes mainly produced by *Trichoderma* genus. They play a significant part in the lysis of cell walls of phytopathogenic fungi, because chitin or fibrils of β-glucan are embedded into the protein matrix.

It was noted that the proteases produced by *Trichoderma* genus may be involved in inactivating extracellular enzymes of phytopathogenic fungi. In this context, it was demonstrated that the protease of *Trichoderma harzianum* isolate 1051 showed antagonistic activity against *Crinipellis perniciosa*, the causal agent of cocoa plant witches' broom disease (De Marco et al. 2002). The hydrolytic enzymes produced by *B. cinerea*, endo and exo-polygalacturonase were partially deactivated by protease from the *T. harzianum* isolates (Elad and Kapat 1999).

5.4 Cellulases

Cellulases are enzymes that degrade cellulose. These enzymes are produced by the genus *Trichoderma* (Roussos and Raimbault 1982, Roussos 1985). The known function of fungal cellulases is to degrade cellulose-rich plant cell walls, and these enzymes are being studied for their role in conversion of biomass into energy, in order to provide sustainable solutions in bio refinery (De la Cruz Quiroz et al. 2015). However only a few studies have reported the basic description of cellulose-induced defense, but there is currently no research on the mechanism of cellulose-induced plant defense.

The *Trichoderma* cellulases play essential roles in the specificity, recognition, and adhesion of certain symbiotic fungi. These enzymes can cause a series of defense reactions in tobacco leaves, such as the cytoplasmic contraction and nuclear accumulation (Yano et al. 1998).

In 2016, Saravanakumar et al. identified the role of *Trichoderma* cellulases in the induced systemic resistance in maize plants to protect them against invading pathogens. Furthermore, Gajera and Vakharia (2012) suggested that percentage growth inhibition of *Aspergillus niger* decreased with the increasing concentration of cellulase produced by *T. viride* (Saravanakumar et al. 2016).

6. Active Secondary Metabolites Produced by *Trichoderma* Strains

Trichoderma genus typically produces SMs that can effect the interactions of plants with their pathogens. These bioactive molecules may also have antifungal and antibiotic properties that are used world-wide for crop protection and bio-fertilization (Demain et al. 2000). In addition to direct toxic activity against phytopathogens, biocontrol-related metabolites may also increase disease resistance by triggering systemic plant defense activity or enhance root and shoot growth. Thereby, more than 100 metabolites with antibiotic and antifungal activity including pyrones, peptaibols, terpenes, polyketides, metabolites derived from amino acids, and polypeptides were detected in *Trichoderma* species culture (Vinale et al. 2014). In Fig. 4, a classification has been reported by Vinale et al. (2014) for several SMs.

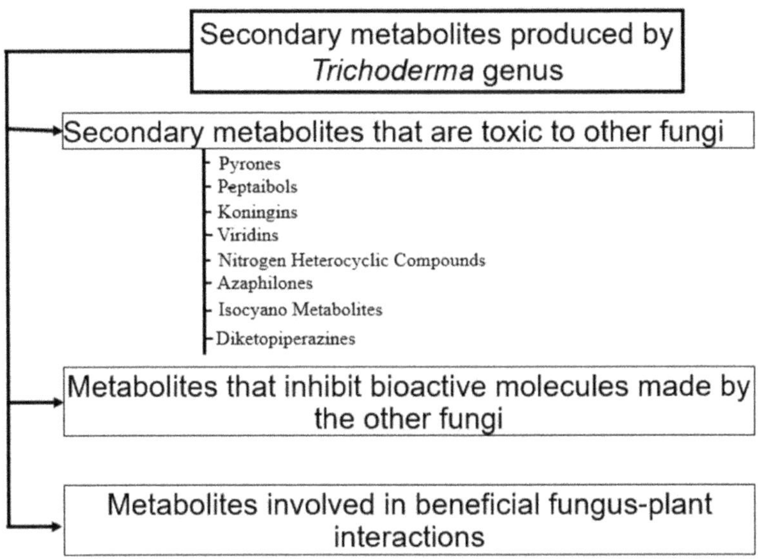

Figure 4. Secondary metabolites production by *Trichoderma* strains.

Another point is that important mycotoxin producing fungal species which affect plants belong mostly to the species *Aspergillus, Alternaria* and *Fusarium.* Examples of these are *A. flavus, A. carbonarius* and *A. alternata* occurring on different food commodities and *Fusarium oxysporum* occurring on different vegetables, fruits and crops. These fungi are all able to produce important mycotoxins such as aflatoxin, ochratoxin, alternariol or trichothecenes. These mycotoxin cause several health issues leading in the worst cases to cancer in humans and animals (Gromadzka et al. 2008).

Besides substantially reduced growth rates, a transcriptional based inhibition of mycotoxin biosynthesis in the competed *Aspergillus* and *Fusarium* by *Trichoderma* genus was demonstrated recently (Braun et al. 2018). Authors suggested that the SMs produced by *Trichoderma* may support this fact.

The following paragraphs report the most significant SMs isolated from *Trichoderma* species that have shown a significant antifungal activity.

6.1 Pyrones

The pyrone **6-pentyl-2H-pyran-2-one** (6-pentyl-□-pyrone or 6-PP) (Fig. 5.1) is a metabolite commonly purified from the culture of different *Trichoderma* species (*T. asperellum, T. viride, T. atroviride, T. harzianum, T. koningii*). It is a volatile compound, responsible for the coconut aroma released by axenically developed colonies (Sarhy-Bagnon et al. 2000, Oda et al. 2009).

In literature, there are many studies available aiming at 6-PP production in solid and liquid fermentation. Ladeira et al. (2010) developed a 6-PP extraction fermentation processes (Oda et al. 2009) and scaled-up an extraction fermentation process (Rocha-Valadez et al. 2006). Hamrouni et al. (2019d) reported a comparison of 6-PP production under two types of cultures (solid and liquid culture) and obtained a significant difference on 6-PP biosynthesis with a higher 6-PP production under SSF.

6-PP has shown both *in vivo* and *in vitro* antifungal activities towards several plant pathogenic fungi and a strong relationship has been found between the biosynthesis of this metabolite and the biocontrol ability of the producing fungi. Walter et al. (2000) observed that high 6-PP concentration significantly affected germination, germ tube growth, colony formation and colony size ($P<0.001$) of 14 *Botrytis cinerea* isolates tested.

Furthermore, it has been indicated that the addition of 0.3 mg/mL of 6-PP to agar medium caused a 31.7% reduction in *F. oxysporum* and a 69.6% growth reduction in *Rhizoctonia solani* after 2 days of culture (Reino et al. 2008).

Cytosporone S (Fig. 5.2) is another pyrone, recently isolated from a *Trichoderma* strain. It has been reported to have *in vitro* antibiotic activity against several bacteria and fungi (Tamás et al. 2018).

6.2 Peptaibols

The name "peptaibol" is formed from the names of the components: Peptide, Aib and Amino alcohol. A wide variety of peptaibols were identified in *Trichoderma* cultures (Vinale et al. 2014). These compounds are amphipathic and linear oligopeptides (5–20 amino acids). They are rich in non-proteinogenic amino acids (α-aminoisobutyric acid and isovaline), the N-terminal group of the peptide is usually acetylated and the C- terminus is an amino alcohol (phenylalaninol, isoleucinol, valinol, leucinol).

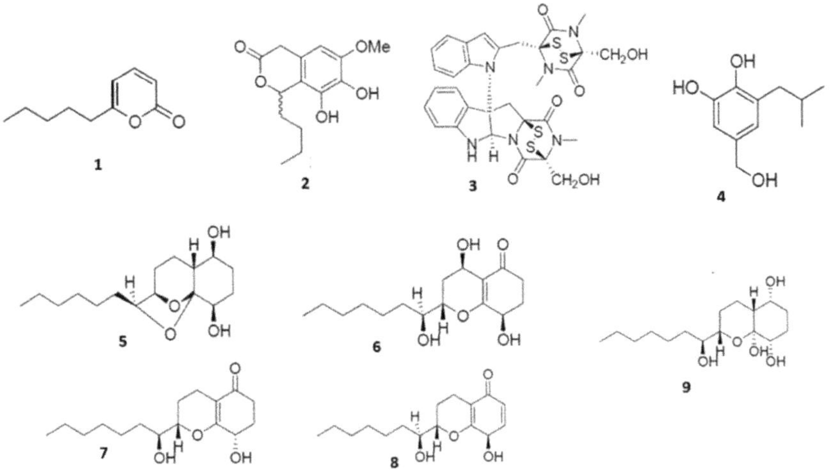

Figure 5. Chemical structure of 6-PP (1), cytosporone S (2), chaetomin (3), bigutol (4), koninginins A, B, D, E and G (5, 6, 7, 8, 9).

The information available in the literature about the peptaibol profiles of *Trichoderma* strains is limited. Efficient production of peptaibols predominantly occurred in solid state cultivation and correlated with sporulation (Tisch and Schmoll 2010). Tamás et al. (2018) produced peptaibols by solid state cultivation, and they identified the peptaibols composition of *T. koningiopsis* and *T. gamsii* using spectrometric methods, such as HPLC-ESI-MS measurements, which revealed a total of 30 peptaibol sequences.

In fact, the culture medium most frequently used for the fermentations of *Trichoderma* species for peptaibols production needs to contain carbon, nitrogen and a phosphate source, and calcium, potassium, magnesium, iron, manganese, copper and zinc as the main trace elements.

Previous reports described the conformational properties, and biological activities of the peptaibols of the *Trichoderma* species. It has been demonstrated that peptaibols inhibited β-glucan synthase activity in the host fungus, this inhibition prevented the reconstruction of the pathogen cell walls, but it acted in a synergistic manner with *T. harzianum* β-glucanases (Degenkolb et al. 2008). In addition, peptaibols have proven activity against mycotoxigenic fungi like *Aspergillus* and *Fusarium* species but are ineffective against *Alternaria* species (Daniel et al. 2007). Based on the chain lengths of the amino acid sequences peptaibols can be classified into three groups:

1. The short-sequence peptaibols with 11–16 amino acid residues (example the harzianins),
2. The long-sequence peptaibols with 18–20 residues (the trichorzianins) and
3. The lipopeptaibols with 6 or 10 residues (the trichogins) (Stoppacher et al. 2010).

Examples of some peptaibols described in literature are included in Table 2 with their respective references.

Table 2. Examples of peptaibols isolated from *Trichoderma* species

Species	*Peptaibols*	*Reference*
T. viride	Suzukacillin A	Krause et al. (2006)
	Trichotoxin A40	Brueckner et al. (1985)
	Trichovirins II	Jaworski et al. (1999)
	Trichorovins	Fujita et al. (1994)
	Trichodecenins I, II	Fujita et al. (1994)
T. koningii	Trichokonins V–VIII	Huang et al. (1995)
	Trikoningin KA, KB	Auvin-Guette et al. (1993)
T. polysporum	Polysporins A–D	New et al. (1996)
	Trichosporin B–V	Iida et al. (1993)
T. harzianum	Trichorzianines A, B	El Hajji et al. (1987),
	Trichokindins I–VII	Iida et al. (1994)
	Harzianins HC	Rebuffat et al. (1995)
	Trichorozins I–IV	Iida et al. (1995)
T. atroviride	Atroviridins A–C	Oh et al. (2000)
	Trichofumin A–D	Berg et al. (2003)

Figures 5.3 and 5.4 show the structures of two residues of peptaibols "chaetomin" and "bigutol" isolated from *Trichoderma viride* culture.

6.3 Koninginins

Koninginins are complex pyranes isolated from *T. harzianum, T. koningii* and *T. aureoviride*. (Ghisalberti et al. 1993). Different authors (Almassi et al. 1991, Ghisalberti et al. 1993, Reino et al. 2008, Vinale et al. 2014) reported that *T. harzianum* produces five residues of Koninginins: Koninginins A, B, D, E and G (Figs. 5.5-5.9). The *in vitro* antifungal activities of these SMs are demonstrated against several phytopathogenic agents such as *Fusarium oxysporum, Rhizoctonia solani, Phytophthora cinnamomi, Pythium middletonii, and Bipolaris sorokiniana* (Vinale et al. 2014).

6.4 Viridins

Viridins (Fig. 6.10) are steroidal metabolites isolated from diverse *Trichoderma* species (*T. viride, T. koningii, T. virens*). Viridins molecules were first described in 1945 as antifungal metabolites of the fungus *Trichoderma virens* (Brian and McGowan 1945).

These antifungal compounds prevent spore germination of *Botrytis cinerea, Fusarium caeruleum, Penicillium expansum, Colletotrichum lini* and *Aspergillus niger* (Reino et al. 2008). *T. viride* and *T. hamatum* produce viridiol, a similar antifungal and phytotoxic metabolite for which the *in vivo* activity has been demonstrated (Fig. 6.11).

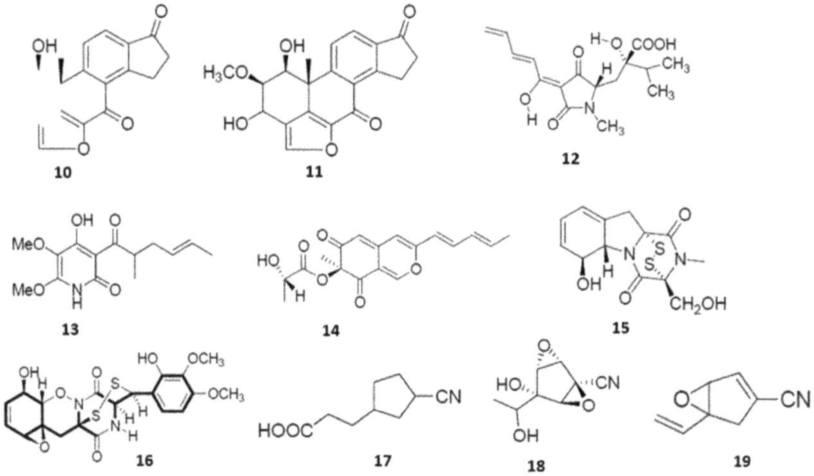

Figure 6. Chemical structure of Viridins (10), Viridol (11), Harzianic acid (12), Harzianopyridone (13), Azaphilones (14), Gliotoxin (15), Gliovirin (16), Dermadin (17), Isonitrile Trichoviridin (18), Homothallins (19).

Results of a transcriptional comparison of viridiol, wild type and a viridin deficient *T. virens* mutant led to the identification that six genes are similar to those involved in secondary metabolism of other fungi. Four of these genes are located as a cluster that is associated with the production of viridin (Mukherjee et al. 2006).

6.5 Nitrogen Heterocyclic Compounds

Harzianic acid (Fig. 6.12) is a *T. harzianum* metabolite characterized by the presence of a pyrrolidinedione ring system. The *in vitro* antifungal activities of these compounds were demonstrated against *Rhizoctonia solani, Pythium irregulare*, and *Sclerotinia sclerotiorum* (Vinale et al. 2006).

Harzianopyridone (Fig. 6.13) a new compound isolated from different *Trichoderma* strains. It is characterized by the presence of penta-substituted pyridine ring system with a 2,3-dimethoxy-4-pyridinol pattern. It has been reported that harzianopyridone isolated from the liquid culture of *T. harzianum* T22 inhibits the growth of several plant pathogens like *Botrytis cinerea, R. solani* and *Pythium ultimum* (Vinale et al. 2009).

6.6 Azaphilones

The azaphilones (Fig. 6.14) have been isolated from different genus of *Trichoderma.* They contain a highly oxygenated bicyclic core and a chiral quaternary center. Two azaphilone-type compounds, **harziphilone** (belong to the class of hydrogenated azaphilones) and **fleephilone**, (seem to be structurally related to azaphilones with a 3-hydroxy butanoyl moiety) were isolated from the butanol–methanol extract of the fermentation broth of *T. harzianum* by bioassay-guided fractionation (Gallos et al. 2001). Harziphilone and fleephilone have demonstrated inhibitory activity against the growth of *Rhizoctonia solani* and *Pythium ultimum* (Vinale et al. 2006). Recently, T22 azaphilone, a new compound was isolated from culture filtrates of commercial strains *T. harzianum* T22. These natural products showed *in vitro* antifungal activity against *R. solani, P. ultimum* and *Gaeumannomyces graminis* (Tamás et al. 2018).

6.7 Diketopiperazines

Two antifungal metabolites named **gliotoxin** (Fig. 6.15) and **gliovirin** (Fig. 6.16) are the two most important *Trichoderma* secondary metabolites belonging to the class of diketopiperazines. Most studies indicated the potential role of antifungal and antibiotic production in the biocontrol mechanism of the gliotoxin/gliovirin producers. Strains producing gliotoxin are antagonistic to *Rhizoctonia solani*, whereas those producing gliovirin are effective against *Pythium ultimum* (Vinale et al. 2006).

6.8 Isocyano Metabolites

Isocyano molecules are other compounds produced by *Trichoderma* strains; they are characterized by a 5-membered ring. In fact, due to their instability, the isolation and separation of these compounds are very difficult (Reino et al. 2008). The first reports on isocyano cyclopentenes in *Trichoderma* species were published 54 years ago (Meyer 1966).

In 1971, two compounds **dermadin** (Fig. 5.17), and **isonitrile trichoviridin** (Fig. 5.18) from *T. viride, T. koningii* and *T. hamatum* (Tamura et al. 1975) were patented. These metabolites showed remarkable fungicidal activities. The ability of several *Trichoderma* strains to produce dermadin is correlated with the ability to suppress *Botrytis cinerea* germination. Activity of dermadin on *B.cinerea* is 10 times higher than that of the *Trichoderma* trichoviridin.

Finally, *T. koningii* produces remarkable fungicidal metabolites such as cyclopentenes isocyano metabolites named **homothallins** (Fig. 6.19) that affect the morphology growth of *Botrytis allii*, but a significant phytotoxicity in the etiolated wheat coleoptile bioassay (100% inhibition at 10^{-3} M) has been mentioned.

7. Conclusions and Future Prospects

From the investigations described above, it is clear that *Trichoderma* species have great potential for the future, due to their ability to produce a wide range of antifungal compounds and for their ability to parasitize other fungi including mycotoxigenic ones. In addition, to these direct effects on fungal phytopathogens, recent evidence indicates that many *Trichoderma* species, especially *Trichoderma harzianum, Trichoderma viride, Trichoderma asperellum* and *Trichoderma virens* can induce both localized and systemic resistance in a range of plants to a variety of plant pathogens, and certain strains can also have substantial influence on plant growth and development.

As mentioned in this chapter, great advances have been made in studies on the biosynthesis, biological activity and structural determination of antifungal metabolites. *Trichoderma* species produce at least three classes of compounds that elicit plant defense responses: peptides (peptaibols...), proteins (lytic enzymes) and low-molecular weight compounds (polyketides such as pyrones).

The solid state fermentation (SSF) system is used preferably rather than synthetic liquid media for antifungal compounds production of biological control agents (BCA)'s microorganisms used as biopesticides. It facilitates the development of biopesticide formulations used in field crops that will permit a decrease in process costs due to the possibility of the use of by-products (Roussos et al. 2020).

Finally, there is important information about *Trichoderma* species as a source of biologically active metabolites topics. It remains necessary to continue to investigate them for years to come in order to enhance antibiotic compound production at greater scales through the evaluation of several reactor types, and different kinds of agro-industrial wastes. The correlation between these compounds and their effectiveness as biological control agents has to be established. The products have to be formulated in order to establish biological indicators to control biopesticides virulence. Tools for traceability have to be developed to monitor biopesticides in soil.

References

Almassi, F., Ghisalberti, E.L., Narbey, M.J. and Sivasithamparam, K. 1991. New antibiotics from strains of *Trichoderma harzianum*. Journal of Natural Products (54): 396–402. DOI: 10.1021/np50074a008.

Auvin-Guette, C., Rebuffat, S., Vuidepot, I., Massias, M. and Bodo, B. 1993. Structural elucidation of trikoningins KA and KB, peptaibols from *Trichoderma koningii*. Journal of the Chemical Society, Perkin Transactions 1(2): 249–255. DOI: P19930000249.

Berg, A., Grigoriev, P.A., Degenkolb, T., Neuhof, T., Haertl, A., Schlegel, B. and Graefe, U. 2003. Isolation, structure elucidation and biological activities of trichofumins A, B, C and D, new 11 and 13mer peptaibols from *Trichoderma* sp. HKI 0276. Journal of Peptide Science 9: 810–816. DOI: 10.1002/psc.498

Brand, D. 2006. Utilization of solid-state fermentation for the production of fungal biological control agents: Case study on *Paecilomyces lilacinus* against root-knot nematodes. Thèse Doctorat de l'Université de Provence, Marseille, France. p. 188.

Braun, H., Woitsch, L., Hetzer, B., Geisen, R., Zange, B. and Schmidt-Heydt, M. 2018. *Trichoderma harzianum*: Inhibition of mycotoxin producing fungi and toxin biosynthesis at a transcriptional level. International Journal of Food Microbiology 18: 30172–30177. DOI: 10.1016/j.ijfoodmicro.2018.04.021.

Brian, P.W. and McGowan, J.C. 1945. Viridin: A highly fungistatic substance produced by *Trichoderma viride*. Nature 156: 144–145. DOI: 10.1038/156144a0.

Brueckner, H., Koenig, W.A., Aydin, M. and Jung, G. 1985. Trichotoxin A40. Purification by counter-current distribution and sequencing of isolated fragments. Biochimica et Biophysica Acta 827: 51–62. DOI: 10.1016/0167-4838(85)90100-1.

Carboué, Q., Perraud-Gaime, I., Tranier, M.S. and Roussos, S. 2017. Production of microbial enzymes by solid-state fermentation for food applications. pp. 437–451. *In*: Ray, R.C., Rosell, C.M. (eds.). Microbial Enzyme Technology for Food Applications. CRC Press, Boca Raton.

Chen, L.L., Liu, L.J., Shi, M., Song, X.Y., Zheng, C.Y., Chen, X.L. and Zhang, Y.Z. 2009. Characterization and gene cloning of a novel serine protease with nematicidal activity from *Trichoderma pseudokoningii* SMF2. FEMS Microbiology Letters 299: 135–142. DOI: 10.1111/j.1574-6968.2009.01746.x.

Dalla-Santa, H.S., Sousa, N.J. and Brand, D. 2004. Conidia production of *Beauveria* sp. by solid-state fermentation for biocontrol of Ilex paraguariensis Caterpillars. Folia Microbiologica 49: 418–422. DOI: 10.1007/bf02931603.

Daniel, J.F. and Filho, E.R. 2007. Peptaibols of *Trichoderma*. Natural Product Reports 24: 1128–1141. DOI: 10.1039/b618086h.

De la Cruz Quiroz, R., Roussos, S., Hernández, D., Rodríguez, R., Castillo, F. and Aguilar, C.N. 2015. Challenges and opportunities of the bio-pesticides production by solid-state fermentation: Filamentous fungi as a model. Critical Reviews in Biotechnology 35: 326–333. DOI: 10.3109/07388551.2013.857292.

De la Cruz-Quiroz, R., Robledo-Padilla, F., Aguilar, C.N. and Roussos, S. 2017a. Forced aeration influence on the production of spores by *Trichoderma* strains. Waste and Biomass Valorization 8: 2263–2270. DOI: 10.1007/s12649-017-0045-4.

De la Cruz-Quiroz, R., Roussos, S., Hernandez-Castillo, D., Rodríguez-Herrera, R., López, L.I.L., Castillo, F. and Aguilar, C.N. 2017b. Solid-state fermentation in a bag bioreactor: Effect of corn cob mixed with phytopathogen biomass on spore and cellulase production by *Trichoderma asperellum*. Ch. 03. *In*: A.F. Jozala (ed.). Fermentation Processes. InTech. DOI: 10.5772/64643.

De La Cruz, J., Rey, M., Lorca, J.M., Hidalgo-Gallego, A., Dominguez, F., Pintor-Toro, J.A., Uobell, A. and Bentiez, T. 1992. Isolation and characterization of three chitinases from *Trichoderma harzianum*. European Journal of Biochemistry 206: 859–867. DOI: 10.1111/j.1432-1033.1992.tb16994.x.

De Marco, J.L., Valadares-Inglis, M.C. and Felix, C.R. 2002. Production of hydrolytic enzymes by *Trichoderma* isolates with antagonistic activity against *Crinipellis perniciosa,* the causal agent of witches' broom cocoa. Brazilian Journal of Microbiology 34: 33–38. DOI: 10.1590/S1517-83822003000100008.

Degenkolb, T., von Dohren, H., Nielsen, K.F., Samuels, G.J. and Bruckner, H. 2008. Recent advances and future prospects in peptaibiotics, hydrophobin, and mycotoxin research, and their importance for chemotaxonomy of *Trichoderma* and Hypocrea. Chemistry & Biodiversity 5: 671–680. DOI: 10.1002/cbdv.200890064.

Demain, A.L. and Fang, A. 2000. The natural functions of secondary metabolites. Advances in Biochemical Engineering/Biotechnology 69: 1–39. DOI: 10.1007/3-540-44964-7_1.

Druzhinina, I.S., Kopchinskiy, A.G. and Kubicek, C.P. 2006. The first 100 *Trichoderma* species characterized by molecular data. Mycoscience 47: 55–64. DOI: 10.1007/S10267-006-0279-7.

Durand, A. 2003. Bioreactor designs for solid state fermentation. Biochemical Engineering Journal 13: 113–125. DOI: 10.1016/S1369-703X(02)00124-9.

El Hajji, M., Rebuffat, S., Lecommandeur, D. and Bodo, B. 1987. Isolation and sequence determination of trichorzianines: A antifungal peptides from *Trichoderma harzianum*. International Journal of Peptide and Protein Research 29: 207–215. DOI:10.1111/j.1399-3011.1987.tb02247.x.

Elad, Y. and Kapat, A. 1999. The role of *Trichoderma harzianum* protease in the biocontrol of *Botrytis cinerea*. European Journal of Plant Pathology 105: 177–189. DOI: 10.1023/A%3A1008753629207.

Fujita, T., Wada, S., Iida, A., Nishimura, T., Kanai, M. and Toyama, N. 1994. Fungal metabolites. XIII. Isolation and structural elucidation of new peptaibols, trichodecenins-I and -II, from *Trichoderma viride*. Chemical and Pharmaceutical Bulletin 42: 489–494.

Gajera, H. P. and Vakharia, D.N. 2012. Production of lytic enzymes by Trichoderma isolates during in vitro antagonism with *Aspergillus niger*, the causal agent of collar rot of peanut. Brazilian Journal of Microbiology 43: 43–52. DOI: 10.1590/S1517-83822012000100005.

Gallos, J.K., Damianou, K.C. and Dellios, C.C. 2001. A new total synthesis of pentenomycin. Tetrahedron Letters 42: 5769–5771. DOI: 10.1016/S0040-4039(01)01099-1.

Ghisalberti, E.L. and Rowland, C.Y. 1993. Antifungal metabolites from *Trichoderma harzianum*. Journal of Natural Products 56: 1799–1804. DOI:10.1021/np50100a020.

Gromadzka, K., Waskiewicz, A., Chelkowski, J. and Golinski, P. 2008. Zearalenone and its metabolites: Occurrence, detection, toxicity and guidelines. World Mycotoxin Journal 1: 209–220. DOI: 10.3920/WMJ2008.x015.

Grondona, I., Rodríguez, A., Gómez, M.I., Rafael Camacho, R., Llobell, A. and Monte, E. 2004. TUSAL®, a commercial biocontrol formulation based on *Trichoderma*. Management of plant diseases and arthropod pests by BCAs. IOBC/wprs Bulletin 27(8): 285–288.

Hamrouni, R., Molinet, J., Dupuy, N., Taieb, N., Carboue, Q., Masmoudi, A. and Roussos, S. 2019a. The effect of aeration for 6-pentyl-alpha-pyrone, conidia and lytic enzymes production by *Trichoderma asperellum* strains grown in solid-state fermentation. Waste and Biomass Valorization 11: 5711–5720. DOI: 10.1007/s12649-019-00809-4.

Hamrouni, R., Molinet, J., Mitropoulou, G., Kourkoutas, Y., Dupuy, N., Masmoudi, A. and Roussos, S. 2019b. From flasks to single used bioreactor: Scale-up of solid-state fermentation process for metabolites and conidia production by *Trichoderma asperellum*. Journal of Environmental Management 252, 109496. DOI: 10.1016/j.jenvman.2019.109496.

Hamrouni, R., Claeys-Bruno, M., Molinet, J., Masmoudi, A., Roussos, S. and Dupuy, N. 2019c. Challenges of enzymes, conidia and 6-pentyl-alpha-pyrone production from Solid-State-Fermentation of agroindustrial wastes using experimental design and *T. asperellum* strains. Waste and Biomass Valorization 11: 5699–5710. DOI: 10.1007/s12649-019-00908-2.

Hamrouni, R., Molinet, J., Miche, L., Dupuy, N., Masmoudi, A. and Roussos, S. 2019d. Production of coconut aroma in solid-state cultivation: Screening and identification of potential strains of *Trichoderma* for 6-pentyl-alpha-pyrones and spores production. Journal of Chemistry. Article Number: 8562384. DOI: 10.1155/2019/8562384.

Harman, G.E. 2000. Myths and dogmas of biocontrol: Changes in perceptions derived from research on *Trichoderma harzianum* T-22. Plant Disease 84: 377–393. DOI: 10.1094/PDIS.2000.84.4.377.

Harman, G.E., Howell, C.R., Viterbo, A., Chet, I. and Lorito, M. 2004. *Trichoderma* species –opportunistic, avirulent plant symbionts. Nature Reviews Microbiology 2: 43–56. DOI: 10.1038/nrmicro797.

Holker, U. and Lenz, J. 2005. Solid-state fermentation are there any biotechnological advantages? Current Opinion in Microbiology 8: 301–306. DOI: 10.1016/j.mib.2005.04.006.

Howell, C.R. 2003. Mechanisms employed by *Trichoderma* species in the biological control of plant diseases: The history and evolution of current concepts. Plant Disease 87: 4–10. DOI: 10.1094/PDIS.2003.87.1.4.

Huang, Q., Tezuka, Y., Kikuchi, T., Nishi, A., Tubaki, K. and Tanaka, K. 1995. Studies on metabolites of mycoparasitic fungi. II. Metabolites of *Trichoderma koningii*. Chemical and Pharmaceutical Bulletin 43: 223–239. DOI: 10.1248/cpb.43.223.

Iida, A., Uesato, S., Shingu, T., Nagaoka, Y., Kuroda, Y. and Fujita, T. 1993. Fungal metabolites. Part 7: Solution structure of an antibiotic peptide, trichosporin B-V, from *Trichoderma polysporum*. Journal of the Chemical Society, Perkin Transactions 1(3): 375–379. DOI: 10.1039/P19930000375.

Iida, A., Sanekata, M., Fujita, T., Tanaka, H., Enoki, A., Fuse, G., Kanai, M., Rudewicz, P.J. and Tachikawa, E. 1994. Fungal metabolites. XVI: Structures of new peptaibols, trichokindins I-VII, from the fungus *Trichoderma harzianum*. Chemical and Pharmaceutical Bulletin 42: 1070–1075. DOI: 10.1248/cpb.42.1070.

Iida, A., Sanekata, M., Wada, S., Fujita, T., Tanaka, H., Enoki, A., Fuse, G., Kanai, M. and Asami, K. 1995. Fungal metabolites. XVIII: New membrane-modifying peptides, trichorozins I-IV, from the fungus *Trichoderma harzianum*. Chemical and Pharmaceutical Bulletin 43: 392–397. DOI: 10.1248/cpb.43.392.

Jackson, M.A., Dunlap, C.A. and Jaronski, S.T. 2010. Ecological considerations in producing and formulating fungal entomopathogens for use in insect biocontrol. Biological Control 55: 129–145. DOI: 10.1007/978-90-481-3966-8_10.

Jaworski, A., Kirschbaum, J. and Bruckner, H. 1999. Structures of trichovirins II, peptaibol antibiotics from the mold *Trichoderma viride* NRRL 5243. Journal of Peptide Science 5: 341–351. DOI: 10.1002/(SICI)1099-1387(199908)5:8%3C341::AID-PSC204%3E3.0.CO;2-0

Jin, X. and Custis, D. 2011. Microencapsulating aerial conidia of *Trichoderma harzianum* through spray drying at elevated temperatures. Biological Control 56: 202–208. DOI:10.1016/j.biocontrol.2010.11.008.

Kalogeris, E., Iniotaki, F., Topakas, E., Christakopoulos, P., Kekos, D. and Macris, B.J. 2003. Performance of an intermittent agitation rotating drum type bioreactor for solid-state fermentation of wheat straw. Bioresource Technology 86(3): 207–213. DOI: 10.1016/S0960-8524(02)00175-X

Keswani, C., Bisen, K., Singh, V., Sarma, B.K. and Singh, H.B. 2016. Formulation technology of biocontrol agents: Present status and future prospects. pp. 35–52. *In*: Arora, N.K., Mehnaz, S. and Balestrini, R. (eds.). Bioformulations: For Sustainable Agriculture. Springer, New Delhi.

Krause, C., Kirschbaum, J., Jung, G. and Brueckner, H. 2006. Sequence diversity of the peptaibol antibiotic suzukacillin-A from the mold *Trichoderma viride*. Journal of Peptide Science 12: 321–327. DOI: 10.1002/psc.728.

Kullnig, C.M., Krupica, T., Woo, S.L., Mach, R.L., Rey, M., Benítez, T., Lorito, M. and Kubicek, C.P. 2001. Confusion abounds over identities of *Trichoderma* biocontrol isolates. Mycological Research 105: 769–772. DOI: 10.1017/S0953756201229967.

Kumar, D.P., Singh, R.K., Anupama, P.D., Solanki, M.K., Kumar, S., Srivastava, A.K., Singhal, P.K. and Arora, D.K. 2012. Studies on Exo-Chitinase Production from *Trichoderma asperellum* UTP-16 and Its Characterization. Indian Journal of Microbiology 52: 388–395. DOI: 10.1007/s12088-011-0237-8.

Ladeira, N.C., Peixoto, V.J., Penha, M.P., de Paula Barros, E.B. and Leite, S.G.F. 2010. Optimization of 6-pentyl-alpha-pyrone production by solid state fermentation using sugarcane bagasse as residue. Bioresources 5: 2297–2306. DOI: 10.15376/biores.5.4.2297-2306.

Lehner, S.M., Atanasova, L. and Neumann, N.K. 2013. Isotope-assisted screening for iron-containing metabolites reveals high diversity among known and unknown siderophores produced by *Trichoderma* spp. Applied and Environmental Microbiology 79: 18–31. DOI: 10.1128/AEM.02339-12.

Loera-Corral, O., Porcayo-Loza, J., Montesinos-Matias, R. and Favela-Torres, E. 2016. Production of conidia by the fungus *Metarhizium anisopliae* using solid-state fermentation. Microbial-based biopesticides: Methods and protocols. pp. 61–69. *In*: Glare, T.R. and Moran-Diez, M.E. (eds.). Springer, New York. DOI: 10.1007/978-1-4939-6367-6-6.

Marcello, C.M., Steindorff, A.S., da Silva, S.P., Silva Rdo, N., Mendes Bataus, L.A. and Ulhoa, C.J. 2010. Expression analysis of the exo-beta-1,3-glucanase from the mycoparasitic fungus *Trichoderma asperellum*. Microbiological Research 165(1): 75–81. DOI: 10.1016/j.micres.2008.08.002.

Marra, R., Ambrosino, P., Carbone, V., Vinale, F., Woo, S.L., Ruocco, M., Ciliento, R., Lanzuise, S., Ferraioli, S., Soriente, I., Gigante, S., Turra, D., Fogliano, V., Scala, F. and Lorito, M. 2006. Study of the three-way interaction between *Trichoderma atroviride*, plant and fungal pathogens by using a proteomic approach. Current Genetics 50: 307–321. DOI: 10.1007/s00294-006-0091-0.

Meyer, C.E. 1966. A new antibiotic. II: Isolation and characterization. Journal of Applied Microbiology 14: 511–512.

Motta, F.L. and Santana, M.H.A. 2014. Solid-state fermentation for humic acids production by a *Trichoderma reesei* strain using an oil palm empty fruit bunch as the substrate. Biotechnology and Applied Biochemistry 172: 2205–2217. DOI: 10.1007/s12010-013-0668-2.

Muiño, B.L., Botta, E., Pérez, E., Moreno, D. and Fernández, E. 2006. Uso de *Trichoderma* como alternativa al bromuro de metilo en los cultivos protegidos, flores y ornamentales en Cuba. Fitosanidad 10(2): 179–180.

Mukherjee, M., Horwitzn, B.A., Sherkhane, P.D., Hadar, R. and Mukherjee, P.K. 2006. A secondary metabolite biosynthesis cluster in *Trichoderma virens*: Evidence from analysis of genes underexpressed in a mutant defective in morphogenesis and antibiotic production. Current Genetics 50: 193–202. DOI: 10.1007/s00294-006-0075-0.

Mukherjee, P.K. and Kenerley, C.M. 2010. Regulation of morphogenesis and biocontrol properties in *Trichoderma virens* by a VELVET protein, Vel 1. Applied and Environmental Microbiology 76: 2345–2352. DOI: 10.1128/AEM.02391-09.

Nakazawa, H., Okada, K., Onodera, T., Ogasawara, W., Okada, H. and Morikawa, Y. 2009. Directed evolution of endoglucanase III (Cel12A) from *Trichoderma reesei*. Applied Microbiology and Biotechnology 83: 649–657. DOI: 10.1007/s00253-009-1901-3.

New, A.P., Eckers, C., Haskins, N.J., Neville, W.A., Elson, S., Hueso-Rodriguez, J.A. and Rivera-Sagredo, A. 1996. Structures of polysporins A-D, four new peptaibols isolated from *Trichoderma polysporum*. Tetrahedron Letters 37: 3039–3042. DOI: 10.1016/0040-4039(96)00463-7.

Oda, S., Isshiki, K. and Ohashi, S. 2009. Production of 6-pentyl-α-pyrone with *T. atroviride* and its mutant a novel extractive liquid-surface immobilization (EXT-LSI) system. Process Biochemistry 44: 625–630. DOI: 10.1016/j.procbio.2009.01.017.

Oh, S.U., Lee, S.J., Kim, J.H. and Yoo, I.D. 2000. Structural elucidation of new antibiotic peptides, atroviridins A, B and C from *Trichoderma atroviride*. Tetrahedron Letters 41: 61–64. DOI: 10.1016/S0040-4039(99)02000-6.

Olson, S. 2015. An analysis of the biopesticide market now and where it is going. Outlooks on Pest Management 26(5): 203–206. DOI: 10.1564/v26_oct_04.

Pandey, A. 2003. Solid-State Fermentation. Biochemical Engineering Journal 13: 81–84. DOI: 10.1016/S1369-703X(02)00121-3.

Pandey, A., Soccol, C.R. and Larroche, C. 2008. Current Developments in Solid State Fermentation. Springer, Asiatech Publishers Inc. DOI: 10.1016/j.bej.2013.10.013.

Persoon, C.H. 1794. Disposita methodica fungorum. Römer's Neues Mag Bot 1: 81–128.

Raimbault, M. and Alazard, D. 1980. Culture method to study fungal growth in solid fermentation. Applied Microbiology and Biotechnology 9: 199–209. DOI: 10.1007/BF00504486.

Rebuffat, S., Goulard, C. and Bodo, B. 1995. Antibiotic peptides from *Trichoderma harzianum:* Harzianins HC, proline-rich 14-residue peptaibols. Journal of the Chemical Society, Perkin Transactions 1(14): 1849–1855. DOI: 10.1039/P19950001849.

Reino, J.L., Guerrero, R.F., Hernández-Galán, R. and. Collado, I.G. 2008. Secondary metabolites from species of the biocontrol agent *Trichoderma*. Phytochemistry Reviews 7: 89–123. DOI: 10.1007/s11101-006-9032-2.

Reithner, B., Brunner, K. and Schuhmacher, R. 2005. The G protein alpha subunit Tga1 of *Trichoderma atroviride* is involved in chitinase formation and differential production of antifungal metabolites. Fungal Genetics and Biology 42: 749–760. DOI: 10.1016/j.fgb.2005.04.009.

Renshaw, J.C., Robson, G.D. and Trinci, A.P.J. 2002. Fungal siderophores: Structures, functions and applications. Mycological Research 106: 1123-1142. DOI: 10.1017/s0953756202006548.

Rifai, M.A. 1969. A revision of the genus *Trichoderma*. Mycol Pap 116: 1–56. DOI: 10.1139/b91-298.

Rocha-Valadez, J.A., Estrada, M., Galindo, E. and Serrano-Carreon, L. 2006. From shake flasks to stirred fermentors: Scale-up of an extractive fermentation process for 6-pentyl-α-pyrone production by *Trichoderma harzianum* using volumetric power input. Process Biochemistry 41: 1347–1352. DOI: 10.1016/j.procbio.2006.01.013.

Rodríguez-Fernández, D.E., Rodríguez-León, J.A., De Carvalho, J.C., Karp, S.G., Sturm, W., Parada, J.L. and Soccol, C.R. 2012. Influence of air flow intensity on phytase production by solid-state fermentation. Bioresource Technology 18: 603–606. DOI.org/10.1016/j.biortech.2012.05.032

Rodríguez-Urra, A.B., Jiménez, C., Nieto, M.I., Rodríguez, J., Hayashi, H. and Ugalde, U. 2012. Signaling the induction of sporulation involves the interaction of two secondary metabolites in *Aspergillus nidulans*. ACS Chemical Biology 7: 599–606. DOI: 10.1021/cb200455u.

Roussos, S. and Raimbault, M. 1982. Cellulose hydrolysis by fungi. 2: Cellulase production by *Trichoderma harzianum* in liquid-medium fermentation. Annales de Microbiologie B133: 465–474.

Roussos, S. 1985. Croissance de *Trichoderma harzianum* par Fermentation en Milieu Solide: Physiologie, Sporulation et Production de Cellulases. Thèse de doctorat d'Etat, Université de Provence, Marseille, p. 188.

Roussos, S., Lonsane, B.K., Raimbault, M. and Viniegra Gonzalez, G. 1997. Advances in Solid State Fermentation. Kluwer Academic Publishers, Dordrecht, p. 631.

Roussos, S., Tranier, M.S. and Hamrouni, R. 2020. Biocontrôle: Pour une agriculture compétitive et durable. Techniques de l'Ingénieur BIO 2500: 1–19.

Rubio, M.B., Hermosa, R., Reino, J.L., Collado, I.G. and Monte, E. 2009. Thctf1 transcription factor of *Trichoderma harzianum* is involved in 6-pentyl-2H-pyran-2-one production and antifungal activity. Fungal Genetics and Biology 46: 17–27. DOI: 10.1016/j.fgb.2008.10.008.

Ruffner, B., Pechy-Tarr, M., Hofte, M., Bloemberg, G., Grunder, J., Keel, C. and Maurhofer, M. 2015. Evolutionary patchwork of an insecticidal toxin shared between plant-associated pseudomonads and the insect pathogens *Photorhabdus* and *Xenorhabdus*. BMC Genomics 16: 609. DOI: 10.1186/s12864-015-1763-2.

Ruocco, M., Lanzuise, S., Vinale, F., Marra, R., Turra, D., Woo, S.L. and Lorito, M. 2009. Identification of a new biocontrol gene in *Trichoderma atroviride*: The role of an ABC transporter membrane pump in the interaction with different plant–pathogenic fungi. Molecular Plant-Microbe Interactions 22: 291–301. DOI: 10.1094/MPMI-22-3-0291.

Saravanakumar, K., Yu, C., Dou, K., Wang, M., Li, Y. and Chen, J. 2016. Synergistic effect of *Trichoderma*-derived antifungal metabolites and cell wall degrading enzymes on enhanced biocontrol of *Fusarium oxysporum* f. sp. *Cucumerinum*. Biological Control 94: 37–46. DOI: 10.1016/j.biocontrol.2015.12.001.

Sarhy-Bagnon, V., Lozano, P., Saucedo-Castañeda, G. and Roussos, S. 2000. Production of 6-pentyl-α-pyrone by *Trichoderma harzianum* in liquid and solid-state cultures. Process Biochemistry 36(1-2): 103–109. DOI: 10.1016/S0032-9592(00)00184-9.

Shah, P.A. and Pell, J.K. 2003. Entomopathogenic fungi as biological control agents. Applied Microbiology and Biotechnology (5-6): 413–423. DOI: 10.1007/s00253-003-1240-8.

Sharma, P. and Pandey, R. 2009. Biological control of root-knot nematode; *Meloidogyne incognita* in the medicinal plant; *Withania somnífera* and the effect of biocontrol agents on plant growth. African Journal of Agricultural Research 4(6): 564–567. DOI: 10.1080/09583157.2010.487935.

Solanki, M.K., Singh, N., Singh, R.K., Singh, P., Srivastava, A.K., Kumar, S., Kashyap, P.L. and Arora, D.K. 2010. Plant defense activation and management of tomato root rot by a chitin-fortified *Trichoderma/Hypocrea* formulation. Phytoparasitica. DOI: 10.1007/s12600-011- 0188-y.

Stoppacher, N., Kluger, B., Zeilinger, S., Krska, R. and Schuhmacher, R. 2010. Identification and profiling of volatile metabolites of the biocontrol fungus *Trichoderma atroviride* by HS-SPME-GCMS. Journal of Microbiological Methods 81(2): 187–193. DOI: 10.1016/j.mimet.2010.03.011.

Tamás, M., Chetna, T., Gordana, R., Dávid, R., András, S., Csaba, V. and László, K. 2018. New 19-Residue Peptaibols from *Trichoderma* Clade *Viride*. Microorganisms 6: 85. DOI: 10.3390/microorganisms6030085.

Tamura, A., Kotani, H. and Naruto, S. 1975. Trichoviridin and dermadin from *Trichoderma* sp. TK-1. Journal of Antibiotics 28: 161–162. DOI: 10.7164/antibiotics.28.161.

Tanaka, T., Fukui, T. and Imanaka, T. 2001. Different cleavage specificities of the dual catalytic domains in chitinase from the hyperthermophilic archaeon *Thermococcus kodakaraensis* KOD1, Journal of Biological Chemistry 276: 35629–35635. DOI: 10.1074/jbc.M105919200.

Thines, E., Anke, H. and Weber, R.W. 2004. Fungal secondary metabolites as inhibitors of infection-related morphogenesis in phytopathogenic fungi. Mycological Research 108: 14–25. DOI: 10.1017/S0953756203008943.

Thomas, L., Larroche, C. and Pandey, A. 2013. Current developments in solid-state fermentation. Biochemical Engineering Journal 81: 146–161. DOI: 10.1016/j.bej.2013.10.013.

Tisch, D. and Schmoll, M. 2010. Light regulation of metabolic pathways in fungi. Applied Microbiology and Biotechnology 85: 1259–1277. DOI: 10.1007/s00253-009-2320-1.

Tranier, M.S., Pognant-Gros, J., Quiroz, R.D.L.C., González, C.N.A., Mateille, T. and Roussos, S. 2014. Commercial biological control agents targeted against plant-parasitic root-knot nematodes. Brazilian Archives of Biology and Technology 57: 831–841. DOI: 10.1590/S1516-8913201402540.

Van Breukelen, F.R., Haemers, S., Wijffels, R.H. and Rinzema, A. 2011. Bioreactor and substrate selection for solid-state cultivation of the malaria mosquito control agent *Metarhizium anisopliae*. Process Biochemistry 46: 751–757. DOI: 10.1016/j.procbio.2010.11.023.

Verma, M., Brar, S.K. and Tyagi, R.D. 2007. Antagonistic fungi, *Trichoderma* spp: panoply of biological control. Biochemical Engineering Journal 37: 1–20. DOI: 10.1016/j.bej.2007.05.012.

Viccini, G., Mannich, M. and Fontana-Capalbo, D.M. 2007. Spore production in solid-state fermentation of rice by *Clonostachys rosea*: A biopesticide for gray mold of strawberries. Process Biochemistry 42: 275–278. DOI: 10.1016/j.procbio.2006.07.006.

Vinale, F., Sivasithamparam, K., Ghisalberti, E.L., Woo, S.L., Nigro, M., Marra, R. and Manganiello, G. 2014. *Trichoderma* secondary metabolites active on plants and fungal pathogens. The Open Mycology Journal 8: 127–139. DOI: 10.2174/1874437001408010127.

Vinale, F., Flematti, G. and Sivasithamparam, K. 2009. Harzianic acid, an antifungal and plant growth promoting metabolite from *Trichoderma harzianum*. Journal of Natural Products 72: 2032–5. DOI: 10.1021/np900548p.

Vinale, F., Marra, R., Scala, F., Ghisalberti, E.L., Lorito, M. and Sivasithamparam, K. 2006. Major secondary metabolites produced by two commercial *Trichoderma* strains active against different phytopathogens. Letters in Applied Microbiology 43: 143–148. DOI: 10.1111/j.1472-765X.2006.01939.x.

Walter, M., Boyd-Wilson, K.S.H., Perry, J.H. and Hill, R.A. 2000. *Botrytis* tolerance to 6-pentyl-alpha-pyrone and massoialactone. New Zealand Plant Protection 53: 375–381. DOI: 10.30843/nzpp.2000.53.3612.

Weindling, R. 1932. *Trichoderma lignorum* as a parasite of other soil fungi. Phytopathology 22: 837–845. DOI: 10.4236/oalib.1101706.

Yang, L., Li, G.Q. and Long, Y.Q. 2010. Effects of soil temperature and moisture on survival of *Coniothyrium minitans*. Biological Control 55: 27–33. DOI: 10.1016/j.biocontrol.2010.06.010.

Yano, A., Suzuki, K., Uchimiya, H. and Shinshi, H. 1998. Induction of hypersensitive cell death induced by a fungal protein in cultures of tobacco cells. Molecular Plant-Microbe Interactions 11: 115–123. DOI: ORG/10.1094/MPMI.1998.11.2.115.

Yedidia, I.I., Benhamou, N. and Chet, II. 1999. Induction of defense responses in cucumber plants (*Cucumis sativus* L.) by the biocontrol agent *Trichoderma harzianum*. Applied and Environmental Microbiology 65: 1061–1070. DOI: 10.1128/AEM.65.3.

Yedidia, I., Shoresh, M., Kerem, Z., Benhamou, N., Kapulnik, Y. and Chet, I. 2003. Concomitant induction of systemic resistance to pseudomonas syringae pv. lachrymans in cucumber by *Trichoderma asperellum* (T-203) and accumulation of phytoalexins. Applied and Environmental Microbiology 69: 7343–7353. DOI: 10.1128/AEM.69.12.

Zachow, C., Berg, C., Müller, H., Monk, J. and Berg, G. 2016. Endemic plants harbour specific *Trichoderma* communities with an exceptional potential for biocontrol of phytopathogens. Journal of Biotechnology 235: 162–170. DOI: 10.1016/j.jbiotec.2016.03.049.

Zeilinger, S. and Omann, M. 2007. *Trichoderma* biocontrol: Signal transduction pathways involved in host sensing and mycoparasitism. Gene Regulation and Systems Biology 1: 227–234. DOI: 10.4137/GRSB.S397.

CHAPTER

2

Food Processing and Decontamination Approaches to Control Mycotoxins

Oluwatosin A. Ijabadeniyi*, Titilayo A. Ajayeoba and Omotola F. Olagunju

Department of Biotechnology and Food Science
Durban University of Technology, South Africa

1. Introduction

Mycotoxigenic fungi, majorly from *Aspergillus, Penicillium* and *Fusarium* genera, contaminate agricultural commodities and have the potential to produce harmful secondary metabolites known as mycotoxins, which are detrimental to the health of humans and animals (Achaglinkame et al. 2017). The growth of mycotoxigenic fungi and subsequent accumulation of mycotoxins in food commodities are influenced by several factors, including high temperature and high relative humidity conditions, poor harvesting practices, improper drying, handling, packaging, storage, and transportation (Bhat et al. 2010). A wide range of food and food products have been reported as an excellent substrate for the growth of mycotoxicological pathogens. In addition, mycotoxin contamination of beverages, especially those made from tropical products have been reported (Granados-Chinchilla et al. 2018).

Mycotoxins are heat stable and not significantly volatile, and are possibly carried over into processed foods due to their stability (Bhat et al. 2010, Anfossi et al. 2016). For example, under most conditions of storage, handling and processing of foods or feeds, aflatoxins remain extremely durable (Vijayanandraj et al. 2014). The subsequent ingestion in high quantities or over a long period of time of mycotoxins from mold-contaminated foods may become harmful, with deleterious effects such as growth faltering, intestinal dysfunction, hemorrhage, cancers, and immune system suppression (Marroquín-Cardona et al. 2014, Okeke et al. 2018). The emerging mycotoxin prevalence in agricultural food commodities have become a concern due to their vast incidences in several food commodities. However, some of these are

*Corresponding author: oluwatosini@dut.ac.za

significant to the health of humans and animals and have a higher prevalence in food matrices. Mycotoxins consist mainly of aflatoxins (AFs), ochratoxin A (OTA), trichothecenes (deoxynivalenol (DON) and T-2 toxin), zearalenone (ZEN), fumonisins (FBs) and patulin (PAT) (Dalié et al. 2010, Afsah-Hejri et al. 2013, Ashiq 2015) and frequency of occurrence in food systems and severity on humans and animals have high economic impact. These mycotoxins as well as ergot alkaloids produced mainly by fungi belonging to *Aspergillus, Penicillium, Fusarium* and *Claviceps* generally pose a threat to food safety (Abrunhosa et al. 2016).

The level of contamination depends on the raw material used and continuous surveillance is encouraged, especially in low-income countries, where there are no standard regulations for toxin analysis. Under appropriate tropical conditions, harvest and post-harvest fungi colonization and mycotoxin production are feasible. This is particularly crucial for beverages made from tropical products such as different varieties of tea, coffee, cocoa, and fruits. The continuous demand of consumers for safe quality processed foods has necessitated the technological advances in food processing to reduce/ control the mycotoxin in foods because the consumption of these mycotoxin-contaminated foods can cause severe toxic effects on human and animal health due to their mutagenicity, teratogenicity, carcinogenicity, nephrotoxicity and immunosuppression (Man et al. 2017).

Due to the stable nature of mycotoxins, it is almost practically impossible to eliminate them from processed foods. Various pre-harvest and post-harvest techniques have therefore been explored as decontamination strategies to control contamination in the field and during storage. Pre-harvest measures include growing resistant crop varieties, crop rotation, soil tillage, chemical and biological control of plant diseases, and insect control (Karlovsky et al. 2016). Unfortunately, pre-harvest measures do not guarantee the absence of mycotoxins in food or feed as fungi are ubiquitous in nature. There is therefore a need for decontamination methods for food products that could be safe, effective, environmentally friendly and presenting as a cost-benefit (Corassin et al. 2013). Effective decontamination should be irreversible, modified forms of mycotoxins should be affected together along with parent compounds, the products should be non-toxic, and the food should retain its nutritive value and remain palatable (Karlovsky et al. 2016).

Post-harvest food processing technologies are increasingly being investigated for their ability to reduce aflatoxins in contaminated products (Mutungi et al. 2008), and in general, reduce mycotoxins. Processing of agricultural products into semi-finished or finished products contributes to increased shelf life, stability of colour and flavour, improved bio-accessibility of nutrients, increased economic value and facilitates the preparation of raw food ingredients (Decker et al. 2014). Mycotoxins in foods may be reduced by many commonly used processing methods, for example, the processing of cereals usually reduces mycotoxin contamination in the final product (Pascale et al. 2011, Ademola et al. 2018). Various studies have also reported the application of different food processing methods such as milling, roasting, cooking, frying, etc., to reduce mold and mycotoxin contamination in foods (Siwela et al. 2005, Yazdanpanah et al. 2005, Bullerman and Bianchini 2007, Matumba et al. 2009, Assohoun et al. 2013, Matumba et al. 2015). However, the effect of processing as a decontamination strategy is influenced by the initial condition of the grain (size distribution and moisture content), the combined harvester settings, the type and extent of contamination, and the cleaning/sorting method (Tibola et al. 2016).

Bio-preservation is a method of preserving food products that is based on the principle of employing the use of one organism to control another. In recent years, it has received much attention for the control of spoilage in foods (Dalié et al. 2010), as consumers demand for safe products devoid of chemical preservatives, while also offering good shelf life (Roger et al. 2015). Lactic acid bacteria belonging to the genera *Lactococcus, Lactobacillus, Leuconostoc* and *Pediococcus*, are autochthonous in food systems, and are traditionally used in food and feed as preservative agents, to prevent spoilage and extend shelf life (Dalié et al. 2010, Roger et al. 2015, Ananthi et al. 2016). The use of chemical compounds such as ammonia, hydrogen peroxide, sodium bisulfite, ozone and organic acids, and aqueous plant extracts have also been reported in mycotoxin decontamination strategies. This chapter takes a critical look at the applications of various food processing operations that can effectively decontaminate mycotoxins in foods.

2. Physical Decontamination Methods

Physical decontamination methods of mycotoxins in foods include sieve-cleaning, sorting, milling, steeping, and extrusion (Karlovsky et al. 2016). Cheli et al. (2013) documented some cleaning/sorting operations that have been used to achieve a reduction in mycotoxin content in wheat, which includes sifting, optical sorting, sieving, scouring and polishing. Other physical methods for mycotoxin decontamination include mechanical sorting and separation, thermal inactivation, density segregation, irradiation, ultrasound and adsorption (Yazdanpanah et al. 2005). These methods are largely being utilized as decontamination strategies, as physical decontamination processes do not produce toxic products (Matumba et al. 2015).

2.1 Cleaning

Cleaning is a unit operation in which contaminating materials such as dirt, debris, fragmented grains and grains showing fungal development are removed from the grain lot, and thereby reduces mycotoxin contamination (Pacin and Resnik 2012, Ismail et al. 2018). For example, raw maize may be contaminated with stones, metals and microbial products such as fungi and mycotoxins (Pacin and Resnik 2012). Prior to milling, sorting and cleaning may be used to reduce mycotoxin contamination in wheat as these processes remove kernels with extensive mold growth, broken kernels, fine materials and dust in which most of the toxins accumulate (Cheli et al. 2013). In the cleaning of wheat, shape, size, relative density and air resistance are some properties of wheat kernels that may determine the selection of the cleaning equipment (Cheli et al. 2013).

The nature of product to be cleaned and the type of contaminants to remove define the cleaning procedure to adopt. Basically, cleaning can be on a dry basis which is commonly used for grains with low moisture content but with great mechanical strength and includes separation by air, magnetism, or physical methods. Wet cleaning include soaking, spraying, floatation washing, and ultrasonic cleaning (Pacin and Resnik 2012). Dry cleaning employs the use of classifiers, magnetic separators, and separators based on screening (Pacin and Resnik 2012).

Sieve-cleaning is a mechanical process that involves the use of a net made from metal wire with well-defined spacing between the wires (sieve). Only particles with a dimension smaller than the openings can pass through. Mycotoxins are usually concentrated in the bran and outer layers of grains and are present at reduced levels in the endosperm, thus the sieving process reduces mycotoxin contamination (Pacin and Resnik 2012).

Pacin and Resnik (2012) tested the efficiency of sieve-cleaning to remove aflatoxins B_1, B_2, G_1 and G_2, zearalenone, deoxynivalenol and fumonisins B_1, B_2 and B_3 in maize with seven sieves with sizes ranging between 10 to 14 mm. Although the levels of contamination by aflatoxin and zearalenone in the original maize samples was low, mean levels of aflatoxin and zearalenone were lower in cleaned maize (over sieve) than in uncleaned maize. Fumonisin detected in all the samples was at a higher level in unsieved maize samples than in cleaned maize. Concentration of fumonisin reduced from 7,847.2 µg/kg in uncleaned maize (using 7-mm sieve) to 3,638.1 µg/kg in the clean fraction. Similarly, Pascale et al. (2011) reported a significant reduction of toxins (T-2 and HT-2) after cleaning using a sifter that removed foreign seeds, broken seeds, and glumes. Ten durum wheat samples that were either artificially inoculated (nine samples) or naturally contaminated (one sample) were used in the experiment. T-2 toxin present in uncleaned wheat ranged from 35-785 µg/kg and in cleaned wheat at a lower concentration of 13-184 µg/kg.

2.2 Sorting

Following harvesting, the primary processing of agricultural goods may involve sorting, washing or milling. Sorting originally employed centrifugation force and floatation in air flow before the advent of optical sorting (Karlovsky et al. 2016). Due to the heterogenous nature of aflatoxin contamination, separating damaged kernels effectively reduces the contamination. Grain sorting using UV light illumination is commonly used. It operates based on the development of a bright greenish-yellow fluorescence, generated not from aflatoxin but from the reaction with endogenous peroxidase (Karlovsky et al. 2016). This method, however, is not entirely reliable. For example, in dried commodities peroxidase is inactivated, leading to false positive and false negative results. The application of optical sorting in decontamination of mycotoxin where there are no visible symptoms may however not be effective. For example, Mutiga et al. (2014) reported that there was no reduction of aflatoxin content by sorting whereas Pearson et al. (2004) recorded some success with the method.

Although sorting and cleaning may reduce mycotoxin concentration in commodities, contamination may not be completely removed. The initial condition of the grain, the type and extent of the contamination, and the type of cleaning processes affect the cleaning efficiency (Bullerman and Bianchini 2007, Cheli et al. 2013).

2.3 Dehulling

Dehulling of grains as a form of milling involves the use of abrasion in the removal of the outer layer of grains (Siwela et al. 2005). The hull, underlying aleurone layer and part of the germ, which are the more highly contaminated parts of the grain, are removed during dehulling (Mutungi et al. 2008). Its successful application has been based on the limitation of fungal colonization and accumulation of mycotoxin to the surface layers of kernels (Karlovsky et al. 2016). Matumba et al. (2009) reported about 29% reduction in AFB_1 in maize by removal of bran. In a similar study, Siwela et al. (2005) reported about 89 and 91% reduction in AFB_1 and total aflatoxin, respectively in maize after dehulling, with a higher level of aflatoxins in the bran, than the grits. Mutungi et al. (2008) also reported 57.3 ng/g mean reduction in aflatoxin after dehulling contaminated maize grains. While dehulling may portend an effective decontamination strategy, the complete removal of the bran in grains may not be achievable except where sophisticated commercial dehulling systems are utilized (Matumba et al. 2009). Furthermore, the removal of the bran is independent of the mycotoxin contamination level in the grain. Matumba et al. (2009) did not obtain a significant relationship between initial AFB_1 concentration and AFB_1 percentage reduction during dehulling of maize. There is also the possibility of nutrient loss as a result of dehulling. Moeser et al. (2002) reported a loss of fibres and some essential amino acids during dehulling. In the application of dehulling as a mycotoxin decontamination process, the efficiency of the dehuller and extent of fungal and mycotoxin penetration, which may lead to differences in results, should be considered (Matumba et al. 2009).

2.4 Heat Treatment (Roasting and Extrusion)

Thermal inactivation of foods for mycotoxin decontamination include heating, extrusion and microwaving (Ismail et al. 2018). In the application of thermal treatment for mycotoxin decontamination, the efficiency of the process generally depends on the initial level of contamination, heating temperature, time of exposure, type and the moisture content of the food (Yazdanpanah et al. 2005, Ismail et al. 2018). Heat treatment as a mycotoxin mitigatory measure has recorded success in food matrices. Rastegar et al. (2017) reported about 50.2% reduction in AFB_1 content of naturally contaminated (contamination level of 268 ng/g) pistachio nuts roasted at 120°C for 60 min, after treatment with lemon juice and stored for four days at room temperature. In the same study, pistachio nuts contaminated with 383 ng/g of AFB_1 showed 76.6-93.1% reduction in AFB_1 concentration after the nuts were treated with water, lemon juice and citric acid and then roasted at 120°C for 60 min. Roasting at 150°C for 30 min reduced AFB_1 concentration in naturally contaminated whole pistachio kernels to about 63% of the initial level (Yazdanpanah et al. 2005).

In the application of heat treatment to reduce mycotoxins in foods, the temperature and time regime is of importance. Usually, a high temperature treatment (about 150°C) will be required as reported by these studies. Although at such high temperature, degradation of toxins occurs, there may also be undesirable changes in the physical and nutritional qualities of the food so treated. According to Yazdanpanah et al. (2005), roasting of pistachio nuts at 150°C achieved considerable reduction of aflatoxin, but the physical appearance of the nuts was compromised, resulting in 'burned' nuts with impaired taste. It also appears that decrease in mycotoxin concentration by application of heat may also be dependent on the initial concentration of mycotoxin in the food sample. In the same study, Yazdanpanah et al. (2005) recorded AFB_1 residual concentration of 19% in naturally contaminated pistachio samples with an initial concentration of 235 ppb roasted at 150°C for 30 min. With a similarly treated sample having an initial concentration of 144 ppb AFB_1, a residual AFB_1 of 63% was recorded. This has been attributed to the availability of more active materials for destruction during roasting (Yazdanpanah et al. 2005).

There is also a possibility of resistance to heat by mycotoxins naturally present in a food substance compared to artificially introduced mycotoxins. Yazdanpanah et al. (2005) also reported that even after roasting pistachio nuts at 150°C, the level of aflatoxin remaining in naturally contaminated pistachio nuts was higher compared to the artificially contaminated samples. Aflatoxins occurring naturally in food systems, have often proved more refractory to degradation treatments than artificially spiked aflatoxins because they are more bound to the food macromolecules and are protected from degradation (Saalia and Phillips 2011). Another consideration in the application of heat is the formation of toxic decomposition products and the loss of heat-labile nutrients in the food matrices.

Extrusion, a high temperature-short time processing, used in the manufacture of breakfast cereals, snack and textured foods, and animal feedstuffs may involve very high temperatures (> 150°C), high pressure, humidity and severe shear forces (Bullerman and Bianchini 2007, Zheng et al. 2015, Ismail et al. 2018). The technology involves pushing a granular food material down a heated barrel and through an orifice by a rotating, tight fitting Archimedean screw (Saalia and Phillips 2011). A combination of frictional, compressive and pressure forces as well as shear forces created by the rotating action of the screws, enable rapid cooking and transforming of the food into visco-elastic melt (Saalia and Phillips 2011). During extrusion, raw materials are modified, new shapes and structures are formed and there is development of different functional and nutritional characteristics (Ismail et al. 2018).

In addition to modifying the texture and increasing the digestibility of the processed product, extrusion cooking can also affect the mycotoxin content of extruded products (Zheng et al. 2015). Several factors govern mycotoxin reduction in extruded products including extruder temperature, screw speed, moisture content of the extrusion mixture, and residence time in the extruder (Bullerman and Bianchini 2007). Using a barrel temperature of 150°C, material moisture of 40 g/100 g, feed rate of 17 g/min, and screw speed of 152 rpm, Zheng et al. (2015) reported about 77.6% degradation of AFB_1 in naturally contaminated defatted peanut meal. The barrel temperature and material moisture significantly influenced the rate of degradation of AFB_1. A reduction of 59% aflatoxins was also recorded in naturally contaminated peanut meal and 91% reduction in artificially spiked samples after extrusion using a die of 2.5 mm internal diameter (Saalia and Phillips 2011).

In addition to high temperature and high shear required in the application of extrusion cooking to inactivate or decontaminate aflatoxins, a right pH is also essential (Saalia and Phillips 2011). Since most nutrients in foods are heat sensitive and can be denatured by high temperature, the nutritional quality of the food products manufactured under this condition may become compromised. Protein cross-linking, isopeptide bonding and amino acid racemization are some transformations that may have an impact on the nutritional quality of these products, leading to a reduction in protein digestibility and amino acid bioavailability (Saalia and Phillips 2011). Alkaline treatment, which aims at increasing pH, may also be applied during extrusion to facilitate aflatoxin removal from foods, also leading to amino acid racemization and lysinoalanine formation (Saalia and Phillips 2011).

2.5 Steeping/Washing

Steeping involves the soaking of grains in water. The soluble nature of some mycotoxins in water has been used efficiently in their reduction from the surface of grains. Alkaline solutions such as sodium carbonate has been used for mycotoxins that are not soluble in water such as ZEN. As a mycotoxin decontamination method, steeping or washing enables water-soluble toxins to migrate from grains to steeping water and facilitate mycotoxin reduction (Ademola et al. 2018). Ademola et al. (2018) observed the steeping pattern of ogi processors from Ibadan, Lagos, and Abeokuta, Nigeria, and the effect of steeping on aflatoxin reduction. Steeping was carried out for a period of 2 to 3 days. The study reported that there was no statistical difference in the mean level of aflatoxin reduction due to the length of steeping in Ibadan and Lagos, but reported a statistically different mean reduction for only AFG_1 among processors in Abeokuta who steeped for two days and those who steeped for three days.

3. Microbial and Enzymatic Processing Methods (Fermentation, Sprouting and Binding with LAB)

The use of microorganisms as mycotoxin decontamination agents may be attributed to their specificity, high efficiency and cost-effectiveness, where the organisms do not release undesirable compounds into the food matrix or utilize the food material for growth (Ismail et al. 2018).

Lactic acid bacteria (LAB) including *Lactococcus, Lactobacillus, Leuconostoc* and *Pediococcus*, are naturally present in food systems and are part of the human diet (Dalié et al. 2010). They are traditionally used in food and feed as preservative agents to prevent spoilage and extend shelf life. They are certified safe for use in food applications and many species have been granted the status of Generally Regarded as Safe and Qualified Presumption of Safety by the Food and Agriculture Organization of the United Nations and the European Food Safety Authority, respectively (Ananthi et al. 2016). LAB are able to produce lactic acid during the fermentation of different types of sugars (Ismail et al. 2018). *Lactobacillus* and *Bifidobacterium* are used in fermented foods such as cheese, meat and cereals as preserving agents (Ismail et al. 2018).

In a study by Oluwafemi and Da-Silva (2009), the activity of LAB to reduce aflatoxin was observed during the fermentation of maize to produce Ogi. Over the 16 h fermentation period, a progressive and significant reduction of aflatoxins from 80 to 4 ng/g by heat-treated *L. plantarum* was recorded as fermentation progressed. This study also showed that binding of aflatoxins by LAB is strain-specific, as *L. plantarum* bound more aflatoxins than *L. brevis, L. acidophilus, L. casei* and *L. deibruekii*. Kachouri et al. (2014) reported a decrease in AFB_1 level from 11 to 5.9 µg/kg at day 0 in naturally contaminated olives inoculated with *L. plantarum*. Assohoun et al. (2013) also reported a reduction in AFB_1 concentration (about 80%) in the processing of Doklu, a fermented maize-based food. Roger et al. (2015) similarly reported a decrease in concentration of AFB_1 in 'Kutukutu' fermented with *L. fermentum* N33 (21.8%), *L. brevis* G25 (19.1%) after 24 h of incubation, and with *L. buchneri* M11 (64.2%) and *L. brevis* G25 (63%) after 120 h. The action of three LAB isolated from curd, *S. lactis, S. cremoris* and *L. acidophilus* was reported by Ananthi et al. (2016). The strains were able to degrade AFB_1 within 24 h of incubation and achieved 54-94% reduction after fermentation. Total AFB_1 reduction was most effective by *S. lactis*, followed by *S. cremoris* and *L. acidophilus*. In another study by Kachouri et al. (2014), naturally contaminated olives harvested in Tunisia for the 2009 and 2010 seasons were inoculated with *L. plantarum* and stored for 16 days at ambient temperature. That study revealed a reduction in AFB_1 level from 11 to 5.9 µg/kg at day 0, with a complete removal of AFB_1 during storage.

Various studies have reported the ability of LAB to reduce mycotoxins, particularly aflatoxins. However, the mechanism of aflatoxin removal by LAB remains unclear. It has been suggested to depend on fractions of the cell wall skeletons of LAB, which are able to bind with mutagens through non-covalent bonds (Roger et al. 2015). This theory has been supported by other researchers.

According to Bueno et al. (2007), in vitro removal of AFB_1 by LAB occurs via adhesion to cell wall components rather than covalent binding or metabolic degradation, and probably involves weak Van der Waals bonds, hydrogen bonds or hydrophobic interactions. The cell wall of LAB such as *Leuconostoc* and *Streptococcus* are able to bind mutagens, which include amino acid pyrolysates and heterocyclic amino acids produced during cooking (Dalié et al. 2010). Another theory suggests that LAB fermentation opens the AFB_1 lactone ring resulting in its complete detoxification (Roger et al. 2015).

While the use of starter cultures of various LAB has been used in mycotoxin decontamination, the development of the natural mycobiota of food matrices, via fermentation have also been effective in decontamination of mycotoxins. For example, the effectiveness of microbial decontamination processes has been demonstrated in the production of Ogi, a maize-gruel which serves as a complementary food for infants and young children. It is convenient for the sick, convalescent and elderly, and provides a quick breakfast for low-income rural dwellers (Okeke et al. 2015). In a study by Ademola et al. (2018), a decrease in mean total aflatoxin level was observed in maize samples collected in Ibadan from 9.10 μg/kg to 5.55 μg/kg in the 'ogi' samples after fermentation. A similar reduction was observed in maize samples from Abeokuta with a mean aflatoxin reduction from 18.35 μg/kg to 5.62 after fermentation. The same study also reported a decrease in the mean total fumonisin level in maize samples collected from Ibadan from 495 μg/kg to 187.50 μg/kg in the fermented product (ogi), showing that lactic acid fermentation significantly lowered the level of fumonisin in the maize samples. In a study by Okeke et al. (2015), steeping/fermentation of maize for 48 h (fermentation) reduced fumonisin (FB_1, FB_2 and FB_3) in maize varieties (white maize: 65.7-80.1%; yellow maize: 78.7-88.8%) as well as aflatoxin (AFB_1: 60.8%; AFB_2: 82.8%) and ZEN (99.2%) in the yellow variety. A remarkable reduction of all mycotoxins especially after 48 h of steeping was achieved by fermentation, mediated by the natural maize flora (Okeke et al. 2015). Significant reductions were observed in kunu-zaki, a fermented non-alcoholic beverage from maize, millet or sorghum as single or mixed grains. The concentrations of DON, FB_1, total FBs, ZEN and some other non-legislated mycotoxins were drastically reduced in the fermented drink. For example, 98.9% reduction was noted in DON, 99.4% in FB_1, 99.5% in tFB, 98.3% in MON and 76.2% in ZEN (Ezekiel et al. 2015).

Although lactic acid fermentation offers a mycotoxin decontamination process which results in a significant loss of aflatoxin and fumonisin. It also shows the limitation of this method as the residual aflatoxin concentration in the fermented samples were still higher than the maximum acceptable level of 4 μg/kg, indicating that locally fermented foods such as ogi may not be safe for consumption, as prolonged exposure to aflatoxin, even at low doses may have deleterious effects on consumers.

An improvement over LAB fermentation is the use of LAB in combination with other organisms, such as yeast cells in aflatoxin decontamination e.g.

Saccharomyces cerevisae. *S. cerevisae* is widely used in the production of alcoholic beverages and possesses the potential to bind aflatoxin B_1 (Dalié et al. 2010). Corassin et al. (2013) studied the removal of AFM_1 by three strains of heat-killed LAB and *S. cerevisae*, singly and in combination, in UHT skim milk spiked with 0.5 μg/kg for 30 and 60 min contact time. *S. cerevisiae* bound 90.3 and 92.7% of AFM_1 content in UHT skim milk for 30 and 60 min respectively. When used in combination with LAB, the efficiency of removal was greater with 91.7 and 100% for 30 and 60 min incubation periods, respectively.

The successful decontamination of AFM_1 by heat-killed cells of LAB and *S. cerevisae* suggests that bacteria viability may not be a prerequisite for mycotoxin removal in foods. This also confirms the hypothesis that the binding of aflatoxin by bacterial cells is partly due to the occurrence of a physical union with the bacterial cell wall components, mainly to polysaccharides and peptidoglycans, instead of through a covalent binding or degradation by the microorganisms metabolism (Corassin et al. 2013). Like in LAB mycotoxin reduction, the mechanism remains unclear in yeast cells but it is also currently accepted that yeast cell wall possess the ability to adsorb the toxin (Corassin et al. 2013). Of course, it remains necessary to test the toxicity of yeasts or LAB linked to mycotoxins.

While biological methods of mycotoxin decontamination may provide a safe alternative, various limitations have also been recorded with these methods. These include a long degradation time ($\geq$ 72 h), incomplete degradation, reversibility of the process, non-adaptability to typical food systems, culture pigmentation, and odour production (Oluwafemi and Da-Silva 2009). The identification and utilization of other beneficial microbes such as yeast cells in combination with LAB as a mycotoxin decontamination strategy, may portend a more efficient mycotoxin mitigatory measure.

4. Chemical Processing

The use of chemical compounds such as ammonia, hydrogen peroxide, sodium bisulfite, ozone and organic acids, and aqueous plant extracts as mycotoxin decontamination agents have been reported (Chen et al. 2014, Vijayanandraj et al. 2014, Hontanaya et al. 2015). Peanuts with an average aflatoxin content of 200 μg/kg and moisture content of 5%, were treated with ozone concentration of 6.0 mg/L for 30 min. Detoxification rates of 65.8% and 65.9% were recorded in total aflatoxin and AFB1, respectively (Chen et al. 2014). In another study, aqueous extracts from Vasaka (*Adhatodavasica* Nees) leaves showed a reduction of over 98% of AFB_1 during incubation at 37°C for 24 h (Vijayanandraj et al. 2014).

Although the effectiveness of application of chemicals in mycotoxin decontamination has been reported, usage of chemical processing is limited by loss of nutritional and organoleptic qualities, undesirable health effects of the treatments, high cost of equipment, amongst others (Oluwafemi and Da-Silva 2009). Furthermore, chemicals may transform mycotoxins to other compounds. Chemical treatment for mycotoxin decontamination is still not authorized within the EU for commodities destined for human food (Karlovsky et al. 2016). A promising method of decontamination is by the application of aqueous extracts of plants. These extracts contain compounds that are biodegradable, environmentally friendly, potentially low-cost, renewable and biologically safe (Ismail et al. 2018). However, their applications in mycotoxin decontamination have been limited due to issues in reproducibility of their activities, and their strong aroma which may restrict food applications (Ismail et al. 2018).

5. Electromagnetic Processing: Microwave and Gamma Ray

Electromagnetic energy is a kind of wave that can selectively and efficiently deliver heat to food. Typically, their spectrum can be divided into two broad categories: nonionizing radiation and ionizing radiation, for example, microwave and gamma rays. Electromagnetic energies are also classified according to wave frequency (Calado et al. 2014), for example, radio wave, microwave, infrared, visible light, ultraviolet radiation, X-ray, and gamma-ray. According to Shi (2016), microwaves frequency ranges from 0.3 GHz to 300 GHz with wavelengths from 1 m – 1 mm. Two frequency bands are allocated in the USA by the Federal Communications Commissions (FCC) for industrial, scientific, and medical applications. The 915 M band is used for industrial heating only, while 2450 M band is used both in the industry and domestic heating. Microwaves systems between 10 and 200 kW heating capacities are used in the food industries

Microwave is a robust method for supplying energy to foods that come into contact with water in order to heat predominantly portions that are wet via polymeric packaging, delivering possible short-term in-package sterilization and/or pasteurization procedures (Clark 2013), while gamma rays are also known as ionizing radiation that interact with molecules as they pass through food to form positively and negatively charged ions that can rapidly change into highly reactive free radicals, which in turn react with each other and with unchanged molecules (Kunstadt 1997). Depending on the wavelength of the microwave system, the extent of exposure and the type of food, non-ionizing (microwave) can partially reduce mycotoxins in food. In addition, microwave treatment may partly reduce the concentrations of Deoxynivalenol (DON) in naturally infected maize, but better results can be obtained at the highest temperatures. Temperatures of 150–175°C, for example, can achieve a decrease of 40% (Karlovsky et al. 2016). In another analysis by Herzallah et al. (2008), test samples subjected to γ-irradiation and microwave heating induced a substantial ($P < 0.05$) decrease in aflatoxin B_1 compositions by 42.7 and 32.3% for γ-irradiation and microwave heating (T3 of 25 kGy and 10 min of microwave heating), respectively. In contrast, it was observed that mycotoxin reductions (aflatoxin B_1 and OTA) in rice samples were 72.5

and 82.4% microwave cooking, 84.0 and 83.0% normal cooking while the rice cooked in excess water had the highest reduction of both mycotoxins (87.5 and 86.6%) (Kaushik 2015) but wet heating at 80°C for 1 h may result in 73% of AFB_1 degraded. Wet heating degradation of AFB_1 involves furofuran moiety hydrolysis and lactone ring along with additional decarboxylation. Microwave heating generated the same degradation products as conventional heating, indicating that microwave heating degradation is due solely to its thermal effects (Shi 2016).

Aflatoxin is destroyed when exposed to microwaves at rates proportional to microwave power and exposure time (Temba et al. 2016). Mycotoxins are extremely stable compounds that appear almost unchanged after food processing and are slightly heat-resistant. The impact of irradiation on the amount of mycotoxin in food depends on five major factors: mycotoxin type, level of mycotoxin, moisture content in food material, synergetic impact with other mycotoxins, synergetic effect with other compounds, and toxic degradation items (Calado et al. 2014). Increasing microwave energy, treatment time, and heating temperature increases the percentage reduction. Temperature is a crucial factor and internal temperature above 150 °C is generally required in as reported to achieve a high percentage (> 90%) decrease in aflatoxin (Shi 2016). However, some of these physical methods may be expensive depending on geographic location and may also remove or destroy important nutrients in commodities.

The gamma rays frequency is above 1019 Hz, implying wavelengths below 10–12 m. Radiation equipment consists of a single source of high-energy radiation (isotope origin) to generate gamma rays, or less commonly, by a system that produces high-energy electron beams (Lima et al. 2018). It has been widely recognized that ionizing radiation is a means for decontaminating meat. Exposing food to radiation treatment retards spoilage, enhances safety through the elimination or reduction of pathogenic microorganisms and partially reduces mycotoxins (Karlovsky et al. 2016). Gamma-based food radiation treatment is based on electromagnetic radiation characterized by high-frequency waves using a gamma source such as cobalt-60 (^{60}Co) or electrons generated by high-energy electron beam accelerators, although electron beams and gamma rays differ greatly in their ability to penetrate matter (Pillai 2016). Compared to electron beams, gamma rays generally exhibit higher penetration into food. The advantage of gamma radiation is the high penetrability and uniformity of the dose expressed (according to the International System Units) by Gray (Gy) which represents an absorbed radiation dose of ionizing radiation that allows the treatment of products of various sizes and shapes (Pillai 2016). Gamma radiation treatment adds to the value of stored foods, minimizing economic losses (as a result of food deterioration) as well as food safety, thus encouraging the adoption of goods exported by developing countries. The Joint FAO / IAEA / WHO Expert Committee for the assessment of toxicological, nutritional, chemical and physical aspects of foods treated with ionizing radiation deduced that foods treated with concentrations up to 10 kGy (kilogray) are safe and nutritionally adequate as long as they can be produced as per good manufacturing practices (Aquino 2011).

Gamma ray is usually achieved by Cobalt 60 although small proportion of today's irradiators are achieved by isotope of cesium (^{137}Cs). The use of these gamma rays is characterized by high penetration energy and short treatment times. Gamma irradiation (at 5 and 10 kGy) has been reported to have an impact on both, AFB_1 and OTA molecules. Radiolytic products formed by gamma irradiation of AFB_1 and OTA are found to be less toxic to cells (Pk15, HepG2, SH-SY5Y) than the parent compounds (non-irradiated mycotoxins). AFB_1 was more susceptible to gamma irradiation than OTA. AFB_1 radiolytic products are probably results of addition reaction on double bound in the terminal furan ring (Domijan et al. 2019). However, increasing doses of radiation could destroy certain mycotoxins until its total destruction at 20 kGy (Aquino 2011, Calado et al. 2014, Karlovsky et al. 2016), solar radiation was more effective in AFB_1 reduction when compared with γ-irradiation and microwave heating (Herzallah et al. 2008). For gamma irradiation, its effect on AFB_1 and OTA molecules is dependent on chemical structure of mycotoxin. Since low-dose gamma irradiation reduced the initial level of both mycotoxins and gamma-irradiated mycotoxins had lower toxicity compared to non-irradiated mycotoxins, it can be concluded that gamma irradiation could be used as a method of decontamination (Domijan et al. 2019).

While the presence of water plays an important role in gamma-energy destruction of aflatoxins (AFs), however, fumonisin molecules are very stable, and their destruction is likely to be difficult because of its relatively high thermal and light stability. According to Aquino (2011), the results of the removal of fumonisin were unsatisfactory when ammonia was used to remove it.

The efficacy of gamma radiation in mycotoxins degradation is significantly dependent on radiation dose and the presence of water plays an important role in gamma radiation destruction of AFs since water radiolysis leads to the formation of highly reactive free radicals. These free radicals can easily attack AFs on the terminal furan ring and yield less biologically active products (Jalili 2016).

6. Novel Technologies: Cold Plasma Treatment and Pulsed Electric Field Technology

Plasma can be described as an ionized gas consisting of several reactive species (e.g., ions, electrons, and free radicals) and ultraviolet (UV) radiation produced at various combinations of pressure and temperature, which can be loosely categorized into thermal and non-thermal (cold) plasma (Gavahian and Cullen 2019). Plasmas produced under high pressure (0.1 kPa) and high power (50,000 kW) conditions are considered to be thermal (equilibrium) plasmas with high and uniform distribution of temperature with their electrons in thermodynamic equilibrium with certain species while cold (non-equilibrium) plasmas are produced under reduced pressure and low power conditions and have lower temperatures (Gavahian and Cullen 2019). Cold (non-thermal) plasma treatment technologies is a promising decontamination approach that has near ambient temperature of 30-60 °C, and could be obtained at atmospheric or reduced pressures (vacuum). Plasma generation at atmospheric pressure is of interest for the food industry because this does not require extreme conditions and equipment. Typical approaches for atmospheric cold plasma generation include the cold plasma jet, dielectric barrier discharges (DBD), corona discharge, plasma jet, microwave radio frequency (Shi 2016). Recent developments have been made in cold atmospheric pressure plasma (CAP) sources and the ability to tailor discharges to produce highly reactive species in high concentrations, but at temperatures close to room temperature, there are more biological applications (Hojnik et al. 2017). The mechanism of decontamination from various CAP sources include inhibition of the cell membrane function for short plasma treatment times (ca. 30 s); subsequent treatment results in complete inactivation, membrane and increase in its permeability, intracellular nano-structural changes (Misra et al. 2019), apoptosis of fungal cells at low doses (Hashizume et al. 2014).

The reactive oxygen species from plasma causes oxidation of intracellular organelles; particularly, lipid phosphates to lipid peroxide through a chain reaction that can result in the oxidation of genomic DNA and cellular proteins (Lu et al. 2014).

However, there is depleting information on the roles and contributions of reactive nitrogen species (RNS) and ultraviolet radiation in cold plasma because fungal spores contain the protective pigment melanin, in the cell wall layers, which confer resistance to external stresses, including UV, hence, fungal interactions with UV and RNS from plasma sources is required (Misra et al. 2019). Non-thermal plasmas are used in low-pressure arc discharges, e.g. fluorescent lamps, in dielectric barrier discharges, e.g. ozone tubes, and in plasma jets. Just as diverse as the discharge devices are the possibilities of electronic control so that, together with pressure, gas flow, and gas type, a wide range of adjustable parameters is provided (Schlüter et al. 2013).

Plasma technology has mostly been used for the mycotoxin treatment of seeds, cereals, crops and fresh products. *A. flavus* and *A. parasiticus* contaminated maize have been reduced (by 5.48 log) with atmospheric pressure fluidized bed plasma system with air and nitrogen as a feed gas. An argon CAP source have been used to inhibit fumonisin B2 and ochratoxin A after the exposure of *A. niger* on date palm fruits and oxygen CAP achieved a 90% reduction of *C. cladosporioides* and *P. citrinum* on the surface of dried filefish fillets. Moreover, argon and oxygen CAP proved to be efficient against *A. brasiliensis* contaminating pistachios (Hojnik et al. 2017). A microwave-induced atmospheric pressure plasma system used with argon as a carrier gas to treat three different mycotoxins, AFB_1, DON, NIV dried on glass coverslips resulted in the complete decontamination of all three mycotoxins after 5 sec of plasma exposure. The cytotoxicity was completely eliminated as tested on mouse macrophage and low-temperature radiofrequency plasma achieved 88.3% AFB_1 degradation after 10 min of treatment (Hojnik et al. 2017, Yang 2019).

Non-thermal methods typically affect the chemical structure of the mycotoxins leading to their degradation. However, the extent of decontamination depends on the presence of water in the treated food products, the extent of mycotoxin contamination, and the intensity of exposure (Hojnik et al. 2017).

In comparison with the other methods, CAP mycotoxin decontamination of food has several advantages such as being environmentally benign, requiring a low energy input, economically favourable, negligible effect on the quality of many types of treated food (Hojnik et al. 2017). These advantages are based on the reactivity of the plasma species which enable the high decontamination efficiency in a very short time compared to alternative decontamination methods (Hojnik et al. 2017, Yang 2019). However, the limitations of CAP technology lies in its inability to precisely control the gas phase chemistry when using ambient air, given that it varies with conditions in the surrounding atmosphere (for example increases in humidity). Furthermore, it is not suitable for the treatment of high-fat food products because it contains ROS, and additional safety measures are required as well as systems for the destruction and exhaust of potentially harmful gaseous pollutant such as O_3 and NO_2, especially when using very high voltages (Hojnik et al. 2017).

The removal/decontamination of mycotoxin i.e. OTA, ZEN, DON or AFB1 with pulsed light in duration with a broad spectrum of light and light flux almost completely removed mycotoxins from the food materials with no toxic effect (Hojnik et al. 2017, Yang 2019). Pulsed electric field (PEF) is an alternative non-thermal pasteurization technique used for the preservation of food products. The possible application of PEF on liquid foods has been well ascertained, wherein its application on solid matrix has become the current area of research with more prospective applications such as solid–liquid extraction, mass transfer in plant and meat processing, for desired textural changes and for inhibition of fungal growth and toxin production in synthetic media and corn grains (Vijayalakshmi et al. 2018).

PEF is classified into two distinct but related areas: reversible electroporation and inactivation and food conservation of microorganisms (Vijayalakshmi et al. 2018). Microbial inactivation requires high electric field pulses (>18 kV/cm) for a relatively short period of time (Khan et al. 2017). Pathogenic and spoilage bacteria, yeasts and certain food-related enzymes are inhibited, but this treatment does not eliminate bacterial spores, but other methods such as high-pressure processing technologies are used in combination with PEF for microorganism inactivation (Woldemariam and Emire 2019). With regard to the attributes of food quality, PEF is considered superior to conventional heat treatment because it prevents or reduces changes in sensory and physicochemical properties, especially in order to preserve the micronutrient content of foods (Vijayalakshmi et al. 2018). There are many factors influencing the outcome of microorganisms with the use of the pulsed electric field. These variables are related to the system parameters (time, field frequency, temperature and number of pulses), the product characteristics and the characteristics of the microorganisms present in the sample influence the PEF result (Lima et al. 2018).

To compare the impact of pulses, single pulse of high hydrostatic pressure (HHP) treatment was also applied with the same pressure/temperature combinations and holding time and results indicated that pressure treatment in combination with mild heat and pulses reduced the levels of patulin in clear apple juice up to 62.11%. Pulse HHP was found to be more effective in low patulin concentrations, whereas HHP was more effective for high patulin concentrations (Avsaroglu et al. 2015).

7. High Pressure Processing (HPP)

The guiding principles of HPP are based on the assumption that foods experiencing HPP in a vessel obey the isostatic law regardless of the food's size or shape. The HPP inhibition of microorganisms results from a combination of factors including changes in cell membranes, cell wall, proteins, and cellular functions regulated by enzymes. Cell membranes are the primary sites of pressure-induced damage, with consequent changes in cell permeability, transportation systems, loss of osmotic sensitivity, disruption of organelles, and inability to maintain intracellular pH (Misra et al. 2019, Woldemariam and Emire 2019). In a model system of protein and lipid membrane, HPP above 200 MPa inactivates vegetative bacteria, yeast, and mold while pressures up to 700 MPa and treatment times from a few seconds to several minutes are used to inactivate microbial cells in practice (Woldemariam and Emire 2019).

Kalagatur et al. (2018a) also observed and concluded that complete fungal spore inactivation in peptone water at 380 MPa and 60°C for 30 min occurred by cell membrane disruption and colony forming unit, DON, and ZEA were completely inhibited at 550 MPa and 45°C for 20 min in maize, suggesting that pressure, temperature, and holding time were determining factors for HPP.

8. Nanotechnology

Nanotechnology works with nanomaterials in the range of 1–100 nm (e.g. nanoparticles). Different chemical, physical and biological techniques have been effectively used to synthesize metal nanoparticles such as gold, silver, palladium, iron, platinum, rhodium, and that it is expected that nanotechnology will pave the way for the development of nanosensors that would be helpful to detect such mycotoxins (Rai et al. 2015). The term 'nanomaterial' is defined as 'material with any external dimension in the nanoscale or having internal structure or surface structure in the nanoscale,' according to the International Organization for Standardization (ISO). In addition, nanoparticles naturally occur as pigments, emulsions, volcanic ashes or powder, are created anthropogenically as with the instance of exhaust fumes, or are purposely manufactured for use as additives in cosmetics, pharmaceuticals and food

(Kotzybik et al. 2016). In the design of biosensors, nanotechnology plays an increasingly important role. By using nanomaterials to build them, the sensitivity and performance of biosensors could be improved. There is a wide range of nanoscale materials available. Using these nanomaterials makes it possible to introduce many new technologies for signal transduction in biosensors. In the fields of chemical and biological research, nanosensors and other nanosystems are very important due to their size (Milanova 2015, El Saadani et al. 2020).

There are three main strategies: mold inhibition, mycotoxin adsorption, and reducing the toxic effect.

Mold inhibition: In practice, the prevention of the occurrence of mycotoxin could be facilitated through antifungal nanoparticles, which can be easily produced on a large scale. The antifungal approach is directed in two methods via antifungal compounds encapsulated in a polymeric nanocage and a nanoparticles-only inhibition effect (Horky et al. 2018). The disadvantage of the former method is air instability, although nanopolymers allow cargo release under suitable enzyme presence conditions, higher temperature, pH change while the latter are strongly influenced by stable metal nanoparticles, they act immediately and offer the possibility of green synthesis, showing less toxicity and improved main characteristics (Adelere and Lateef 2016, Horky et al. 2018).

The potentials of nanoparticles in removing pollutants via adsorption rely on the high surface area, high affinity to organic compounds and the fact that nanomaterials can be modified specifically to enhance selectivity to specific target pollutants. The type of nanoparticle material depends on the structure of mycotoxins involved as it influences the chemical and physical properties (polar or nonpolar molecules) (Horky et al. 2018).

Activated charcoal, a type of nanoparticles used in the treatment of mycotoxins has superior stability, inertness, strong adsorptive properties, large surface area by weight and colloidal stability at different pHs (Chen et al. 2007). Chemically, the covalent carbon–carbon bonding and crystalline framework have unique characteristics such as strength, elasticity, and high conductivity. Graphene, graphene oxide, nano-diamonds, completeness, fibers and nanotubes have enormous potential to become modern mycotoxin adsorbents (Horky et al. 2018). The use of carbon nanotubes (CNT) in a biosensor have been reported and it may be single-walled (SWL) or multi-walled (MWL) structures (Milanova 2015). For an example of SWL, cadmium telluride-CNT were used to form signal tag in sensor development by attaching AFM_1 antibodies on their surface for the detection of AFM_1 in milk (Gan et al. 2013). Conversely, by the use of SWCNT, single-walled carbon nanotubes in combination with aptamer can selectively identify ochratoxin A where Ochratoxin A induces the switching of the aptamer to antiparallel G-quadruplex (Guo et al. 2011). Ahmadou et al. (2019) had also used a form of carbon called biochar to fix ochratoxin A in food. Furthermore, an electroanalytical method to analyse citrinin mycotoxin was carried out and the technique uses carbon paste electrodes filled with multi-walled carbon nanotubes embedded in a mineral oil, horseradish peroxidase, and ferrocene as a redox mediator (Zachetti et al. 2013).

Chitosan (CS) is a natural cationic polysaccharide produced from chitin. CS contains hydroxyl groups, acetylamine, or free amino groups and are nontoxic, biodegradable, and possesses low immunogenicity. CS has shown promising results for mycotoxin elimination from different raw materials with approximately 70% efficacy (Horky et al. 2018). The antifungal and anti-mycotoxin activities of *Cymbopogon martinii* essential oil (CMEO) and encapsulated CMEO nanoparticles (Ce-CMEO-NPs) against *F. graminearum* were assessed in maize grains under laboratory conditions over a storage period of 28 days. Ce-CMEO-NPs presented efficient and enhanced antifungal and antimycotoxin activities by controlled release of antifungal constituents from Ce-CMEO-NPs (Kalagatur et al. 2018b). CS nanoparticles are able to encapsulate various compounds such as glutaraldehyde and tripolyphosphate. Glutaraldehyde adsorption ability for AFB_1 (73%), OTA (97%), ZEN (94%), and FUM 1 (99%) have been reported but glutaraldehyde is considered to be toxic while tripolyphosphate is a non-toxic, anionic chelating agent forming stable CS nanoparticles. Other nanoclay binders such as montmorillonite, bentonite, zeolite, or hydrated sodium (calcium) aluminosilicate are used for the detoxification of mycotoxins from food and feed (Horky et al. 2018). Unlike nanocomposites, halloysite ($Al_2Si_2O_5(OH)_4$) naturally occurs as an efficient absorbent for both cations and anions. Moreover, halloysite nanotubes could be modified by various surfactants to enhance their sorption properties and specificity (Horky et al. 2018).

The influence of particles based on silica and silver, exhibiting nominal sizes between 0.65 nm and 200 nm, on the physiology of the mycotoxigenic filamentous fungus *Penicillium verrucosum* is influenced by the applied concentration and time-point, the size and the chemical composition of the particles (Kotzybik et al. 2016). In a report, green synthesis of zinc oxide nanoparticles from Syzygiumaromaticum (SaZnO NPs) flower bud extract reduced the growth and production of deoxynivalenol and zearalenone of *F. graminearum* in broth culture. Further analysis revealed that treatment of mycelia with SaZnO NPs enhanced lipid peroxidation, depleted ergosterol content, and caused detrimental damage to the membrane integrity of fungi, suggesting the potential application in agriculture and food industries due to their potent antifungal activity (Lakshmeesha et al. 2019).

Silver nanoparticles may change the metabolism and toxicity of molds. AgNPs have shown to reduce mycotoxin production of *Aspergillus* spp. (81–96%) and mold cytotoxicity (50–75%) (Pietrzak et al. 2015). Similarly, the reduction of mycelial growth and production of aflatoxin and ochratoxin were evaluated by synthesizing AgNPs using Egyptian honey as reducing and capping agents (El-Desouky and Ammar 2016). Results obtained indicated that 3 mg-100 mL media of honey derived AgNPs have reduced the aflatoxin G_1, G_2, B_1 and B_2 production by *A. parasiticus* to 77.55, 62.91, 58.76 and 66.56%, respectively and ochratoxin A (OTA) by *A. ochraceus* to 79.85% with significantly inhibitory effect on mycelial growth (El-Desouky and Ammar 2016).

Conclusion

Certain food unit operations and diverse food processing techniques are able to reduce the level of mycotoxin in food. However, some of these techniques discussed in this chapter may modify the chemical structures of masked mycotoxins to forms not detectable by conventional analytical methods, while their toxic effects are maintained. It is therefore important that whatever method of decontamination is applied to foods in the control of mycotoxins, the methods should take into consideration loss of food materials and/or nutrients as well as the possible release or bioavailability of masked mycotoxins. Furthermore, importance of having a strategy to prevent contamination with pathogenic fungi and their mycotoxins in the food pipeline cannot be overemphasized since prevention is best safeguard against the need for removal in the first instance. An adequate and reliable food safety system should be mandatory for facilities producing food that are high risk for mycotoxin contamination.

Acknowledgements

This work is based on the research supported in part by the National Research Foundation, South Africa for the grant, Unique Grant no. 118910 and the Durban University of Technology RFA-Food and Nutrition Security Grant.

References

Abrunhosa, L., Morales, H., Soares, C., Calado, T., Vila-Cha, A.S., Pereira, M. and Venancio, A. 2016. A review of mycotoxins in food and feed products in Portugal and estimation of probable daily intakes. Critical Reviews in Food Science and Nutrition 56: 249–265. Doi.10.1080/10408398.2012.720619.

Achaglinkame, M.A., Opoku, N. and Amagloh, F.K. 2017. Aflatoxin contamination in cereals and legumes to reconsider usage as complementary food ingredients for Ghanaian infants: A review. Journal of Nutrition & Intermediary Metabolism 10: 1–7. Doi.org/10.1016/j.jnim.2017.09.001.

Adelere, I.A. and Lateef, A. 2016. A novel approach to the green synthesis of metallic nanoparticles: The use of agro-wastes, enzymes, and pigments. Nanotechnology Reviews 5: 567–587. Doi.10.1515/ntrev-2016-0024.

Ademola, O., Liverpool-Tasie, L.S.O., Obadina, A., Saha Turna, N. and Wu, F. 2018. The Effect of Processing Practices on Mycotoxin Reduction in Maize Based Products: Evidence from Lactic Acid Fermentation in Southwest Nigeria. https://ageconsearch.umn.edu/record/279860/files/FSP%20Research%20Paper%20116.pdf. Accessed April 9, 2020.

Afsah-Hejri, L., Jinap, S., Hajeb, P., Radu, S. and Shakibazadeh, S. 2013. A review on mycotoxins in food and feed: Malaysia case study. Comprehensive Reviews in Food Science and Food Safety 12: 629–651. Doi.org/10.1111/1541-4337.12029.

Ahmadou, A., Brun, N., Napoli, A., Durand, N. and Montet, D. 2019. Effect of pyrolysis temperature on ochratoxin A adsorption mechanisms and kinetics by cashew nut shell biochars. Journal of Food Science and Nutrition 4–7: 877–888. Doi: 10.25177/JFST.4.7.RA.565

Ananthi, V., Sankara Subramanian, R.K. and Arun, A. 2016. Detoxification of Aflatoxin B1 using Lactic acid bacteria. International Journal of Biosciences and Technology 9: 40–45. https://www.ijbst. org/papers-published/ijbst-2016-volume-9-Issue-7.

Anfossi, L., Giovannoli, C. and Baggiani, C. 2016. Mycotoxin detection. Current Opinion in Biotechnology 37: 120–126. Doi.org/10.1016/j.copbio.2015.11.005.

Aquino, S. 2011. Gamma radiation against toxigenic fungi in food, medicinal and aromatic herbs. pp. 272–281. *In*: Mendez-Vilas, A. (ed.). Science Against Microbial Pathogens: Communicating Current Research and Technological Advances. Brazilian Journal of Microbiology. ISSN 1517-8382. On-line version: ISSN 1678-4405 doi: https://doi.org/10.1590/S1517-83822005000400009

Ashiq, S. 2015. Natural occurrence of mycotoxins in food and feed: Pakistan Perspective. Comprehensive Reviews in Food Science and Food Safety 14: 159–175. Doi.org/10.1111/1541-4337.12122.

Assohoun, M.C.N., Djeni, T.N., Koussemon-Camara, M. and Kouakou, B. 2013. Effect of fermentation process on nutritional composition and aflatoxin concentrations of Doklu: A fermented maize based food. Food and Nutrition Sciences 4: 1120–1127. https://m.scirp.org/papers/37561.

Avsaroglu, M., Bozoglu, F., Alpas, H., Largeteau, A. and Demazeau, G. 2015. Use of pulsed-high hydrostatic pressure treatment to decrease patulin in apple juice. High Pressure Research 35: 214–222. Doi.org/10.1080/08957959.2015.1027700.

Bhat, R., Rai, R.V. and Karim, A.A. 2010. Mycotoxins in food and feed: Present status and future concerns. Comprehensive Reviews in Food Science and Food Safety 9: 57–81. Doi.org/10.1111/j.1541-4337.2009.00094.

Bueno, D.J., Casale, C.H., Pizzolitto, R.P., Salvano, M.A. and Oliver, G. 2007. Physical adsorption of aflatoxin B1 by lactic acid bacteria and *Saccharomyces cerevisiae*: A theoretical model. Journal of Food Protection 70: 2148–2154. Doi.org/10.4315/0362-028x-70.9.2148.

Bullerman, L.B. and Bianchini, A. 2007. Stability of mycotoxins during food processing. International Journal of Food Microbiology 119: 140–146. Doi.org/10.1016/j.ijfoodmicro.2007.07.035.

Calado, T., Venâncio, A. and Abrunhosa, L. 2014. Irradiation for mold and mycotoxin control: A review. Comprehensive Reviews in Food Science and Food Safety 13: 1049–1061. Doi.org/10.1111/1541-4337.12095.

Cheli, F., Pinotti, L., Rossi, L. and Dell'Orto, V. 2013. Effect of milling procedures on mycotoxin distribution in wheat fractions: A review. LWT-Food Science and Technology 54: 307–314. Doi.org/10.1016/j.lwt.2013.05.040.

Chen, R., Ma, F., Li, P.W., Zhang, W., Ding, X.X., Zhang, Q., Li, M., Wang, Y.R. and Xu, B.C. 2014. Effect of ozone on aflatoxins detoxification and nutritional quality of peanuts. Food Chemistry 146: 284–288. Doi.10.1016/j.foodchem.2013.09.059.

Chen, W., Duan, L. and Zhu, D. 2007. Adsorption of polar and nonpolar organic chemicals to carbon nanotubes. Environmental Science & Technology 41: 8295–8300. Doi.org/10.1021/es071230h

Clark, J.P. 2013. Electromagnetic energy in food processing. Food Technology 67(4): 72–74. http://microwaveheating.wsu.edu/news/files/Electromagnetic%20 Energy.pdf" Electromagnetic Energy.pdf (wsu.edu)

Corassin, C.H., Bovo, F., Rosim, R.E. and Oliveira, C.A.F.D. 2013. Efficiency of *Saccharomyces cerevisiae* and lactic acid bacteria strains to bind aflatoxin M1 in UHT skim milk. Food Control 31: 80–83. Doi.org/10.1016/j.foodcont.2012.09.033.

Dalié, D.K.D., Deschamps, A.M. and Richard-Forget, F. 2010. Lactic acid bacteria – potential for control of mould growth and mycotoxins: A review. Food Control 21: 370–380. Doi.10.1016/j.foodcont.2009.07.011.

Decker, E.A., Rose, D.J. and Stewart, D. 2014. Processing of oats and impact of processing operations on nutritional and health benefits. British Journal of Nutrition 112: S58–S64. Doi.10.1017/s000711451400227x.

Domijan, A.M., Marjanović Čermak, A.M., Vulić, A., Tartaro Bujak, I., Pavičić, I., Pleadin, J., Markov, K. and Mihaljević, B. 2019. Cytotoxicity of gamma irradiated aflatoxin B1 and ochratoxin A. Journal of Environmental Science and Health Part B, 54: 155–162. Doi.10.1080/03601234.2018 .1536578.

El-Desouky, T.A. and Ammar, H. 2016. Honey mediated silver nanoparticles and their inhibitory effect on aflatoxins and ochratoxin A. Journal of Applied Pharmaceutical Science 6: 083–090. Doi.10.7324/JAPS.2016.60615.

Elsaadani, M., Durand, N., Sorli, B., Guibert, B., Alter, P. and Montet, D. 2020. Aptamer assisted ultrafiltration cleanup with high performance liquid chromatography – Fluorescence Detector for the determination of OTA in green coffee. Food Chemistry 125851. Doi.10.1016/j.foodchem.2019.125851.

Ezekiel, C.N., Abia, W.A., Ogara, I.M., Sulyok, M., Warth, B. and Krska, R. 2015. Fate of mycotoxins in two popular traditional cereal-based beverages (kunu-zaki and pito) from rural Nigeria. LWT-Food Science and Technology 60: 137–141. Doi. org/10.1016/j.lwt.2014.08.018.

Gan, N., Zhou, J., Xiong, P., Hu, F., Cao, Y., Li, T. and Jiang, Q. 2013. An ultrasensitive electrochemiluminescent immunoassay for Aflatoxin M1 in milk, based on extraction by magnetic graphene and detection by antibody-labeled CdTe quantumn dots-carbon nanotubes nanocomposite. Toxins 5: 865–883. Doi.10.3390/toxins5050865.

Gavahian, M. and Cullen, P. 2019. Cold plasma as an emerging technique for mycotoxin-free food: Efficacy, mechanisms, and trends. Food Reviews International. Doi.org/10.1080/87559129.2019.1630638.

Granados-Chinchilla, F., Redondo-Solano, M. and Jaikel-Víquez, D. 2018. Mycotoxin contamination of beverages obtained from tropical crops. Beverages 4: 83. Doi.org/10.3390/beverages4040083.

Guo, Z., Ren, J., Wang, J. and Wang, E. 2011. Single-walled carbon nanotubes based quenching of free FAM-aptamer for selective determination of ochratoxin A. Talanta 85: 2517–2521. Doi.10.1016/ j.talanta.2011.08.015.

Hashizume, H., Ohta, T., Takeda, K., Ishikawa, K., Hori, M. and Ito, M. 2014. Quantitative clarification of inactivation mechanism of *Penicillium digitatum* spores treated with neutral oxygen radicals. Japanese Journal of Applied Physics 54 (1S) Doi.10.7567/jjap.54.01ag05.

Herzallah, S., Alshawabkeh, K. and Fataftah, A.A. 2008. Aflatoxin decontamination of artificially contaminated feeds by sunlight, γ-radiation, and microwave heating. Journal of Applied Poultry Research 17: 515–521. Doi: 10.3382/japr.2007-00107.

Hojnik, N., Cvelbar, U., Tavčar-Kalcher, G., Walsh, J. and Križaj, I. 2017. Mycotoxin decontamination of food: Cold atmospheric pressure plasma versus "classic" decontamination. Toxins 9. Doi.10.3390/ toxins9050151.

Hontanaya, C., Meca, G., Luciano, F., Mañes, J. and Font, G. 2015. Inhibition of aflatoxin B1, B2, G1 and G2 production by *Aspergillus parasiticus* in nuts using yellow and oriental mustard flours. Food Control 47: 154–160. Doi.10.1016/j.foodcont.2014.07.008.

Horky, P., Skalickova, S., Baholet, D. and Skladanka, J. 2018. Nanoparticles as a solution for eliminating the risk of mycotoxins. Nanomaterials (Basel, Switzerland) 8: 727. Doi.10.3390/nano 8090727.

Ismail, A., Gonçalves, B.L., de Neeff, D.V., Ponzilacqua, B., Coppa, C.F.S.C., Hintzsche, H., Sajid, M., Cruz, A.G., Corassin, C.H. and Oliveira, C.A.F. 2018. Aflatoxins in foodstuffs: Occurrence and recent advances in decontamination. Food Research International 113: 74–85. Doi.10.1016/ j.foodres.2018.06.067.

Jalili, M. 2016. A review on aflatoxins reduction in food. Iranian Journal of Health, Safety and Environment 3: 445–459. http://www.ijhse.ir/index.php/IJHSE/article/view/136/pdf_58.

Kachouri, F., Ksontini, H. and Hamdi, M. 2014. Removal of aflatoxin B1 and inhibition of *Aspergillus flavus* growth by the use of *Lactobacillus plantarum* on olives. Journal of Food Protection 77: 1760–1767. Doi.10.4315/0362-028x.jfp-13-360.

Kalagatur, N.K., Kamasani, J.R., Mudili, V., Krishna, K., Chauhan, O.P. and Sreepathi, M.H. 2018a. Effect of high pressure processing on growth and mycotoxin production of *Fusarium graminearum* in maize. Food Bioscience 21: 53–59. Doi.org/10.1016/j.fbio.2017.11.005.

Kalagatur, N.K., Nirmal Ghosh, O.S., Sundararaj, N. and Mudili, V. 2018b. Antifungal activity of chitosan nanoparticles encapsulated with Cymbopogon martini essential oil on plant pathogenic fungi *Fusarium graminearum*. Frontiers in Pharmacology 9: 610. Doi.10.3389/fphar.2018.00610.

Karlovsky, P., Suman, M., Berthiller, F., DeMeester, J., Eisenbrand, G., Perrin, I., Oswald, I.P., Speijers, G., Chiodini, A., Recker, T. and Dussort, P. 2016. Impact of food processing and detoxification treatments on mycotoxin contamination. Mycotoxin Research 32: 179–205. Doi.10.1007/ s12550–016-0257-7.

Kaushik, G. 2015. Effect of processing on mycotoxin content in grains. Critical Reviews in Food Science and Nutrition 55: 1672–1683. Doi. 10.1080/10408398.2012.701254.

Khan, I., Tango, C.N., Miskeen, S., Lee, B.H. and Oh, D.H. 2017. Hurdle technology: A novel approach for enhanced food quality and safety – A review. Food Control 73: 1426–1444. Doi.org/10.1016/j.foodcont.2016.11. 010.

Kotzybik, K., Gräf, V., Kugler, L., Stoll, D.A., Greiner, R., Geisen, R. and Schmidt-Heydt, M. 2016. Influence of different nanomaterials on growth and mycotoxin production of *Penicillium verrucosum*. PloS One 11. Doi.10.1371/journal.pone.0150855.

Kunstadt, P. 1997. Food irradiation: Gamma processing facilities. https://inis.iaea.org/search/search.aspx?orig_q=RN:30031635. Accessed April 1 2020.

Lakshmeesha, T.R., Kalagatur, N.K., Mudili, V., Mohan, C.D., Rangappa, S., Prasad, B.D., Ashwini, B.S., Hashem, A., Alqarawi, A.A., Malik, J.A., Abd-Allah,

E.F., Gupta, V.K., Siddaiah, C.N. and Niranjana, S.R. 2019. Biofabrication of zinc oxide nanoparticles with syzygium aromaticum flower buds extract and finding its novel application in controlling the growth and mycotoxins of *Fusarium graminearum*. Frontiers in Microbiology 10. Doi.10.3389/fmicb.2019.01244.

Lima, F., Vieira, K., Santos, M. and de-Souza, P.M. 2018. Effects of radiation technologies on food nutritional quality. Descriptive Food Science 137. Doi.10.5772/intechopen.80437.

Lu, Q., Liu, D., Song, Y., Zhou, R. and Niu, J. 2014. Inactivation of the tomato pathogen *Cladosporium fulvum* by an atmospheric-pressure cold plasma jet. Plasma Processes and Polymers 11: 1028–1036. Doi.10.1002/ppap.201400070.

Man, Y., Liang, G., Li, A. and Pan, L. 2017. Recent advances in mycotoxin determination for food monitoring via microchip. Toxins 9(10): 324. Doi.10.3390/toxins9100324.

Marroquín-Cardona, A., Johnson, N., Phillips, T. and Hayes, A. 2014. Mycotoxins in a changing global environment – A review. Food and Chemical Toxicology 69: 220–230. Doi.10.1016/ j.fct.2014.04.025.

Matumba, L., Monjerezi, M., Chirwa, E., Lakudzala, D. and Mumba, P. 2009. Natural occurrence of AFB_1 in maize and effect of traditional maize flour production on AFB_1 reduction in Malawi. African Journal of Food Science 3: 413–425. https://www.researchgate.net/publication/228687721_Natural_occurrence_of_AFB_in_maize_and_effect_of_traditional_maize_flour_production_on_AFB_reduction_in_Malawi

Matumba, L., Poucke, C., Ediage, E.N., Jacobs, B. and De Saeger, S. 2015. Effectiveness of hand sorting, flotation/washing, dehulling and combinations thereof on the decontamination of mycotoxin-contaminated white maize. Food Additives and Contaminats Part A, 32: 960-969. Doi.10.1080/19440049.2015.1029535.

Milanova, S.N. 2015. Application of nanotechnology in detection of mycotoxins and in agricultural sector. Journal of Central European Agriculture 16(2): 117–130. https://doi.org/10.5513/jcea.v16i2.3472

Misra, N., Yadav, B., Roopesh, M. and Jo, C. 2019. Cold plasma for effective fungal and mycotoxin control in foods: Mechanisms, inactivation effects, and applications. Comprehensive Reviews in Food Science and Food Safety 18: 106-120. Doi.org/10.1111/1541-4337.12398.

Moeser, A., Kim, I., Van Heugten, E. and Kempen, T. 2002. The nutritional value of degermed, dehulled corn for pigs and its impact on the gastrointestinal tract and nutrient excretion. Journal of Animal Science 80: 2629–2638. Doi.10.1093/ansci/80.10.2629.

Mutiga, S.K., Were, V., Hoffmann, V., Harvey, J.W., Milgroom, M.G. and Nelson, R.J. 2014. Extent and drivers of mycotoxin contamination: Inferences from a survey of Kenyan maize mills. Postharvest Pathology and Mycotoxins 104: 1221–1231. Doi.10.1094/PHYTO-01-14-0006-R.

Mutungi, C., Lamuka, P., Arimi, S., Gathumbi, J. and Onyango, C. 2008. The fate of aflatoxins during processing of maize into muthokoi – A traditional Kenyan food. Food Control 19: 714–721. Doi.10.1016/j.foodcont.2007.07.011.

Okeke, C.A., Ezekiel, C.N., Nwangburuka, C.C., Sulyok, M., Ezeamagu, C.O., Adeleke, R.A., Dike, S.K. and Krska, R. 2015. Bacteria diversity and mycotoxin reduction during maize fermentation (steeping) for Ogi production. Frontiers in Microbiology. Doi. 10.3389/fmicb.2015.01402.

Okeke, C.A., Ezekiel, C.N., Sulyok, M., Ogunremi, O.R., Ezeamagu, C.O., Šarkanj, B., Warth, B. and Krska, R. 2018. Traditional processing impacts mycotoxin levels and nutritional value of ogi – A maize-based complementary food. Food Control 86: 224–233. Doi.org/10.1016/j.foodcont.2017.11. 021.

Oluwafemi, F. and Da-Silva, F.A. 2009. Removal of aflatoxins by viable and heat-killed *Lactobacillus* species isolated from fermented maize. Journal of Applied Biosciences 16: 871–876. http://www. m.elewa.org/JABS/2009/16/4.pdf. Accessed April 10 2020.

Pacin, A.M. and Resnik, S.L. 2012. Reduction of mycotoxin contamination by segregation with sieves prior to maize milling. pp. 219–234. *In*: Novel Technologies in Food Science. Springer. Doi.10.1007/ 978-1-4419-7880-6_10.

Pascale, M., Haidukowski, M., Lattanzio, V.M.T., Silvestri, M., Ranieri, R. and Visconti, A. 2011. Distribution of T-2 and HT-2 toxins in milling fractions of durum wheat. Journal of Food Protection 74: 1700–1707. Doi.10.4315/0362-028X.JFP-11-149.

Pearson, T.C., Wicklow, D.T. and Pasikatan, M.C. 2004. Reduction of aflatoxin and fumonisin contamination in yellow corn by high-speed dual-wavelength sorting. Cereal Chemistry 81(4): 490–498.

Pietrzak, K., Twarużek, M., Czyżowska, A., Kosicki, R. and Gutarowska, B. 2015. Influence of silver nanoparticles on metabolism and toxicity of moulds. Acta Biochimica Polonica 62: 851–857. Doi.10. 18388/abp.2015_1146.

Pillai, S.D. 2016. Introduction to electron-beam food irradiation. Chemical Engineering Progress 112: 36–44. https://www.aiche.org/resources/publications/cep/2016/november/introduction-electron-beam-food-irradiation. Accessed April 10 2020.

Rai, M., Jogee, P.S. and Ingle, A.P. 2015. Emerging nanotechnology for detection of mycotoxins in food and feed. International Journal of Food Sciences and Nutrition 66: 363–370. Doi.org/10.3109/09634 86.2015.1034251.

Rastegar, H., Shoeibi, S., Yazdanpanah, H., Amirahmadi, M., Khaneghah, A.M., Campagnollo, F.B. and Sant'Ana, A.S. 2017. Removal of aflatoxin B1 by roasting with lemon juice and/or citric acid in contaminated pistachio nuts. Food Control 71: 279–284. Doi.10.1016/j.foodcont.2016.06.045.

Roger, T., Léopold, T.N. and Mbofung, C.M.F. 2015. Effect of selected lactic acid bacteria on growth of *Aspergillus flavus* and aflatoxin B1 production in Kutukutu. Journal of Microbiology Research 5: 84–94. Doi.10.5923/j. microbiology.20150503.02.

Saalia, F.K. and Phillips, R.D. 2011. Degradation of aflatoxins by extrusion cooking: Effects on nutritional quality of extrudates. LWT-Food Science and Technology 44: 1496–1501. Doi.10.1016/ j.lwt.2011.01.021.

Schlüter, O., Ehlbeck, J., Hertel, C., Habermeyer, M., Roth, A., Engel, K., Holzhauser, T., Knorr, D. and Eisenbrand, G. 2013. Opinion on the use of plasma processes for treatment of foods. Molecular Nutrition & Food Research 57: 920-927. Doi.10.1002/mnfr.201300039.

Shi, H. 2016. Investigation of methods for reducing aflatoxin contamination in distillers grains. Purdue University 29th August, 2019). https://docs. lib. purdue.edu/cgi/viewcontent.cgi?article=2215&context=open_access_dissertations Accessed April 1 2020.

Siwela, A.H., Siwela, M., Matindi, G.S.D. and Nziramasanga, N. 2005. Decontamination of aflatoxin-contaminated maize by dehulling. Journal of the Science of Food and Agriculture 85: 2535–2538. Doi.org/10.1002/jsfa.2288.

Temba, B.A., Sultanbawa, Y., Kriticos, D.J., Fox, G.P., Harvey, J.J.W. and Fletcher, M.T. 2016. Tools for defusing a major global food and feed safety risk: Nonbiological postharvest procedures to decontaminate mycotoxins in foods and feeds. Journal of Agricultural and Food Chemistry 64: 8959–8972. Doi.org/10.1021/acs.jafc.6b03777.

Tibola, C.S., Fernandes, J.M.C. and Guarienti, E.M. 2016. Effect of cleaning, sorting and milling processes in wheat mycotoxin content. Food Control 60: 174–179. Doi.org/10.1016/j.foodcont. 2015.07.0310956-7135.

Vijayalakshmi, S., Nadanasabhapathi, S., Kumar, R. and Kumar, S.S. 2018. Effect of pH and pulsed electric field process parameters on the aflatoxin reduction in model system using response surface methodology: Effect of pH and PEF on aflatoxin reduction. Journal of Food Science and Technology 55: 868–878. Doi.10.1007/s13197-017-2939-3.

Vijayanandraj, S., Brinda, R., Kannan, K., Adhithya, R., Vinothini, S., Senthil, K., Chinta, R.R., Paranidharan, V. and Velazhahan, R. 2014. Detoxification of aflatoxin B1 by an aqueous extract from leaves of Adhatodavasica Nees. Microbiological Research 169: 294–300. Doi.10.1016/j.micre s.2013.07.008.

Woldemariam, H.W. and Emire, S.A. 2019. High pressure processing of foods for microbial and mycotoxins control: Current trends and future prospects. Cogent Food & Agriculture 5. Doi.10.1080/23311932.2019.1622184.

Yang, Q. 2019. Decontamination of Aflatoxin B1. *In*: Aflatoxin B1 Occurrence, Detection and Toxicological Effects. Intechopen. Doi.10.5772/itechopen.88774.

Yazdanpanah, H., Mohammadi, T., Abouhossain, A. and Cheraghali, A.M. 2005. Effect of roasting on degradation of aflatoxins in contaminated pistachio nuts. Food and Chemical Toxicology 43: 1135–1139. Doi.10.1016/j.fct.2005.03.004.

Zachetti, V.G.L., Granero, A.M., Robledo, S.N., Zon, M.A. and Fernández, H. 2013. Development of an amperometric biosensor based on peroxidases to quantify citrinin in rice samples. Bioelectrochemistry 91: 3743. Doi.10.1016/j.bioelechem.2012.12.004.

Zheng, H., Wei, S., Xu, Y. and Fan, M. 2015. Reduction of aflatoxin B1 in peanut meal by extrusion cooking. LWT-Food Science and Technology 64: 515–519. Doi.10.1016/j.lwt.2015.06.045.

CHAPTER

3

New Insight of Preventive and Curative Approaches to Reduce Aflatoxin B1 (AFB1) and Ochratoxin A (OTA) Contamination

Isaura Caceres, Selma P. Snini and Florence Mathieu*

Laboratoire de Génie Chimique, Université de Toulouse, CNRS, INPT, UPS, Toulouse, France Avenue de l'Agrobiopole - BP 32607 - Auzeville-Tolosane 31326 Castanet-Tolosan Cedex

1. Introduction

Today, many strategies have been developed to reduce mycotoxin contamination in food and feed. Aflatoxin B1 (AFB1) and Ochratoxin A (OTA) are some of the most studied mycotoxins due to their occurrence, global distribution and hazardous toxicity. AFB1 is produced by many fungal species belonging to the *Flavi* section of the *Aspergillus* genus, such as *Aspergillus flavus*, the most preoccupying specie (Amaike and Keller 2011). AFB1 is the most potent naturally occurring carcinogen and is classified by the International Agency for Research on Cancer (IARC) in group 1A (IARC 1993a) and is responsible for causing human hepatocarcinoma. OTA is produced by several species of *Penicillium* and *Aspergillus*. OTA has nephrotoxic, hepatotoxic, teratogenic and immunotoxic effects on animals and is classified by the IARC in the group 2B as a possible human carcinogen (IARC 1993b). These mycotoxins contaminate different food and feed. Their elimination is very difficult due to their specific properties such as high thermal and chemical stability (Raters and Matissek 2008). Therefore, it is necessary to establish preventive strategies from the early crop stages to prevent mycotoxin accumulation. Good Agricultural Practices (GAP), Good Manufacturing Practices (GMP) and Good Hygienic Practices (GHP) together with the Hazard Analysis and Critical Control Points (HACCP) are the key prevention measures during the different stages of crop transformation

*Corresponding author: florence.mathieu@toulouse-inp.fr

(FAO 2001, Magan and Aldred 2007). Unfortunately, these strategies are nowadays not sufficient to ensure food safety leading to the use of important quantities of fungicides as a remedy. In the last few years, the use of fungicides has been intensified although they present numerous disadvantages such as detrimental effects in mammals, environmental contamination and subsequent strong impact on microbial biodiversity (Ferrer and Cabral 1991, Calhelha et al. 2006, Carvalho 2017, Gupta 2018, Zubrod et al. 2019). The mode of action of these fungicides is mainly to inhibit the fungal growth. In this case, the total or partial elimination of the targeted microorganisms unbalance the natural microbial ecosystem, causing the potential emergence of new microorganisms, which may be even more dangerous (Tano 2011). Moreover, recurring applications of fungicides can also lead to the development of resistance which compromises disease control (Lucas et al. 2015). Azole fungicides are one of three major classes of fungicides used to control fungal disease on cereal crops. They target the CYP51 (sterol 14α-demethylase) an important enzyme involved in the ergosterol biosynthesis which is essential to maintain fungal membrane fluidity and permeability (Mansfield et al. 2010). However, for several years, fungal resistance has arisen against these compounds (Price et al. 2015). Recently, the work of Mateo et al.(2017) showed that azole fungicides proved to be highly efficient in reducing *A. flavus* growth and aflatoxins production. However, stimulation of aflatoxins production was found under particular culture conditions and low-dosage treatments. Besides this, the variation of the hydrothermal conditions under climate change could induce favorable circumstances to increase mycotoxin contamination in areas where they were not previously widespread, raising a new challenge for food safety (Medina et al. 2015). Indeed, warm temperatures and high humidity are key factors for mycotoxigenic fungi, influencing their germination, growth, sporulation and mycotoxin production. Due to the physiological properties of aflatoxigenic strains, it was long admitted that AFB1 contamination occurs in tropical and subtropical regions. Conversely, OTA contamination seems to be more disseminated since its producers belong to *Aspergillus* and *Penicillium*, two fungai generally requiring different conditions for growth and mycotoxin production. However, recently, European countries have been facing the serious emerging threat of the possible occurrence of AFB1 in maize (Battilani et al. 2016) and OTA in grapes produced in Europe, which can lead to serious health and economic problems (Bellí et al. 2004, Battilani et al. 2006). Moreover, in France, a recent study has demonstrated that *Aspergillus* section *Flavi* is frequently present in maize and in the case of favorable environmental conditions, its development can lead to AFB1 contamination (Bailly et al. 2018). Given the proven toxicity of AFB1 and OTA and their widespread distribution, it is necessary to develop methods that prevent their occurrence in feed and food. The toxicity of fungicides on human health, environment, living organisms and on food quality has led researchers to develop alternative strategies. Several techniques can be adopted in fields, during the storage of harvested goods or during industrial processes. These techniques can be classified into two categories. First, preventive methods can be applied in fields or during storage and act directly on fungal development and/or on mycotoxin production. These methods involve regulation mechanisms that lead to the downexpression of genes coding for the enzymes involved in the mycotoxin biosynthesis pathway. Hence, at best, the biosynthesis pathway is completely extinguished but sometimes, the biosynthesis pathway is only interrupted and leads to the accumulation of intermediate compounds that may be more toxic than the mycotoxin itself. As an example, Theumer et al. (2018) have compared the toxicity of AFB1 and it precursors on three human cell lines and have demonstrated that some AFB1 metabolic precursors have genotoxic potential equivalent or greater to AFB1. Among these precursors, the versicolorin A (VerA) presents higher potent cytotoxic effects compared to AFB1. Indeed, VerA induces higher oxidative stress, causes more DNA damages, is a stronger inhibition of cell proliferation and apoptotic cell death (Gauthier et al. 2020). Thus, this must be taken into account in the development of strategies to reduce mycotoxin contamination. These preventive methods

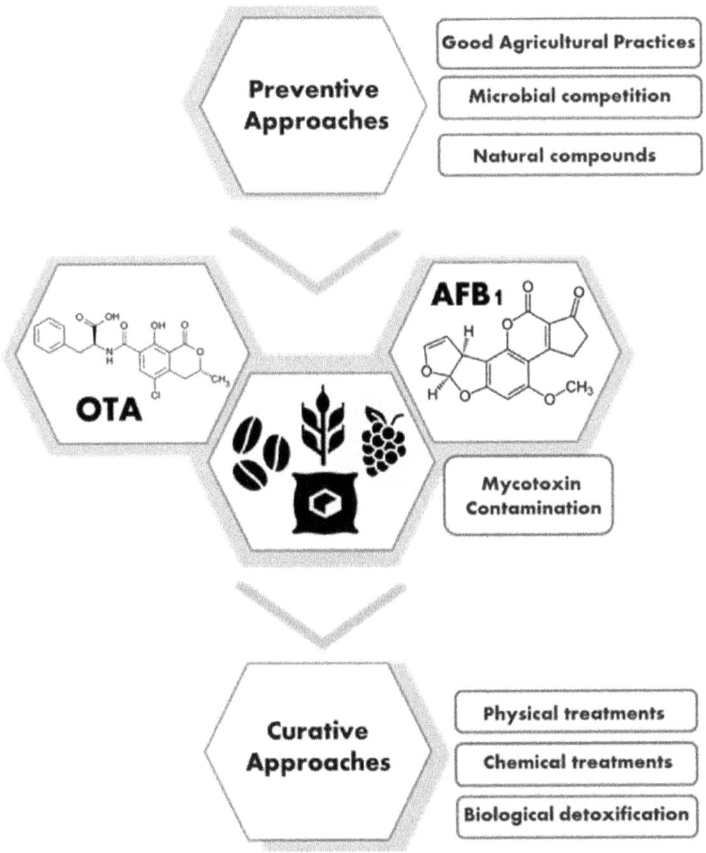

Figure 1. Schematic representation of the preventive and curative approaches proposed to reduce AFB1 and OTA contamination.

can also reduce fungal growth. In the case where mycotoxin production is correlated to the fungal growth, this can lead to the reduction of mycotoxin concentration. However, some studies have demonstrated that the reduction of fungal growth can lead to an increase in mycotoxin production. Conversely, during storage or in industrial production processes, the reduction of mycotoxin accumulation through fungal growth reduction may be desired. Indeed, during storage, fungal growth even without mycotoxin production is responsible for huge food spoilage.

The second category gathered curative methods that aim to detoxify contaminated matrices by eliminating the produced mycotoxin. These curative methods include physical and chemical decontamination, as well as biological detoxification. Some of these strategies can lead to the formation of new toxic substances (degraded products, metabolites, or by-products) whose toxicity must be assessed. However, most of these methods lead to a reduction in the quality of decontaminated foods (Grenier et al. 2014).

This chapter is organized by the approach type i.e. preventive and curative approaches, and within each type a description of each method used for reducing AFB1 and OTA contamination is detailed. Figure 1 shows a schematic illustration describing the preventive and curative approaches and the major categories that will be described in this chapter.

2. Preventive Approaches

2.1 Good Agricultural Practices and HACCP Strategies

Good agricultural practices together with the HACCP system are the first strategies to prevent fungal and mycotoxin contamination at different crop stages. As a general overview, Good Agricultural Practices maintain healthy crops and ensure sustainable agriculture; Good Manufacturing Practices establish basic operational and environmental conditions to ensure food quality and the Hazard Analysis and Critical Control Points system ensures food safety through the identification of critical points of risk during the entire food processing system. The correct management of these practices is extensively described in FAO and IARC reports (FAO 2001, 2002, IARC 2012) and hereby, only several of them will be described with a special focus in AFB1 and OTA management.

Preharvest Practices

Seed Selection. Several seed varieties present a higher and natural resistance to fungal and mycotoxin contamination. In peanuts, frequently contaminated by AFB1, fungal resistance is partially determined by the seed coat composition such as the quantity of lignin, cellulose, proteins, enzymes or phenolic compounds (Cobos et al. 2018). However, a recent study involving the restriction site-associated DNA sequencing (RAD-sequencing)method demonstrated that several peanut genotypes were highly resistant to aflatoxin contamination in different environments (Yu et al. 2020). Similarly, a comparison between certain varieties of Italian grapes demonstrated that several of them presented a better resistance against *A. carbonarius* and thus, to mycotoxin contamination. Indeed, OTA levels were systematically lower in varieties such as 'Pampanuto' and 'Uva di Troia' than others like 'Cabernet Sauvignon' (Battilani et al. 2004).

Soil Conditions. Nutriment and water supplies of soil, as well as pH, are also conditional factors that predispose fungal contamination. Within nutriment impact, nitrogen sources were demonstrated to modulate AFB1 and OTA production (Abbas et al. 2009, Fasoyin et al. 2019). Interestingly, in *A. flavus*, fungal growth and AFB1 production were impacted not only by different nitrogen sources but also, by the ionic charges of the compounds founded in these sources (Wang et al. 2017). The pH of the soil is also a parameter predisposing fungal contamination since it affects microbial communities among others. Experiments varying the pH of soil (4.0 to 8.0) demonstrated that lower levels of the pH of soil induced a decrease in bacterial growth and an increase in fungal growth (Rousk et al. 2009). Indeed, acidic pH conditions have been demonstrated to be more favorable than alkaline ones for AFB1 and OTA production in *A. ochraceus* and *A. flavus* (Cotty 1988, Maor et al. 2017). Moreover, microbial diversity has also a role in mycotoxin contamination since as it will be further described in this chapter, several soil microorganisms are capable of inhibiting OTA and AFB1 production. Elimination of old-crop residues as well as the determination of mycotoxin content in soil, before planting, is also a need to be evaluated. On one side, contaminated residues could serve as an initial inoculum of fungi and on the other side, mycotoxin determination in soil can prevent further contamination in crops. Indeed, it was demonstrated that the roots of plants can gradually take AFB1 content from soil and latter accumulate it in seeds *via* xylem (Snigdha et al. 2015). Drought stress conditions have been related to AFB1 production. In groundnuts, different levels of dryness were tested demonstrating that AFB1 contamination increased under high drought conditions compared to low levels. Moreover, dry stress conditions also impacted the *Aspergillus* population in the soil, showing that the presence of *A. flavus* was higher in prolonged drought conditions (Sibakwe et al. 2017).

Crop Rotation. The Codex Alimentarius (2003) recommended the planning of two-year intervals as one of the strategies to prevent mycotoxin contamination. In crops affected by aflatoxin contamination, a study demonstrated that depending on the sequential alternation of crops, the AFB1 content was modulated. For instance, crop sequences of 'pigeonpea-maize' or 'fallow-groundnuts' reduced AFB1 levels in maize and groundnuts while crop sequences of 'maize-groundnuts' or 'groundnut-maize' had the opposite effect by increasing toxin content (Abraham et al. 2016). Crop rotation also reduces the risk of repetitive insect infection and fungal contamination typical of a monoculture crop (IARC 2012).

Insect Injury. The damage of crops by insect injury allows an available nutrient source for opportunistic fungi and thus for mycotoxin contamination. To date, the use of insecticides is one of the most used remedies. However, several alternatives have been already developed. Such is the case of the genetically modified (GM) Bt maize, a variety with an inserted toxin-producing protein of *Bacillus thuringiensis* that has been demonstrated to effectively reduce insect damage and to present resistance against some AFB1-producer strains (Pruter et al. 2019). The Bt maize is one of the controlled GM crops that are allowed for cultivation in the European Union (Halford 2019). However, its cultivation is controversial due to environmental, ethical or economic reasons among others. Insecticidal and repellent coatings with encapsulated essential oils and plant extracts have been also proposed (Féat et al. 2019). Nevertheless, to our knowledge, these coatings have not already been reported in *in situ* conditions. In preharvest conditions, grape berries are frequently contaminated by OTA due to the insect injury (e.g. *Lobesia botrana*, grape fruit fly) caused when the berries are fully ripe or near to maturity (Zhang et al. 2016).

Technological Methods. Simulation models with changing climate scenarios allow us to now predict AFB1 and OTA contamination. Predictive simulations can be adapted as a complementary tool but they do not replace GAPs. Simulations are not 100% reliable. However, greater percentages of contamination were already demonstrated such as 54% of AFB1 contamination in peanut kernels (Hendrix et al. 2019). Other examples include predicting AFB1 contamination by considering climate changes, drought stress, elevated CO_2 conditions, the impact of A_w and temperature as well as the infectious life cycle of *A. carbonarius* in grapes (Medina et al. 2015, Battilani et al. 2016, Krulj et al. 2019).

Postharvest Practices

Temperature and Moisture Control. Environmental conditions of stock are one of the major influencing parameters to limit fungal infection. Aflatoxins and OTA contents in food commodities have been largely associated with tropical and subtropical countries where climate conditions are optimal for fungal development. For instance, B-type aflatoxin contents in Brazil nuts infected by *A. nomius* were reported to be higher at 25°C, which in fact, coincided with the Amazon region temperature (Yunes et al. 2019). Similarly, fungal and aflatoxins contamination in maize kernels by *A. flavus* were impacted by temperature and relative humidity (RH) conditions. With a RH of 90%, AFB1 content was significantly higher at 30°C than at 20°C while a RH of 60% significantly decreases toxin production independently of the temperature (Muga et al. 2019). In green coffee, A_w activities below of 0.80 were demonstrated to protect grains from OTA contamination by *A. ochraceus* (Suárez-Quiroz et al. 2004). The material used for stock can also influence fungal contamination. While jute and woven bags are highly porous and can easily absorb moisture from the environment, Purdue Improved Crop Storage (PICS) bags have demonstrated to be effective against insect infection, fungal growth and AFB1 contamination (Williams et al. 2014, 2017).

Sorting. Cleaning and sorting are the frontline barriers still relevant today to eliminate visual contamination of fungi and to reduce contaminants including mycotoxins (Matumba et al. 2015). Sorting consists of removing damaged grains before using them in the manufacturing process. Indeed, damaged cereal grains or grape berries are more easily contaminated by toxigenic fungal strains (Raghavender et al. 2007, Cozzi et al. 2013). The traditional sorting method is hand sorting and is based on the visible abnormalities of kernels, like smaller size, mold spots or color changes, which could be associated with a higher possibility of mycotoxin contamination (Kabak et al. 2006, Xu et al. 2017). However, the efficiency of this method depends on the experience of workers; it is time-consuming and can be only performed in small-scale productions. A recent study performed in distinct regions of South Africa demonstrated that most of the farmers were unaware of mycotoxin existence and their effects on human and animal health. However, visual separation of grains from contaminated crops was a very usual practice in farmer communities allowing the reduction of mycotoxin contents (Phokane et al. 2019). Training of women farmers in hand sorting achieved a 42.9% of AFB1 reduction in groundnuts (Xu et al. 2017). Aflatoxin detection by electronic, optical technologies and manual color sorting have been also reported as mycotoxin-reducer methods (Zivoli et al. 2016, Stasiewicz et al. 2017). Moreover, a partial removal of AFB1 can be obtained by using fluorescence sorting from peanuts. Indeed, under UV light illumination the appearance of a bright green-yellow fluorescence (BGY) has been correlated to aflatoxin contamination. However, the BGY fluorescence was due to kojic acid whose presence was strongly correlated with AFB1 contamination (Pelletier and Reizner 1992). AFB1 contamination can also be observed by infrared spectroscopic methods (Levasseur-Garcia 2018). For large-scale productions, the sorting step can be mechanized and combine some of the equipment, including air separators, sieves, gravity separators, and indented cylinders. Whatever the sorting method used, manual or automatic, sorting reduces mycotoxin contamination (Tibola et al. 2016, Zivoli et al. 2016; Chalwe et al. 2019). After the sorting step, washing methods can also reduce mycotoxin contamination by removing mycotoxins from the outer surface of grain (Matumba et al. 2015).

Drying. Before certain industrial processes, grains or foodstuffs must be dried under well-controlled conditions to preserve them. Efficient drying techniques prevent fungal contamination by reducing moisture levels in food commodities. In the treatment of red chili peppers, the application of the HACCP system demonstrated that drying was one of the key steps to ensure the quality and safety of the product and that oven drying could better reduce the aflatoxins contamination risk rather than sun-drying (Ozturkoglu-budak 2017). Indeed, ineffective sun drying conditions allow fungal infection due to higher moisture levels at the beginning of the process. Different drying methods have been reported to reduce AFB1 and OTA contamination. For instance, the Mandela cock method, frequently used in smallholder production, was effective to reduce aflatoxin contamination by *A. flavus* in groundnuts by 75% (Dambolachepa et al. 2019). Similarly, a roasted treatment in robusta coffee beans resulted in lower OTA levels by *A. niger* and *A. ochraceus,* compared to the drying yard method (Barcelo and Barcelo 2018).

2.2 Microbial Competition

Among biocontrol strategies employed to reduce AFB1 and OTA contamination, the use of microorganisms is of great importance. These are called biocontrol agents (BCAs) and generally, they are separated into two types: non-toxigenic or antagonistic microorganisms. BCAs can be applied directly in the field to protect crops as a preharvest strategy, during the storage of raw materials or during the production process as a postharvest strategy.

*Microbial Competition with Non-Toxigenic Fungal Strains.*Concerning the reduction of AFB1 and OTA contamination in crops, several BCAs have been studied and presented different levels of efficiency. The use of non-toxigenic fungal strains (i.e. mycotoxin non-producer fungal strains) to manage mycotoxin contamination was more described for the reduction of AFB1 contamination than for OTA contamination.

The most effective AFB1 reducing strategy is the preharvest application of non-toxigenic *A. flavus* strains (also called atoxigenic strains) (Amaike and Keller 2011, Ojiambo et al. 2018). These atoxigenic strains are often isolated from fields, directly from soil cultures or seeds. Isolation of atoxigenic *A. flavus* strain is not difficult because they represent a high percentage of *A. flavus* strains in the natural environment (Ehrlich 2008, 2014). These atoxigenic strains are applied in fields by using inoculated seeds or spore suspensions directly sprayed on crops (Abbas et al. 2006, 2017). Through competitive exclusion, these non-toxigenic strains exclude AFB1 producers by reducing their development and the production of AFB1 (Torres et al. 2014). Historically, the International Institute of Tropical Agriculture (IITA) and the United States Department of Agriculture (USDA) with other partners have successfully developed several country-specific indigenous aflatoxin biocontrol products. The most well-known is named Aflasafe®, a combination of four *A. flavus* strains isolated from Nigeria which do not produce AFB1 (Atehnkeng et al. 2008). Moreover, two atoxigenic *A. flavus* strains, AF36 (NRRL 18543) and Afla-Guard® (NRRL 21882), were also successfully developed for the reduction of AFB1. These strains have been approved by the U.S. Environmental Protection Agency for biocontrol of aflatoxin contamination and are commercialized in USA. The non-aflatoxigenic strain AF36, isolated from cottonseed in Arizona, has a full-aflatoxin gene cluster with one mutated gene (Cotty 1989, Ehrlich and Cotty 2004). The Afla-Guard® was isolated from a naturally infected peanut in Georgia but unlike the AF36 strain, the entire gene cluster involved in the AFB1 biosynthesis pathway is missing (Dorner 2004, Moore et al. 2009). The use of these native non-toxigenic *A. flavus* strains as BCAs has proved to be very effective in reducing AFB1 contamination in cotton (Cotty and Bhatnagar 1994), peanuts (Hulikunte Mallikarjunaiah et al. 2017, Alaniz Zanon et al. 2018) and maize (Atehnkeng et al. 2008, 2016, Camiletti et al. 2018). Recently, another effective atoxigenic fungal strain, the AF-X1®, has been commercialized in Italy for AFB1 management in maize (Mauro et al. 2018). The effectiveness of such BCAs was based on the fact that these atoxigenic *A. flavus* strains are genetically stable and have an asexual reproduction cycle. However, recent studies have demonstrated that *A. flavus* can have a sexual reproduction cycle under specific conditions (Horn et al. 2009a, 2009b, 2016) leading to an exchange of genetic material by recombination and could result in a displacement of *A. flavus* populations in favor of increasing AFB1 contamination (Olarte et al. 2012). Other studies have demonstrated that the repeated application of atoxigenic *A. flavus* strains could increase the efficacy of biocontrol and thus, leading to a better reduction of AFB1 contamination (Bhatnagar-Mathur et al. 2015, Weaver and Abbas 2019). These contradictory results demonstrate that the use of these atoxigenic strains and their long-term efficacy deserve further evaluation. Recently, some studies have tried to identify factors that can influence the effectiveness of BCAs in reducing AFB1 contamination under field conditions. Apparently, the genetic structure of indigenous soil populations of *A. flavus* is a determining factor in the effectiveness of the applied BCA in reduction of AFB1 contamination (Lewis et al. 2019). Interactions between atoxigenic and aflatoxin-producing strains are complex and deserve more attention before the development of biocontrol solutions. Moreover, it seems that a soil study is unavoidable to characterize the subpopulation of toxigenic strains to improve the effectiveness of BCAs.

Concerning the reduction of OTA contamination, as mentioned above, the use of non-toxigenic fungal strains is not well described. Indeed, unlike the production of AFB1 by *A. flavus*, OTA production is a very constant characteristic of *A. carbonarius*, the main responsible for OTA contamination of grapes and derived products. Thus, the isolation of atoxigenic *A. carbonarius* strains is difficult (Cabañes et al. 2013, Castellá et al. 2018). However, some strains have been isolated and the sequencing of their genome has demonstrated that in *A. carbonarius* atoxigenic strains there is a high accumulation of mutations in genes encoding the PKS and NRPS explaining the lack of OTA production (Cabañes et al. 2015, Castellá et al. 2018). Recently, the use of these non-OTA-producing strains as BCA was evaluated *in vitro* and the obtained results demonstrated that its co-inoculation with an *A. carbonarius* OTA-producing strain decreases the OTA production up to 89%. The co-inoculation of *A. carbonarius* non-OTA producing strains with *A. niger* OTA producing strain results in a complete inhibition of OTA production (Castellá et al. 2020). Two other fungal species, belonging to the black *Aspergilli* family, *A. niger* and *A. tubingensis* are also frequently isolated from grapes. However, unlike *A. carbonarius*, the percentage of OTA-producing strains in *A. niger* is low and *A. tubingensis* is a non-OTA-producing species (Frisvad et al. 2011, Castellá et al. 2018, Choque et al. 2018). The ability of *A. tubingensis* to reduce the production of OTA by *A. carbonarius* and *A. niger* was tested and in both cases, OTA reduction was achieved (Castellá et al. 2020). Thus, the use of such fungal species as BCAs to reduce OTA contamination in the vineyard is promising and needs further investigations.

Microbial Competition with Antagonistic Microorganisms. To protect crops, antagonistic microorganisms can also be used. Several modes of action with more or less direct effects are known for these BCAs. First, the BCA and the toxigenic fungal strain can compete because they share the same ecological niche and nutrient requirements (such as carbon, azote and, iron). This competition can lead to the inhibition of the toxigenic fungal strain growth and/or to the reduction of mycotoxin production. Second, antagonistic microorganisms can also produce molecules and secondary metabolites which can directly act on fungal growth and thus, leading to the reduction of mycotoxin production (Nguyen et al. 2017). Concerning the management of AFB1 contamination, several bacterial species such as *Bacillus* spp., *Pseudomonas* spp. and *Streptomyces* spp., have demonstrated their ability to reduce fungal growth and AFB1 production *in vitro* under laboratory conditions. Numerous research publications have demonstrated the ability of *Bacillus* strains to control aflatoxigenic strains. Indeed, *Bacillus* spp. produce many antimicrobial compounds and are nonpathogenic species, thus making good candidates for the BCA status (Ren et al. 2020). For example, *B. amyloliquefaciens* and *B. subtilis* culture filtrates were able to inhibit *A. parasiticus* growth and aflatoxins production. Moreover, both strains were able to degrade the four aflatoxins (Siahmoshteh et al. 2017, 2018). Another *Bacillus* strain, *B. velezensis* was able to reduce fungal growth of *A. flavus* and *A. parasiticus* and their associated aflatoxins through the production of three kinds of lipopeptides (Liu et al. 2020). Among the *Pseudomonas* genus, *P. fluorescens* and *P. protegens* have demonstrated their ability to reduce *A. flavus* growth and AFB1 production. In addition to the reduction of *A. flavus* growth, *P. fluorescens* strain can inhibit AFB1 biosynthesis and to degrade AFB1 (Yang et al. 2017, Mannaa and Kim 2018). *Streptomyces* spp. is another microbial genus widely found in soil and in plants. Some strains belonging to the *Streptomyces* genus are entophytic bacteria promoting plant defenses. They are also able to produce a wide range of secondary metabolites, hydrolytic enzymes and antimicrobial compounds (Lee et al. 2018, Spasic et al. 2018). Several studies have demonstrated the capacity of *Streptomyces* strains to reduce *A. flavus* and *A. parasiticus* growth and subsequently, reduce aflatoxins accumulation (Zucchi et al. 2008, Shakeel et al. 2018). In recent years, studies have focused on elucidating the mechanism of action of *Streptomyces* strains leading to mycotoxin reduction. A promising approach is the study of volatile compounds as a biofumigant to reduce aflatoxins contamination during storage (Yang et al. 2019, Lyu et al. 2020). In all these studies, the reduction of mycotoxin concentration is achieved through the reduction of fungal growth. However, the fungal growth reduction of mycotoxigenic strains can unbalance the microbial ecosystem leading to the appearance of new microorganisms potently more hazardous. Moreover, the fungal growth reduction can also lead to the stimulation of mycotoxin production (Audenaert et al. 2013, Morcia et al. 2017). Recently, studies have focused on the search for BCAs that do not affect mycotoxinogenic fungal growth. For example, several actinobacteria strains have demonstrated their ability to reduce AFB1 concentration without affecting *A. flavus* or *A. parasiticus* growth (Verheecke et al. 2015, Caceres et al. 2018). Moreover, these actinobacteria strain belonging to the *Streptomyces* genus are spore-forming bacteria, endophytic and rhizosphere bacteria able to colonize hostile environments, a fact highlighting their sustainability as phyto-protective agents (Barka et al. 2016). In conclusion, many microbial strains have demonstrated an ability to inhibit aflatoxigenic fungal strains. Despite all these promising results, all of these studies were conducted under laboratory conditions and none of these antagonistic microbial agents have been already commercialized.

Concerning the management of OTA contamination in a vineyard or in grapes, there is an emerging consensus in the literature on the use of yeast as BCAs. Several yeast species including different genera such as *Saccharomyces* spp., *Candida* spp., *Rhodotorula* spp. *Lachancea* spp., *Picha* spp. and others are considered as potential BCAs towards ochratoxigenic *Aspergillus spp.* (Gil-Serna et al. 2018). Moreover, yeast constitutes the major part of the microbial community on the surface of grape berries. They are adapted to this ecological niche and can colonize grapes to compete for space and nutrients with other microorganisms (Nally et al. 2015). In many studies the efficiency of saprophytic yeast to control ochratoxigenic *Aspergillus* spp. was tested in artificially inoculated grapes (Bleve et al. 2006, Dimakopoulou et al. 2008, Fiori et al. 2014, Zhu et al. 2015, De Souza et al. 2017, Tryfinopoulou et al. 2019, 2020). Ponsone et al. (2011) have isolated from grapes two indigenous yeast strains of *Lachancea. thermotolerans* which were able to control *A. carbonarius* and *A. niger* growth and OTA accumulation. The study of the mechanism of action has demonstrated that these strains were also able to impact the expression of OTA biosynthesis genes (Ponsone et al. 2013). In a more recent study, the same research team has demonstrated the effectiveness of these two yeast strains against black rot and OTA accumulation under greenhouse and field conditions(Ponsone et al. 2016). Moreover, this effect was significantly higher when mechanical damage was applied, suggesting that the yeast rapidly colonized the wounds preventing *Aspergillus* spp. infection. (Ponsone et al. 2016). These results are promising for the development of a biocontrol strategy based on *L. thermotolerans*. However, the authors indicated that the effectiveness of this strain is very depending on environmental factors and decreases when temperature and humidity are optimum for fungal development. Moreover, as biocontrol yeast strains can compete with *S. cerevisiae* a microorganism frequently used in the fermentation of grape must, it is also important to evaluate the influence of such biocontrol agents used in vineyards on grapes and on the winemaking process.

2.3 Natural Compounds

The use of natural compounds is an alternative strategy that has been also reported to fight against mycotoxin contamination. Within this context, essential oils, organic and aqueous extracts and isolated molecules are frequently studied. Most of these natural products have been studied in *A. flavus* and *A. parasiticus* (for AFB1) or in *A. ochraceus* (for OTA) while only a few information is reported in other toxigenic strains also belonging to the *Flavi, Ochraceorosei, Nidulantes, Circumdati* and *Nigri* sections. To date, the mechanisms of action by which these natural products inhibit mycotoxin production are yet to be elucidated. This is a great challenge since it involves the study of a large variety of existing natural compounds as well as the understanding of the complex mechanism of mycotoxin biosynthesis in fungi. To date, a great majority of the studies demonstrate that natural products inhibit mycotoxin production due to a strong fungal inhibition. However, several others effectively reduce mycotoxin content with a minimal impact on fungal development. The study of the impact that natural compounds have on fungal development contribute to their appropriate utilization within the different stages of the food chain. While an effective inhibition of mycotoxin production without a strong impact in fungal growth can be suitable in pre- and harvested stages, in postharvest stage fungal inhibition is a critical point for commercialization. In the last years, the study of the molecular mechanisms of action of natural products also begins to be elucidated due to the development of new analytical techniques. This represents a strategic tool to understand the impact that natural products have on gene modulation providing insights in the identification of possible fungal targets and in their impact within the mycotoxin biosynthetic pathway.

However, the efficacy of the use of natural products against fungal growth and/or mycotoxin inhibition also depends on multiple factors such as the fungal species and test conditions as well as the plant material and its extraction process, among several others. It stays necessary to take care of the toxicity of the natural compounds.

In the next section, a summary concerning different aspects of the use of natural products against AFB1 and OTA contamination will be presented.

Essential Oils (EOs). EOs are a mixture of volatile low molecular weight compounds (<500 Da) mostly including terpenoids, phenylpropanoids, and hydrocarbons as major constituents (Singh et al. 2019). Within this context, the EOs of several aromatic plants have received special attention such as rosemary, thyme, basil, cinnamon, citronella, cumin, oregano or cloves. Depending on the nature of the EO, its impact on fungal growth is variable. For instance, the *in vitro* use of the EOs of Holy Basil (0.2 µL per mL) and cumin seed (0.5 µL per mL) in *A. flavus*, completely inhibited AFB1 production with a fungal reduction of 90.2% and 91.3% respectively (Kumar et al. 2010, Kedia et al. 2014). On the other hand, the EO of lemongrass (0.2 mg per mL) also inhibits AFB1 in *A. flavus* with only a fungal inhibition of 3.3% (Paranagama et al. 2003). A similar effect was also observed in *A. carbonarius*, where the EOs of cinnamon, taramira, oregano and thyme (5 µL per mL) completely inhibited fungal growth and consequently, OTA production (El Khoury et al. 2016). Nevertheless, at the same concentration, the EOs of fennel, rosemary, cardamom, celery, anise and chamomile inhibited OTA production from ranges of 70% to 89% with a fungal development from 11.8% to 33% (El Khoury et al. 2016).

As previously mentioned, great progress in the study of the molecular mechanisms of action of natural compounds have been observed in the last few years. However, while the AFB1 biosynthetic pathway is completely elucidated, the OTA pathway is still under study. Studies demonstrated that different EOs (e.g. rosemary, lemongrass, clove, mandarin) used as OTA inhibitors in *A. carbonarius,* were capable to down-regulate the genes of the OTA cluster *acOTApks* and *acOTAnrps* regardless of their impact in fungal development (El Khoury et al. 2016, Lappa et al. 2017). For *A. flavus*, the EO of turmeric (*Curcuma longa* L.) also inhibits AFB1 production in a transcriptional manner by down-regulating the expression of seven genes gathered in the AFB1 cluster including the principal regulators *aflR* and *aflS* (Hu et al. 2017). Moreover, a recent study demonstrated that the EOs of rosemary and clove in *A. flavus,* inhibit AFB1 production by modulating the expression of the global regulator *laeA,* as well as *metP* and *lipA,* genes involved in fungal virulence (Oliveira et al. 2020).

Concerning the use of EOs in the food industry, several of them are already commercialized as crop protectors while several others present drawback limitations (Bhavaniramya et al. 2019). For instance, some EOs are highly toxic and can induce dead in mammals and insects (Isman 2000). Moreover, as part of their mechanism of action, several EOs present an alteration of the fungal cell membrane and it has been suggested that the same deleterious effect could be observed in plants or food structures (Isman and Machial 2006). EOs can also modify the organoleptic properties of food due to their high content in volatile compounds but only a few studies have reported their sensory appreciation at doses that inhibit mycotoxin production. For instance, the use of the EO of *Artemisia nilagirica* effectively reduces AFB1 and OTA production in table grapes and sensory analyses resulted in a positive acceptability of consumers (Sonker et al. 2015). Nevertheless, this evaluation was performed only at the beginning of the treatment, limiting the appreciation of the EO's stability and its impact on food properties during the time. Indeed, environmental factors such as light, temperature, pH or humidity can alter the efficacy of the EOs due to the oxidation, isomerization, cyclization or dehydrogenation reactions (Turek and Stintzing 2013). It has been demonstrated that the exposition of the EO of Boldo to UV light can reduce by 29.1% its antifungal effect on *A. flavus* (Passone et al. 2013). On the other side, the EO of *Citrus sinensis* L. can support temperatures from 40 to 100°C and even autoclaving conditions without losing its antifungal activity against *A. niger* (Sharma and Tripathi 2008). These facts highlight the importance not only on the discovery of natural products that inhibit mycotoxin production but also, of the importance to study their stability facing to environmental conditions. Regarding this, several strategies have been proposed such as vapor diffusion or the microencapsulation technique. Indeed, vapor diffusion of EOs of boldo, poleo and clove have demonstrated an effective inhibition of fungal growth and AFB1 content in *A. flavus* and *A. parasiticus* (Passone et al. 2013). Moreover, in *A. flavus,* microencapsulated EOs of *Satureja montana* and *Origanum virens* have been shown to be more effective as antifungal and antiaflatoxigenic agents compared to the non-encapsulated ones (García-Díaz et al. 2019).

Aqueous and Organic Extracts. According to the International Organization for Standardization, extracts are obtained by treating a natural raw material with one or several solvents. Due to their nature, they present different properties from those of EOs and several of them have been demonstrated to have antifungal and antimycotoxin activities that have been strongly related to their content of phenols, flavonoids, tannins and saponins, among others. Aqueous and organic extracts are less reported than EOs but a great variety of leaves, branches, seeds and scopes of plants using different extraction methods have been already reported.

Concerning aqueous extracts, native Indian, Egyptian and American plant extracts are demonstrated as antifungal and antitoxigenic agents in *A. flavus, A. parasiticus* and *A. niger* (Mahmoud 1999, Thippeswamy et al. 2014, Sinha 2018, Juarez-Segovia et al. 2019). For instance, Neem leaf extract greatly inhibits sterigmatocystin and aflatoxins production by *A. parasiticus* and *A. flavus* (Bhatnagar and McCormick 1988, Ghorbanian et al. 2007, Sinha 2018). Nevertheless, the same extract did not inhibit OTA production but instead, it stimulated fungal growth, sporulation and exudates in *P. verrucosum* and *P. brevicompactum* (Mossini et al. 2009), demonstrating that a same extract can have a specific impact on different fungal species.

For organic extracts, *in vitro* tests of the *Thuja orientalis* ethanolic extract resulted in a dose-dependent inhibition on *A. flavus* growth. In addition, detoxification of AFB1 in corn seeds was observed with percentages of 71.35% to 100% for two months of exposure (Amna 2017). The ethanolic extract of *Solanum indicum*, a type of berries currently eaten in many African countries, also showed potential to reduce OTA production in *A. niger, A. carbonarius* and *A. ochraceus* (Ahou Kouadio et al. 2019). Similarly, hexanic and methanolic extracts of several traditional Mexican plants also showed antifungal properties against *A. niger* (Navarro García et al. 2003). Besides, *in vivo* conditions have demonstrated that the supplementation of several ethanolic extracts (e.g. Chamomile, papaya) in animal diet can counteract the toxic effects induced by OTA and AFB1 (El-Nekeety et al. 2017, Nazarizadeh et al. 2019).

Concerning the study of the molecular mechanism of action of natural extracts, little information has been reported. For instance, the aqueous extract of *Micromeria graeca*, greatly inhibits AFB1 in *A. flavus* without a strong impact in fungal growth (El Khoury et al. 2017). The molecular study revealed that the principal aflatoxin gene regulators *aflR* and *aflS* were significantly down-regulated inducing a low expression of the other genes belonging to the aflatoxin gene cluster. As well, this extract had an impact on the expression of other genes involved in the environmental response previously demonstrated to be linked to the production of AFB1.

Compared to the EOs, aqueous and organic extracts present other benefits that can be adapted in the food industry such as a lower impact in the organoleptic changes in food or less toxicity due to the nature of their compounds. However, this assumption needs to be proved by performing sensorial analyses and comparative toxicological studies. The valorization of natural extracts that are considered as industrial wastes can also be a solution. Such is the case of the aqueous phase obtained during the production of

EOs by the hydrodistillation process. Indeed, clove, cinnamon, and oregano aqueous phases presented antifungal properties against *A. niger* and other fungal contaminants (Cáceres et al. 2014). The combination between various natural extracts as well as the combination of natural extracts with EOs also demonstrated to improve their efficacy. In *A. flavus*, a methanolic extract of *Citrullus colocynthis* inhibits fungal growth at 36.2% and total aflatoxins (including AFB1) at 27.3%. However, the combination of the *C. citrullus* extract with the EO of citronella inhibited 85.6% fungal growth and more than 90% of total aflatoxins (Sidhu et al. 2009). The use of natural extracts in food commodities has also been reported. As an example, an artichoke extract inhibits AFB1 production in sesame seeds contaminated by *A. parasiticus* and organic extracts of 12 different Indian plants also showed AFB1 inhibitions from 28.75% to 91.25% in maize seeds contaminated by *A. flavus* (Thippeswamy et al. 2014, Kollia et al. 2017).

Natural Molecules. A great variety of molecules issues from plants and spices have been studied against AFB1 and OTA production. Such is the case for cinnamaldehyde, caffeine, caffeic acid, capsaicin, citral, geraniol as well as piperine and piperine like compounds among several others. In this section, only the study of eugenol and curcumin will be described due to their recurrent study in AFB1- and OTA-producer strains.

Eugenol is a major terpene compound present in basil, cloves and other plants. It was demonstrated to inhibit fungal growth in *A. niger* and *A. ochraceus* with MIC and MFC of 400 and 450 μg per mL and of 350 and 400 μg per mL respectively (Abbaszadeh et al. 2014). However, in *A. carbonarius*, MIC and an MFC were of 2000 ppm or higher demonstrating that the effect of eugenol in OTA-producer strains is fungal dependent (Šimović et al. 2014).

Regarding the aflatoxin-producing strains, eugenol seems to effectively inhibit toxin production without a strong impact on fungal growth. For instance, it inhibits AFB1 production in *A. parasiticus* and in *A. flavus* in a dose-dependent manner with low (11.4% to 34.5%) or null impact in fungal development (Jayashree and Subramanyam 1999, Caceres et al. 2016). It has also been demonstrated that eugenol inhibits AFB1 production in a transcriptional manner. In *A. parasiticus*, eugenol concentrations of 62.5 and 125 μg per mL significantly downregulated the genes *alfM, aflD, aflC, aflP* and *aflR* (Jahanshiri et al. 2015). Similarly, in *A. flavus*, the use of eugenol at 0.80 mM downregulated the expression of the aflatoxin genes *aflP, aflM, aflR, aflD* and *aflT* (Liang et al. 2015) and further studies demonstrated that the *in vitro* use of 0.5 mM of eugenol downregulated the entire gene cluster with the only exception of *aflT* (Caceres et al. 2016). Moreover, the expression of several external regulators linked with AFB1 biosynthesis and belonging to families of oxidative stress response, velvet complex, cellular signaling as well as other global transcription factors was also modulated (Caceres et al. 2016).

Curcumin is a main polyphenol belonging to the ginger family and founded in *Curcuma longa* and other species such as *Curcuma* spp. In *A. parasiticus*, curcumin inhibited in a dose-dependent manner AFB1 production from 22.6% to 94.9% using concentrations from 125 to 2000 μg per mL. This effect was accompanied with a fungal growth inhibition (34% to 60.8%) and a significant reduction on gene expression of the *aflR, alfM, aflD, aflC, aflP* (Jahanshiri et al. 2012). In the presence of curcumin at 0.1% (w/v), fungal growth of *A. Alliaceus* was completely inhibited and thus, not detectable levels of OTA, OTB and citrinin were detected (Lee et al. 2007). However the use of 0.01% (w/v) of curcumin in *A. alliaceus* inhibited fungal growth by 50% and OTA and citrinin production (71.9% and 89% respectively) but surprisingly, the OTB production increased by three times. The authors suggested that the mode of action of curcumin is *via* the inhibition of the chlorination step from OTB to OTA (Lee et al. 2007). *In vivo* assays have also demonstrated that curcumin supplementation has a hepatoprotective effect through the inhibition of a cytochrome P450 involved in the AFB1 bioactivation in broilers and it also alleviates the oxidative stress response pathways induced by OTA in ducks (Muhammad et al. 2017, Zhai et al. 2020).

In order to illustrate the effect that eugenol and curcumin have over the expression of the *aflR* gene, the principal regulator in the aflatoxin gene cluster, Fig. 2 is presented.

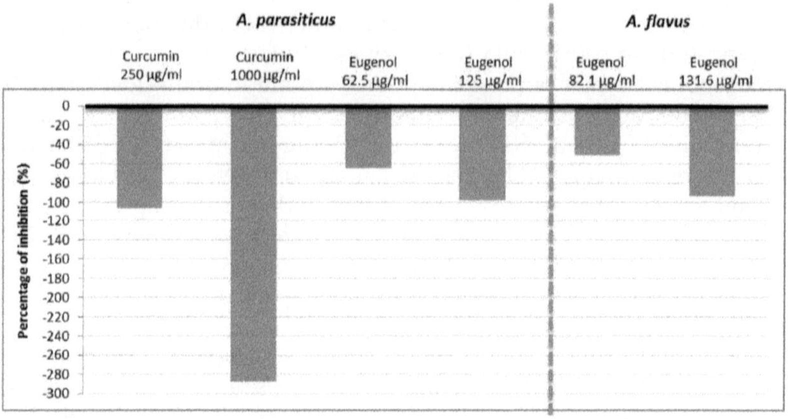

Figure 2. Effect of curcumin and eugenol at different concentrations over the expression of the *aflR* gene in *A. parasiticus* and *A. flavus*. Control level corresponds to zero. This figure is based on the works of Jahanshiri et al. (2012, 2015), Liang et al. (2015), Caceres et al. (2016).

As for EOs, the microencapsulation of molecules has also been reported. For instance, microcapsules of the antioxidants compounds BHA: 2(3)-tert-butyl-4 hydroxyanisole and BHT:2,6-di(tert-butyl)-p-cresol were tested in peanut kernels contaminated with *A. flavus* and *A. parasiticus* demonstrating antifungal and antiaflatoxigenic activity for 45 days (Garcia et al. 2016). Concerning the application of molecules in food, a study demonstrated that a direct application of eugenol and carvacrol in watermelon (*C. lanatus* L) inhibited in a dose-dependent manner the *A. carbonarius* growth and this effect, increased at lower temperatures (25°C *vs* 15°C) (Šimović et al. 2014). Indeed, temperature is a key factor for fungal development and a good strategy to improve the efficacy of natural compounds against mycotoxin inhibition. The application of natural compounds under modified atmospheres has also been reported with promising results. Such is the case of the use of geraniol and citral compounds that were capable to delay fungal infection of *A. flavus* and *A. ochraceus* in rice without harmful effects (Tang et al. 2018).

3. Curative Approaches

Curative approaches can be of two types. On the one hand, decontamination methods aim to remove or degrade the targeted mycotoxin by various physical, chemical or biological processes. However, degradation treatments lead to the appearance of new compounds resulting from the degradation of the parent compound. In this case, it is necessary to evaluate the toxicity of these new compounds before implementing this type of approach. On the other hand, detoxification method corresponds to the transformation of the targeted mycotoxin by physical, chemical or biological processes leading to a reduced toxicity. Curative methods can be applied in addition to the preventive methods mentioned and described above when those treatments have not been effective enough to reduce AFB1 and OTA concentrations to comply with current European regulations. Over the past five years, many original bibliographic reviews have compiled data on the various possible strategies to reduce the accumulation of mycotoxins contaminating food and feed and some focus on AFB1 and OTA (Pfliegler et al. 2015, Verheecke et al. 2016, Chen et al. 2018, Peng et al. 2018, Vila-Donat et al. 2018, Haque et al. 2020).

The great diversity of approaches and applications mentioned in all of those works show that the classification of these decontamination methods is not simple. To sum up, we can say that there are physical, chemical treatments and biological methods. According to these methods, the mycotoxins can be modified, degraded or adsorbed. In any case, all of these curative methods can be controversial when used in food matrices. Thus, their major drawback would be the potential loss of the organoleptic and nutritional qualities of food as well as the presence of known or unknown residues which can also be toxic. Thus, currently, these methods are applied in animal feed but rarely in human food. However, the challenge of 2050 will be to feed more than nine billion people. To reach this objective, decontamination and detoxification approaches could be part of the solution.

In such a context and focusing on AFB1 and OTA contaminating food and beverages, we will enumerate and compare several curative methods existing to date based on bibliographic data from the last five years by addressing successively (1) the physical treatments, such as heat treatments, cold plasma, irradiation, light effects, without also forgetting all the supports which are used as mycotoxin binders; (2) chemical treatments currently used such as ammoniation, the use of ozone, neutral electrolyzed oxidizing water and (3) the biological methods, in particular the use of microorganisms targeting the mechanisms of adsorption and degradation.

3.1 Physical Treatments

Various physical treatments can be applied to food/feed matrices and to raw materials to reduce AFB1 and OTA contamination. Several strategies in these physical treatments can be considered, i.e. (1) denaturation of mycotoxins by heat, irradiation, and nonthermal treatments and (2) mycotoxin elimination by the great diversity of binders existing to date (magnetic carbon, aflatoxin selective clay, carbon and activated carbon…) (Peng et al. 2018). It should be noted that the border between certain physical methods and adsorbing agents is difficult to define. Thus here, these will be considered as physical treatments.

Extrusion Cooking. It is a conventional cooking method widely applied in food industries especially for cereals and cereal foodstuffs to improve the quality of processed products (De Pilli and Alessandrino 2020). Extrusion cooking combines high temperature, mechanical treatment at high pressure in a very short time. Due to the heat resistance of mycotoxins, thermal processes may not be sufficient enough to destroy them without significantly affecting the quality of processed food and feed. Indeed, it is known that industrial heat treatment degrades mycotoxins to a certain extent depending on the initial concentration. For example, to reduce the OTA content in white flour by 76%, it must be subjected for 40 min to a temperature of 250°C and for AFB1, roasted corn from 145 to 165°C reduces AFB1 from 40 to 80% (Scott 1984). Thus, in certain cases, a mycotoxin reduction is obtained through the extrusion process. For example, Zheng et al. (2015) have demonstrated that AFB1 in peanut meal was efficiently degraded by extrusion cooking (up to 77% reduction). They have also demonstrated that the temperature and moisture are two parameters linked to the effectiveness of the process and influence the AFB1 degradation rate (Zheng et al. 2015). Recent studies have shown less promising results for the use of extrusion as a decontamination mean. To improve this process, Mendez-Albores et al. (2008) evaluated the effect of the addition of different acid concentrations in the extrusion-cooking process on the stability of AFB1 in sorghum flour. They used lactic and citric acid at 2M, 4M and 8M and various moisture contents (20% to 30%). Initial AFB1 concentration (140 ppb) decreased when moisture content and the concentration of acids increases. Under some conditions, AFB1 reduction is more effective when using aqueous citric acid (up to 92%) than when using aqueous lactic acid (up to 67%). In corn flour process, extrusion cooking is effective for the inactivation of deoxynivalenol (DON) but is of limited value for AFB1. Besides, in some studies, the use of solvents as an additive was necessary to increase the efficiency of the process. In the corn tortilla process, extrusion cooking is not very effective given a limited reduction of 10 to 25% going up to 46% in the presence of 0.3% of lime or 1.5% of hydrogen peroxide (Peng et al. 2018). As with AFB1, the efficiency in extrusion cooking on OTA reduction depends on various parameters including moisture content, screw speed, barrel temperature, and matrix size. Lee et al.(2017) examined the impact of extrusion processing on the stability of OTA in oat flakes and observed from 0 to 28% of OTA reduction depending on the processing parameters employed. Thus, extrusion is a process widely used for many different processed foods above all cereals, which are matrices particularly contaminated with mycotoxins. Extrusion cooking allows quick detoxification at large-scale production in some cases. Nevertheless, there are still process optimizations to be done for better efficiency to reduce the accumulation of mycotoxins whose initial concentrations on these matrices can be high (De Pilli and Alessandrino 2020).

Irradiation. Food irradiation is a physical method recognized as a reliable and safe method for the preservation of food and feed, promoting food safety and the nutritional value of them. This method has been gradually applied in the food industry. Food irradiation can reduce the amounts of mycotoxins, such as AFB1 and OTA (Calado et al. 2014). Gamma (γ) irradiation is frequently used in the food industry because of its penetration efficiency. The effectiveness of γ irradiation to degrade mycotoxin depends on factors such as matrix composition, water content, radiation dose, types of mycotoxin and its concentration. Moreover, in the literature, γ radiation seems to be a controversial subject. Concerning AFB1, several studies have been conducted on the evaluation of γ irradiation to degrade it in different matrices, such as peanuts, almonds, walnuts, pistachio… (Pankaj et al. 2018a). On the one hand, Zhang et al. (2018) have demonstrated that the application of γ irradiation at 10 kGy in contaminated soybeans results in a significant reduction of AFB1 up to 76%. They also have demonstrated that when the irradiation dose, increased to 30 kGy, no AFB1 was detected in all irradiated samples. On the other hand, Vita et al. (2014) obtained 19% of AFB1 degradation at 15 kGy in almond samples. Concerning OTA, there are lesser studies than for AFB1 but they are also contradictories. Some studies have demonstrated a good effectiveness of γ-irradiation on OTA degradation (reduction rate 60%) in several raw materials (Jalili et al. 2012, Maatouk et al. 2019). There are also studies in which the elimination does not exceed 25% (Vita et al. 2014, Domijan et al. 2015, Calado et al. 2018). As mentioned above, γ-irradiation effectiveness to degrade depends on factors (matrix composition, water content, types of mycotoxin and initial concentration) that may explain these conflicting results. Recently a novel irradiation technology, the Electron Beam Irradiation (EBI) was developed for potential applications for mycotoxin degradation. EBI is more convenient, less costly, and safer than γ-irradiation (Freita-Silva et al. 2015, Liu et al. 2016). Liu et al. (2016) demonstrated the degradation of AFB1 by EBI in aqueous medium. Three different AFB1 initial concentrations were tested (0.5 to 5 ppm) and the treatment by EBI has led to the formation of five degradation products. The toxicity of these degradation products was evaluated and the results indicated that the toxicity of EBI treated samples decreased significantly but did not completely disappear. In addition, it is planned to explore the degradation pathways of AFB1 used during this treatment, which should make it possible to improve the process of applying EBI. Luo et al. (2017) demonstrated that EBI significantly degrade OTA in solution and also that OTA was more easily degraded in corn kernels than in corn flour where the moisture content was lower. However, irradiation has affected corn quality. These recent studies support that EBI method is promising and could become, in the years to come, an effective means of reducing contamination by AFB1 and OTA while ensuring the health of consumers. Degradation of mycotoxins by γ-irradiation and EBI can lead to the appearance of end products which can be toxic. Thus, toxicological assays to evaluate the safety of the radiolytic end products should be conducted. Currently, data on the toxicity of products formed by γ-irradiation of mycotoxins are lacking. Recently, Domijan et al.(2019) have evaluated the toxicity of γ-irradiated mycotoxins AFB1 and OTA *in vitro* on three cell lines. Their results demonstrated that AFB1 and OTA radiolytic products were less toxic than the parent mycotoxins to all of the tested cell lines, thus supporting the use of irradiation to degrade AFB1 and OTA in food matrices.

Cold Plasma. To replace the use of thermal food processing which causes undesirable effects such as color change, texture, and loss of nutrients, cold plasma (CP) technology is considered a non-thermal process, which has generated considerable interest in its use in food processing (Pankaj et al. 2018b). Cold plasma is the fourth state of matter, composed of ions, free electrons, atoms and molecules in their fundamental or excited states (Schmidt et al. 2019). Now, a crucial interest remains to be demonstrated i.e. its effectiveness in reducing mycotoxin concentration. Wielogorska et al. (2019) tested the use of cold atmospheric pressure plasma (CAPP) to treat maize, contaminated with AFB1. With an initial AFB1 concentration of 1.25 ppb, a reduction up to 66% is obtained for AFB1 after 10 min of plasma exposure at a voltage of 6 kV and a continuous frequency of 20 kHz. The majority of AFB1 degradation by-products are obtained by modification terminal furanic cycles without alteration of the lactone cycle, or on the methoxy group corresponding to a less cellular toxicity on HepG2 cells than the initial molecule. However, a potential increase in toxicity must still be studied with other cellular or *in vivo* tests to fully confirm the safety of CAPP, particularly in products intended for consumption. Casas-Junco et al. (2019) investigated the effect of CP on OTA content and also on mycotoxigenic fungi in roasted coffee. From an artificially contaminated roasted coffee with OTA concentrations of 60 to 95 ppb, the CP technology reduces the OTA content by 50% after 30 min of exposure. The authors also explored the effect of the technology on the viability of mycotoxin-producing spores of *Aspergillus* spp. An initial concentration of 5×10^5 spores per g was reduced by 4 log after 6 min of exposure to 850 V. This study did not make it possible to highlight the degradation products obtained. The authors, therefore, recommend carrying out in-depth research to reveal the mechanisms of detoxification. Ultimately, CP is a new physical method with great potential as a postharvest treatment mycotoxin mitigation method (Ten Bosch et al. 2017).

Ultrasound Treatment. Considering the preservation of food by non-thermal processes such as ultrasound represents a challenge for the years to come. This type of treatment must also contribute to the safety of food and feed (Thokchom et al. 2015). Some studies have already tried to demonstrate the effect of this technology on the inactivation of fungal spores (Evelyn and

Silva 2015). They show that ultrasound treatment with an ultrasonic processor (Hielscher UP200S) equipped with a probe of 3 mm (24 kHz - 0.33 W per mL) for 5 min, alone, increases the initial spore concentration by one log. However, the process applied for 15 min leads to 1.8 log spore inactivation. To be even more effective and reduce the time of exposure to ultrasound, the authors demonstrated that a heat treatment at 75°C must be combined with an ultrasound treatment for 10 min to lead inactivation of spores. More research is still needed to design an ultrasonic probe capable of withstanding higher temperatures, then leading to a more efficient process for inactivating fungal spores. Liu et al. (2019a) proposed applying this emerging chemical process to test its efficacy to degrade mycotoxins. An AFB1 aqueous solution (10 ppm) was degraded more than 85% after 80 min of ultrasound exposure in a 550-W power ultrasonic instrument equipped with a 13-mm probe at 20 kHz. This study shows that AFB1 is degraded into eight products whose structure has been identified, implying two degradation pathways, and leading to the rupture of the double bond of AFB1. The reaction products generated from pure AFB1 exposed to ultrasound are complex but the toxicity of these compounds were considerably lower than that of AFB1. Thus, the authors show that ultrasound treatment is effective on aqueous solutions of AFB1 with a reduction in its toxicity under specific conditions, including aqueous solutions. The next logical step that must follow these tests is the application of this treatment in a complex matrix and on several mycotoxins (Liu et al. 2019a, 2019b).

Light Effects: UV Light and PL Treatments. Mycotoxin degradation can be achieved by several light treatments. Several studies report the use of ultraviolet (UV) light and pulsed light (PL) to degrade AFB1. UV radiation has been proved to be efficient in the degradation of AFB1 in water solution. Several UV doses were applied to several AFB1 standard solutions. The obtained results have demonstrated that the increase of UV dosage decreased the aflatoxin-induced cytotoxicity in cells thus suggesting that UV irradiation can be used as an effective technique for the reduction of AFB1 (Patras et al. 2017). However, this experiment was carried out in an aqueous solution allowing a deep penetration of UV rays while the food matrices are more complex and thus UV treatment for AFB1 degradation is limited due to low penetration depth in those matrices (Pankaj et al. 2018a). Some studies have demonstrated that UV can reduce AFB1 concentration in food matrices (Tripathi and Mishra 2010, Liu et al. 2011, Diao et al. 2015, Mao et al. 2016). In these studies, the safety of degradation products was evaluated and results have demonstrated in all cases a lower cytotoxicity that AFB1, thus supporting the use of UV light as a decontamination method. However, according to these studies, the effectiveness of the UV treatment depends on the food matrix, the initial AFB1 concentration and the UV dose used. About PL treatment, it is a relatively novel food processing and preservation technology. It was closely related to UV treatment but with greater effectiveness. PL process consists of the emission of flashes of very intense white light (broad spectrum of wavelengths from UV to near-infrared) by xenon lamps in a very short time (μs to ms). PL has been demonstrated to be an effective method to decontaminate the surface of food and raw material by destroying bacteria, fungi, and viruses (Elmnasser et al. 2007). Over the last ten years, other studies have demonstrated the effectiveness of pulsed light in the degradation of mycotoxins. Moreau et al. (2013) were the first to report that PL can degrade several mycotoxins. They have demonstrated that eight flashes of PL (1 J cm^{-2} during 300 μs) degrade 93% and 98% of AFB1 and OTA in water solution. Moreover, the treatment of AFB1 by PL eliminate its mutagenic potential (Moreau et al. 2013). AFB1 decontamination by PL was tested in food matrices such as peanuts, peanut oil and rice and was efficient in all studied cases with a reduction rate up to 91% depending on the treated matrix (Wang et al. 2016, Abuagela et al. 2018a, 2018b).

Adsorbing Agents. In the bibliographic review of Vilat-Dona et al. (2018), there is a definition of adsorbing agents. The authors point out that these agents are divided into three groups, which are inorganic, synthetic or organic agents, all of them being effective in binding mycotoxins present in foodstuffs and thus decontaminate these matrices. Note that in mycotoxin adsorbents of organic origin, everything related to adsorption *via* microorganisms will be treated in the next part below entitled 'Biological decontamination'. Those adsorbing agents are also called mycotoxins binders, sequestering agents or adsorbents (Vila-Donat et al. 2018). The mode of action of eliminating the toxin by adsorption is not always well studied. However, mycotoxin binders can be of different types and their effectiveness depends on the chemical structure, the dimensions of pores and the available surface. The feed or food composition can also have a major impact on adsorption effectiveness (Dragacci et al. 2009). From a nutritional and/or organoleptic point of view, mycotoxin binding is unfortunately often associated with a degradation of the treated matrix due to the adsorption of beneficial elements on these mycotoxin binders (Kihal et al. 2020).

In some cases, the addition of mycotoxin binders in feed is a strategy to limit the absorption of mycotoxins in the gastrointestinal tract, thus limiting their negative effects on animals (Gallo et al. 2010). The formed complexes (mycotoxin + binder) pass through the animals and are eliminated *via* the feces. The Commission Regulation No. 386/2009 regulates the use of such binders as feed additives in the European Union. These compounds are specified as 'substances for reduction of the contamination of feed by mycotoxins: substances that can suppress or reduce the absorption, promote the excretion of mycotoxins or modify their mode of action' (European Commission (EC) 2009).

As inorganic adsorbing agents are the best known and used as feed additives to reduce exposure of animals to mycotoxins are aluminosilicate which includes several mineral clays such as calcium aluminosilicate sodium hydrate (HSCAS), bentonite, montmorillonite, smectite and zeolite. All of these adsorbents are known to have very good binding efficiency with mycotoxins (Pappas et al. 2016, Bhatti et al. 2017, Vila-Donat et al. 2018, Wang et al. 2018, Zavala-Franco et al. 2018, Rajendran et al. 2019, Haque et al. 2020).

Clays can be used in their natural state or treated through various processes in order to improve their efficiency in binding some mycotoxins. To illustrate this approach, the recent work carried out by Rejeb et al. (2019) describes the use of Tunisian mineral clay as a feed additive. The effect of heat treatment on the clay, i.e. calcination (550°C for 5 h) was tested. During the calcination process, it was possible to observe a decrease in the cation exchange capacity and the specific surface and on the contrary an increase in the size of the pores, which could lead to a modification of adsorption properties on the treated clay. In this study, the authors demonstrated that the two clays (untreated and calcined) were able to bind AFB1 (5 ppb) at pH 3.0 and pH 7.0 to more than 90% in poultry feed with even an improvement in favor of calcined clays probably due to the increase in pore size during the calcination process. The authors conclude that this calcined clay has the potential to be used as a mycotoxin binder in poultry feed and probably for other animal species.

Haque et al. (2020) described the latest advances in the development of detoxification methods and alternative ones to reduce the presence of mycotoxins in food and feed and thus limit human and animal exposure to these deleterious molecules for their health. As innovative methods, they cited in particular the nanobiotechnology with the use of nanomaterials such as silver nanoparticles (AgNPs) which could be effective in binding AFB1 and OTA from maize-based medium (Gómez et al. 2019). González-Jartín et al. (2019) presented a green technology based on the development of 25 magnetic nanostructured materials to remove the main types of mycotoxins from liquid food matrices. These detoxifying agents have the great feature of being magnetically separated from the liquid matrix. The authors showed that these nanoparticles could eliminate up to 87% of mycotoxins from aqueous solutions, with a maximum adsorption capacity of 450 µg of toxin for each g of nanoparticles. To stay in the adsorption of mycotoxins via nanomaterials, Ji and Xie (2020) designed magnetic graphene composite adsorbents, magnetic graphene (MrGO) and magnetic graphene oxide (MGO), to remove AFB1 from a specific substrate, contaminated oil and more specifically rice bran oil. This type of emerging adsorbent is particularly interesting because it offers a large contact surface and it has a porous structure. In addition, the recovery of these types of adsorbents after use is facilitated by magnetic separation. Similarly, its recycling is easy with minimal weight loss due to magnetic attractions. Adsorption experiments are carried out with artificially contaminated rice bran oil with AFB1 concentrations of 5 to 250 ppb in batch conditions with an appropriate dose of 20 mg per mL of adsorbing agent. The results revealed that MGO and MrGO were both able to eliminate AFB1 in 60 min with almost the same efficiency of more than 85%. Since GO is easier to prepare, MGO appears to be a promising new composite adsorbent for the edible oil industry to decontaminate them and make them mycotoxin free (Ji and Xie 2020). Always in the category of synthetic adsorbents, the work of Arak et al. (2019) attempt to assess the adsorption efficiency of a molecularly imprinted polymer (TMU95) at different pH values, in comparison with a commercial binder. The experiments are carried out *in vivo* on AFB1 contaminated feed for ducklings.

The results show that this synthetic polymer (TMU95) is as effective as the commercial adsorbent in adsorbing AFB1 and thus makes it possible to reduce the exposure of ducklings to AFB1 to improve their growth.

As organic mycotoxin binders, the use of activated carbon is widespread. Indeed, the adsorbent capabilities of activated carbon (AC) were recognized early on (Jindal et al. 1994, Var et al. 2008, Espejo and Armada 2009, Olivares-Marín et al. 2009). Activated carbon is often in the form of a water-insoluble black powder obtained by pyrolysis of different types of organic materials. Its adsorbent capacities vary according to its porosity (contact surface), the medium in which it is used (aqueous/non-aqueous, pH ...) and the type of organic material used to obtain it. Activated carbon powder can be used as feed additive binding mycotoxins to inhibit their absorption from the gastrointestinal tract. For example, activated charcoal has been shown to be efficient in detoxifying AFB1 and OTA in contaminated poultry feed (Bhatti et al. 2018). Recently, another form of activated carbon, other than powder, has been tested to adsorb OTA, i.e. in the form of carbon fibers to facilitate its application in particular in winemaking (El Khoury et al. 2018). During this process, the use of oenological fining agents such as chitosan, bentonite, chitin, egg albumin, or potassium caseinate and activated carbon powders (ACPs) are authorized and can reduce OTA concentration (Castellari et al. 2001, Quintela et al. 2013, Balcerek et al. 2017, Sun et al. 2017). Among them, ACPs are the most efficient but it can alter the organoleptic properties and color of wine by adsorbing molecules of interest such as polyphenols (Castellari et al. 2001) and a filtration step is required to remove ACPs after the treatment. Recently, a study has demonstrated that activated carbon fibers (ACFs) can be used to reduce OTA concentration in wine. Indeed, they show a similar adsorption capability that ACPs but they have supplementary advantages. They can be easily removed after treatment and could be used in the form of tissues in the grape juice or winemaking process (El Khoury et al. 2018). Currently, some work tends to develop new organic mycotoxin binders from agricultural by-products (Fernandes et al. 2019, Greco et al. 2019). The advantage of this approach is to recover agricultural waste that would be thrown away and to provide sequestering agents at a very low cost. However, before proposing these innovative binders, it is necessary to study their effectiveness in adsorbing one or more mycotoxins. Greco et al. (2019) compared more than 50 agricultural by-products in sequestering mycotoxins, including AFB1 and OTA. The selected matrices include a large range of food plants and by-products including waste products from industrial and agricultural operations. For example, pea seed and pea seed pod, onion leaf and bulbs, olive tree branches and leaves, broccoli and celery stalks, orange in the form of waste, pulp and peel, lemon pulp and peel. The screening experiments were carried out with 1 and 10 mg per mL of mycotoxin binders in the presence of 1 ppm of AFB1, OTA and a combination of five mycotoxins was tested. As a major result, 66% AFB1 were adsorbed on Sangiovese grape marc at pH 3 and 38% OTA on the almond shell at pH 3. When the five mycotoxins are mixed, AFB1 is the most adsorbed and OTA the least on. Finally, among all those supports, it is grape marc, artichoke waste and almond hulls which are the most promising for adsorbing AFB1 and OTA. The best multi-mycotoxin adsorbent products were those rich in undegradable fibers (lignin, cellulose) and flavonoids, such as grape wastes, artichoke, and almond by-products. A recent study has focused on the adsorption of AFB1 and OTA from by-products of the wine chain and that of olive oil and more particularly on micronized grapes stems and olive pomace (Fernandes et al. 2019). Of these two agro-by-products, the one from the vine is the most effective to bind AFB1 at pH 2.0 and pH 7.0 and to bind OTA at pH 2.0:10 g per L of binder are enough to bind at least 90% of both mycotoxins. Equal efficiency was obtained with olive pomace by increasing the concentration of the binder to 30 g/L. Despite these good adsorption results, the best binder remains activated carbon, which was chosen as a control in this experiment and which shows the maximum adsorption capacity for both mycotoxins and regardless of the pH value.

3.2 Chemical Treatments

Besides this, many studies reported the use of chemical methods to reduce mycotoxin concentrations. Among these methods, there are conventional methods such as acid and base treatments (ammoniation and nixtamalisation), oxidizing agents, organic acids and novel approaches such as neutral electrolyzed oxidizing water (NEW) which can degrade AFB1 (Peng at al. 2018, Schmidt et al. 2019). The application of chemical methods can lead to the appearance of mycotoxin residues on processed food, and for which the safety must be validated.

Ammoniation. It is a chemical process that uses a gaseous ammonia or aqueous ammonium hydroxide to treat food and feed. It has been used since the 90s, and has proved to be effective in treating cultures contaminated with AFB1 or OTA (corn, wheat, barley). For AFB1, ammoniation creates openness of the lactone cycle to form the hydroxy amide that is less toxic than the original molecule, depending on the process parameters: the amount of ammonia applied, the duration of the reaction, and the temperature and pressure levels (Temba et al. 2016). Despite its proven efficiency to reduce AFB1 and OTA concentration, this method is not permitted in the European Community (EC) for human food.

Oxydizing Agents. In 2016, Agriopoulou et al. demonstrated the degradation of aflatoxins in the presence of ozone at concentrations of 8.5 to 40 ppm. Even lowest ozone concentrations were effective in completely degrading 2 or 10 ppb of AFB1 and AFG1, this is not the case with AFB2 and AFG2 which are more stable. On sultanas, 12.5 ppm of ozone led to the reduction of OTA concentration, up to 82% in 240 min without causing a significant decrease in the concentration of phenolic substances (Torlak 2018). In conclusion, chemical methods are effective in reducing mycotoxin concentrations, but the main problem is related to the potential residues which can persist in the treated matrix.

Concerning the neutral electrolyzed oxidizing water (NEW) method, the results are variable according to the matrices treated. Thus, this method would be effective in reducing AFB1 in peanuts but would have no effect on contaminated corn, retaining the toxic effects identical to the untreated control (Peng et al. 2018). Other chemical processes are being studied for their effectiveness in reducing mycotoxins concentration and represent a major challenge for human and animal health.

3.3 Biological Detoxification

As mentioned above, the physical and chemical methods have been so far widely applied in food/feed industries as curative methods to reduce the concentration of mycotoxins. Unfortunately, they have drawbacks such as the destruction of nutritional and organoleptic properties or the production of residues having negative effects on human and animal health. Another approach which is of great concern to the scientific community is the application of biological methods to reduce the accumulation of mycotoxins. These are presented as being healthier, respectful of the nutritional and organoleptic qualities of processed foods and therefore more promising in their application (Chen et al. 2018, Peng et al. 2018). In curative biological methods, two major mechanisms are mentioned, i.e. the adsorption and biodegradation of mycotoxins. In both cases, microorganisms are likely to intervene. They can adsorb mycotoxins on their cell wall; for example, they can be used as biotransformation agents and finally as producers of enzymes used in purified form or not.

Adsorption. The adsorption phenomenon by mycotoxin binders of microbial origin has been well documented in the past five years. Sometimes part of the cell wall is used in this binding process. Bzducha-Wrobel et al. (2019) tested yeast cell walls from *Candida utilis* as mycotoxins trap. Different preparation of cell walls (cell walls or β(1,3)/(1,6)-glucan (β-G)) at 1% and several mycotoxins including AFB1 and OTA at 500 ppb for each one were tested. Depending on the cell wall preparations and mycotoxins, the percentage of mycotoxin adsorption varied. The maximum of 80% was obtained for OTA and AFB1 with cell walls and β-G preparations at pH 3. In other studies, whole cells are considered. For example, yeast cells have been immobilized and evaluated for their capacity to decontaminate a matrix. Farbo et al. (2016) encapsulated *Candida intermedia* yeast cells (2.5 x 10^{10} CFU per mL of living or autoclaved cells) in alginate beads and added them to commercial grape juice spiked with OTA (20 ppb). After 48 h of incubation, a significant reduction (70%–80%) of the total OTA content was achieved with living and autoclaved cells. However, a prolonged stay (from 72 to 140 h) led to the slow release of OTA from the alginate beads in the treated grape juice. The authors proposed an original application of this process. They designed a prototype bioreactor, consisting of a glass chromatography column, filled with these types of beads. Optimizing this process aimed to eliminate OTA traces from liquid matrices while avoiding the phenomenon of desorption. Currently, probiotic bacteria, known for their beneficial effect on human health, are also being tested for their ability to reduce the accumulation of mycotoxins (Sarlak et al. 2017, Azeem et al. 2019). *Lactobacillus acidophilus* has been reported to be effective in the elimination of AFM1 from contaminated milk during the fermentation of a traditional yoghurt-based drink and its storage (Sarlak et al. 2017). Other *Lactobacillus* isolates, *Lb. gallinarum, Lb reuteri, Lb fermentum,* and *Lb paracasei* were tested *in vitro* against toxigenic *A. flavus* (AFB1 producer) by well diffusion assay. In both studies, the authors claim that AFB1 is adsorbed on those probiotic bacteria without giving further details in the mechanism involved. A major concern for the dairy industries is how to manage the presence of AFM1 in milk, following the consumption of AFB1 contaminated feed. One of the solutions is the addition of adsorbents of microbial origin, bacteria and/or yeast, in milk before its use for dairy products or direct consumption (Assaf et al. 2020). Numerous probiotic bacteria and yeast are referred to be particularly efficient to adsorb AFM1. Yeast is the most effective: for example, *Saccharomyces cerevisiae* eliminate nearly 90% of AFM1 (0.5 ppb) after 1 h. Assaf et al. (2020) discussed the interest of studying the stability of the mycotoxin-binder complex formed in order to control the mycotoxin release phenomenon. Thus, for example, the heat treatment of bacteria, yeast or mixtures, the working temperature for binding experiments, the washing steps, etc. are all important parameters to take into account to optimize the binding mechanism and its stability. A new source of effective microbial adsorbents is the use of fungal mycelium with original properties, i.e. pleasant to taste and with nutritional value. Haidukowski et al. (2019) determined the adsorption capacity of AFB1 from the non-viable mycelium of *Pleurotus eryngii*, an edible fungus. The adsorbent material (1000 mg), obtained from lyophilized mycelium, can adsorb up to 85% of AFB1 (50 ppb) at 37°C at pH 5 and 7. The percentage of desorption was 10% ± 4% at pH 3 and 7% ± 4% at pH 7.4. These results indicate good system stability. This study showed that this new microbial adsorbent that is completely different from the materials currently used in the industry could be used as a low cost and effective feed additive for AFB1 decontamination.

Living microorganisms. There is a very large amount of work dealing with the use of living microorganisms in the biodegradation of mycotoxins as a curative method. One potential explanation is that bacteria (De Bellis et al. 2015, Adebo et al. 2017, Ben Taheur et al. 2019), yeast (Chlebicz and Śliżewska 2020) and fungi (Sun et al. 2016, Xiong et al. 2017) showed real abilities due to their great diversity. However, in most cases, authors demonstrated the effectiveness of living microorganisms in degrading mycotoxins and assumed the intervention of extracellular enzymes. Unfortunately, they do not continue the purification works. Pfliegler et al. (2015) highlights the effectiveness of several genera of yeast (*Saccharomyces cerevisiae, Geotrichum candidum, Kluyveromyces marxianus* and *Metschnikowia pulcherrima*) in the degradation of mycotoxins, including OTA degraded in OTa by splitting the amide bond. The authors suggested that the degradation processes could be improved by heterologous production of mycotoxin-degrading enzymes in transgenic yeast. This approach may be particularly suitable in the fermentation industry to increase the efficiency of yeast strains. Pflieger et al. (2015) suggested as future prospects that the most appropriate solution would be to obtain hybrid yeast on the selection of desired characteristics such as the degradation of OTA for the use of these yeast in the winemaking process, in order to comply with European regulation CE N° 123/2005. Verheecke et al. (2016)reported the microbial degradation of AFB1 from *Actinobacteria, Bacillus, γ-Proteobacteria* and *α-Proteobacteria* classes, with particular attention to the mechanisms involved (mostly through extracellular enzymes) as well as to the generated degradation products. They underlined the great diversity of the protocols used during these experiments, which complicates the comparative analysis of the results. They also point out the lack of *in vivo* tests to better assess the toxicity of these curative treatments. Among the *Bacillus* class, Liu et al. (2018a) demonstrated the capabilities of lactic acid bacteria such as *Lactobacillus acidophilus, Lb. plantarum,* and *Enterococcus faecalis* to degrade AFB1 as effective as clays (HSCAS) commonly used in feed for female broiler chicks. Experiments were carried out with artificially contaminated feed rations with 40 ppb of AFB1 and inoculation of LAB at 10^{10} CFU/kg. LAB supplementation feed improved the growth performance of female broiler chicken, and the effect of LAB was greater than HSCAS. Moreover, the reduction of AFB1 residues in tissues and excrement indicates the biodegradation of the mycotoxin by these bacteria. From γ-Proteobacteria, *Pseudomonas putida* is tested for its ability to degrade AFB1 (Singh et al. 2019). Both, culture and supernatant led to a reduction of 80% in AFB1 within 24h. Through various experiments, the authors demonstrated that the active ingredient seems to be of a protein nature, that the growing cells do not seem essential. They concluded that the AFB1 degradation might be an enzymatic process. The class of *Actinobacteria* is of great importance for finding bacterial species able of achieving biological degradation of mycotoxins (Harkai et al. 2016, Prettl et al. 2017, Risa et al. 2018). The ability of *Rhodococcus pyridinivorans* to degrade AFB1 was conducted on by-products during the production of bioethanol from contaminated corn (Prettl et al. 2017). Their results demonstrated that AFB1 content of the stillage significantly decreased (more than 63% in the solid phase and 75% in the liquid phase) between the 3rd and 7th days correlated to the evolution of microbial growth. The authors hope to be able to reduce the concentration of AFB1 below the strict regulatory limit after optimizing the viability of the bacteria. The genus *Streptomyces* is particularly known for its incredible ability to produce bioactive metabolites (Liu et al. 2018b). Harkai et al. (2016) explored the potential of 124 strains to degrade AFB1 (1 ppm) in liquid medium. After the screening, they were able to select one isolate degrading AFB1 by up to 88% and totally eliminating genotoxicity. For the past five years, several studies reported the use of fungi for the degradation of mycotoxin as postharvest treatment. Out of 20 *Aspergillus niger* isolates (10^6 conidia/mL), 14 were able to completely inhibit AFB1 production when they are cocultivated with *A. flavus* (10^6 conidia/mL) in liquid agitated medium at 28°C for 15 days (Xing et al. 2017). The authors claimed that finally, one *A. niger* isolate was able to down-regulate 19 of 20 AFB1 biosynthetic genes when 2% of its culture filtrate was added to the *A.*

flavus growth medium. They demonstrated the good potential of *A. niger* to be used as BCA. Out of 130 fungal isolates from traditional Korean meju, two isolates identified as *A. tubingensis*, a black *Aspergillus* as *A. niger*, were selected for OTA-biodegradation activity (Cho et al. 2016). Both isolates inoculated on OTA supplemented solid medium degraded OTA (40 ppb) by more than 95% after 14 days in OTa. The crude enzymes of the two fungal cultures were able to degrade up to 97% of OTA (40 ppb) in OTa in 24 h at pH 7.0. Those results demonstrated that *A. tubingensis* isolate and their targeted enzymes exerted a great potential to reduce OTA contamination in food and feed.

Degradation by enzymes. The challenge of finding effective enzymes to catabolize, cleave or transform mycotoxins into less or non-toxic metabolites lies in the specificity of the enzyme-substrate bond. The chemical structure of the targeted mycotoxin has an impact on the type of enzyme involved: hydrolase, transferase, epimerase and oxidoreductase. AFB1 has a coumarin structure which is a derivative of the lignin monomeric l-p-coumaryl alcohol. Thus, enzymes such as laccase or manganese peroxidase degrading lignin are good potential candidates for the AFB1 degradation. Wang et al. (2019b) explored the capacity of recombinant produced manganese peroxidases (MnPs), responsible for the oxidative depolymerization of lignin, to degrade four major mycotoxins, including AFB1 and OTA. As expected, AFB1 (5 ppm) was the most rapidly degraded mycotoxin up to 85% in 3 h, while MnPs showed only negligible activity on OTA (50 ppm), which has a different structure. The authors demonstrated that the addition of malonate in the reaction medium improved the degradation of all mycotoxins except that of OTA. The authors suggested that it is the free radicals derived from malonate which must play an important role to improve the mycotoxins degradation by reacting with substrates. In another work, Wang et al. (2019a) use a recombinant bacterial laccase and a fungal one, supplemented by chemical mediators, in particular methyl syringate for the production of highly reactive free radicals which improves the enzymatic reaction. Under their experimental conditions, the degradation rate of AFB1 (5 ppm) was 44.5% after 30 min, reaching 90.8% after 5 h. In these two studies, the authors observed a very reduced toxicity for the degradation products of AFB1. The use of laccase enzyme type supplemented by chemical mediators to improve their efficiency is in line with promising use in food. To adapt the use of such enzymes to liquid matrices in particular, Ren et al. (2019) developed a multifunctional ultrafiltration membrane named metal-organic frameworks (MOFs), a new class of porous inorganic–organic hybrid materials, to immobilize peroxydase-like enzymes. Peroxidase like MOFs such prepared improves the stability of enzymes and are more efficient to remove AFB1. The two mechanisms of adsorption and enzymatic degradation contribute to the reduction of the accumulation of AFB1 thanks to a strong AFB1 elimination at 8 h, related to the strong capacity of adsorption of MOF. Thus, AFB1 could be eliminated from contaminated liquid matrices by treatment on this type of membrane, producing degradation products from which the toxic groups are cleaved. The main advantage would be the potential regeneration of this type of modified membrane for prolonged use over time, with proven effectiveness and significant economic gain. For OTA, studies on its biodegradation date back to 1969 with the use of bovine pancreatic carboxypeptidase A to cleave OTA at the amide bond leading to OTa, which is less toxic than the initial molecule. Since then, several studies have been carried out on carboxypeptidases of various origins and their effectiveness on the OTA degradation (Chen et al. 2018). An enzyme isolated from *A. niger* was able to degrade OTA (1 µg per mL) in OTa up to 99.8% after 25 h of incubation at pH 7.5 and 37°C while at pH 3, the maximum degradation of OTA reaches only 3%. A recombinant ochratoxinase has been shown to be approximately 600 times more effective in degrading OTA at pH 7.5. However, the maximum activity was obtained at 66°C at pH 6 (Chen et al. 2018). It should be noted that the hydrolysis of OTA to OTa by carboxypeptidases is the natural mechanism most often reported to detoxify contaminated feed, directly by the rumen microflora in ruminants (Tao et al. 2018). This hydrolysis mechanism is therefore beneficial for the animal and mainly studied in vitro to detoxify all foodstuffs.

Potential toxicity of degraded or side products and regulation. As presented in this second part, there is a very wide variety of curative methods to decontaminate and/or detoxify AFB1 and OTA in food and feed. Many of them are only allowed for animal feed. European Commission Regulation 2015/786 defines the criteria for the acceptability of detoxification processes applied to products intended for animal feed. The main problem in obtaining authorization to apply all these methods for food is the lack of data on the toxicity of these methods. The degradation of mycotoxins can indeed lead to the formation of degraded or sideproducts which can be less or more toxic than the original compound. Without complete toxicological studies on these newly formed compounds generated during the physical, chemical or biological degradation treatments, the toxicity of these compounds must be presumed to have at most the same toxicity of the parent mycotoxin. Finally, the use of adsorbents, which do not modify the parental mycotoxin, could be the best solution for decontaminating food matrices once the release phenomenon is perfectly managed. In conclusion, none of these treatments led to the total decontamination of contaminated matrices. Thus, a combination of treatments would be the most appropriate strategy to achieve the levels of contamination accepted by the regulations (EU N° 1881/2006, EU N° 1126/2007).

References

Abbas, A., Valez, H. and Dobson, A.D.W. 2009. Analysis of the effect of nutritional factors on OTA and OTB biosynthesis and polyketide synthase gene expression in *Aspergillus ochraceus*. International Journal of Food Microbiology 135: 22–27. doi:10.1016/j.ijfoodmicro.2009.07.014.

Abbas, H., Zablotowicz, R., Bruns, H.A. and Abel, C. 2006. Biocontrol of aflatoxin in corn by inoculation with non-aflatoxigenic *Aspergillus flavus* isolates. Biocontrol Science and Technology 16: 437–449. doi:10.1080/09583150500532477.

Abbas, H.K., Accinelli, C. and Shier, W.T. 2017. Biological control of aflatoxin contamination in U.S. crops and the use of bioplastic formulations of *Aspergillus flavus* biocontrol strains to optimize application strategies. Journal of Agricultural and Food Chemistry 65: 7081–7087. doi:10.1021/acs.jafc.7b01452.

Abbaszadeh, S., Sharifzadeh, A., Shokri, H., Khosravi, A.R. and Abbaszadeh, A. 2014. Antifungal efficacy of thymol, carvacrol, eugenol and menthol as alternative agents to control the growth of food-relevant fungi. Journal de Mycologie Medicale 24: e51–e56. doi:10.1016/j.mycmed.2014.01.063.

Abraham, A., Saka, V., Mhango, W., Njoroge, S. and Brandenburg, R. 2016. Effect of crop rotation on aflatoxin contamination in groundnuts. *In*: RUFORUM Working Document Series (ISSN 1607-9345) 14: 173–178. Available in: http://repository.ruforum.org

Abuagela, M.O., Iqdiam, B.M., Mostafa, H., Gu, L., Smith, M.E. and Sarnoski, P.J. 2018a. Assessing pulsed light treatment on the reduction of aflatoxins in peanuts with and without skin. International Journal of Food Science and Technology 53: 2567–2575. doi:10.1111/ijfs.13851.

Abuagela, M.O., Iqdiam, B.M., Baker, G.L. and MacIntosh, A.J. 2018b. Temperature-controlled pulsed light treatment: Impact on aflatoxin level and quality parameters of peanut oil. Food and Bioprocess Technology 11: 1350–1358. doi:10.1007/s11947-018-2105-6.

Adebo, O.A., Njobeh, P.B., Sidu, S., Adebiyi, J.A. and Mavumengwana, V. 2017. Aflatoxin B1 degradation by culture and lysate of a *Pontibacter* specie. Food Control 80: 99–103. doi:10.1016/j.foodcont.2017.04.042.

Agriopoulou, S., Koliadima, A., Karaiskakis, G. and Kapolos, J. 2016. Kinetic study of aflatoxins' degradation in the presence of ozone. Food Control 61: 221–226. doi:10.1016/j.foodcont.2015.09.013.

Ahou Kouadio, I., Koffi Ban, L. and Bretin Dosso, M. 2019. Prevention of ochratoxin A (OTA) production in coffee beans using natural antifungal derived from *Solanum indicum* L. green berries. Journal of Food Security 7: 63–71. doi:10.12691/jfs-7-3-1.

Alaniz Zanon, M.S., Clemente, M.P. and Chulze, S.N. 2018. Characterization and competitive ability of non-aflatoxigenic *Aspergillus flavus* isolated from the maize agro-ecosystem in Argentina as potential aflatoxin biocontrol agents. International Journal of Food Microbiology 277: 58–63. doi:10.1016/j.ijfoodmicro.2018.04.020.

Amaike, S. and Keller, N.P. 2011. *Aspergillus flavus*. Annual Review of Phytopathology 49: 107–133. doi:10.1146/annurev-phyto-072910-095221.

Amna, M.A. 2017. Activity of Thuja (*Thuja orientalis*) alcoholic extract in inhibition *Aspergillus flavus* growth and detoxification of aflatoxin B1 in contaminated corn seeds. Pakistan Journal of Biotechnology 14: 503–506.

Arak, H., Torshizi, M.A.K., Hedayati, M. and Rahimi, S. 2019. The first *in vivo* application of synthetic polymers based on methacrylic acid as an aflatoxin sorbent in an animal model. Mycotoxin Research 35: 293–307. doi:10.1007/s12550-019-00353-z.

Assaf, J.C., Nahle, S., Louka, N., Chokr, A., Atoui, A. and El Khoury, A. 2020. Assorted methods for decontamination of aflatoxin. Toxins 11: 304. doi: 10.3390/toxins11060304.

Atehnkeng, J., Ojiambo, P.S., Ikotun, T., Sikora, R.A., Cotty, P.J. and Bandyopadhyay, R. 2008. Evaluation of atoxigenic isolates of *Aspergillus flavus* as potential biocontrol agents for aflatoxin in maize. Food Additives and Contaminants – Part A: Chemistry, Analysis, Control, Exposure and Risk Assessment 25: 1264–1271. doi:10.1080/02652030802112635.

Atehnkeng, J., Donner, M., Ojiambo, P.S., Ikotun, B., Augusto, J., Cotty, P.J. and Bandyopadhyay, R. 2016. Environmental distribution and genetic diversity of vegetative compatibility groups determine biocontrol strategies to mitigate aflatoxin contamination of maize by *Aspergillus flavus*. Microbial Biotechnology 9: 75–88. doi:10.1111/1751-7915.12324.

Audenaert, K., Vanheule, A., Höfte, M. and Haesaert, G. 2013. Deoxynivalenol: A major player in the multifaceted response of *Fusarium* to its environment. Toxins 6: 1–19. doi:10.3390/toxins6010001.

Azeem, N., Nawaz, M., Anjum, A.A., Saeed, S., Sana, S., Mustafa, A. and Yousuf, M.R. 2019. Activity and anti-aflatoxigenic effect of indigenously characterized probiotic *Lactobacilli* against *Aspergillus flavus*—A common poultry feed contaminant. Animals 9: 166. doi:10.3390/ani9040166

Bailly, S., El Mahgubi, A., Carvajal-Campos, A., Lorber, S., Puel, O., Oswald, I.P., Bailly, J.D. and Orlando, B. 2018. Occurrence and identification of *Aspergillus* section *Flavi* in the context of the emergence of aflatoxins in french maize. Toxins 10: 525. doi:10.3390/toxins10120525.

Balcerek, M., Pielech-Przybylska, K., Patelski, P., Dziekońska-Kubczak, U. and Jusel, T. 2017. Treatment with activated carbon and other adsorbents as an effective method for the removal of volatile compounds in agricultural distillates. Food Additives and Contaminants – Part A: Chemistry, Analysis, Control, Exposure and Risk Assessment 34: 714–727. doi:10.1080/19440049.2017.1284347.

Barcelo, J. and Barcelo, R. 2018. Post-harvest practices linked with ochratoxin A contamination of coffee in three provinces of Cordillera Administrative Region, Philippines. Food Additives & Contaminants 35: 328–340. doi:10.1080/19440049.2017.1393109

Barka, E.A., Vatsa, P., Sanchez, L., Gaveau-Vaillant, N., Jacquard, C., Klenk, H.-P., Clément, C., Ouhdouch, Y. and Van Wezel, G.P. 2016. Taxonomy, physiology and natural products of *Actinobacteria*. Microbiology and Molecular Biology Reviews 80: 1–43. doi:10.1128/MMBR.00019-15.

Battilani, P., Logrieco, A., Giorni, P., Cozzi, G., Bertuzzi, T. and Pietri, A. 2004. Ochratoxin A production by *Aspergillus carbonarius* on some grape varieties grown in Italy. Journal of the Science of Food and Agriculture 84: 1736–1740. doi:10.1002/jsfa.1875

Battilani, P., Magan, N. and Logrieco, A. 2006. European research on ochratoxin A in grapes and wine. International Journal of Food Microbiology 111: S2–S4. doi:10.1016/j.ijfoodmicro.2006.02.007.

Battilani, P., Toscano, P., Van der Fels-Klerx, H.J., Moretti, A., Camardo Leggieri, M., Brera, C., Rortais, A., Goumperis, T. and Robinson, T. 2016. Aflatoxin B1 contamination in maize in Europe increases due to climate change. Scientific Reports 6: 24328. doi:10.1038/srep24328.

Bellí, N., Pardo, E., Marín, S., Farré, G., Ramos, A.J. and Sanchis, V. 2004. Occurrence of ochratoxin A and toxigenic potential of fungal isolates from

Spanish grapes. Journal of the Science of Food and Agriculture 84: 541–546. doi:10.1002/jsfa.1658.

Ben Taheur, F., Mansour, C., Kouidhi, B. and Chaieb, K. 2019. Use of lactic acid bacteria for the inhibition of *Aspergillus flavus* and *Aspergillus carbonarius* growth and mycotoxin production. Toxicon 166: 15–23. doi:10.1016/j.toxicon.2019.05.004.

Bhatnagar-Mathur, P., Sunkara, S., Bhatnagar-Panwar, M., Waliyar, F. and Sharma, K.K. 2015. Biotechnological advances for combating *Aspergillus flavus* and aflatoxin contamination in crops. Plant Science 234: 119–132. doi:10.1016/j.plantsci.2015.02.009.

Bhatnagar, D. and McCormick, S.P. 1988. The inhibitory effect of neem (*Azadirachta indica*) leaf extracts on aflatoxin synthesis in *Aspergillus parasiticus*. Journal of the American Oil Chemists' Society 65: 1166–1168. doi:10.1007/BF02660575.

Bhatti, S.A., Khan, M.Z., Kashif, M., Saqib, M., Khan, A. and Ul-Hassan, Z. 2017. Protective role of bentonite against aflatoxin B1- and ochratoxin A-induced immunotoxicity in broilers. Journal of Immunotoxicology 14: 66–76. doi:10.1080/1547691X.2016.1264503.

Bhatti, S.A., Khan, M.Z., Hassan, Z.U., Saleemi, M.K., Saqib, M., Khatoon, A. and Akhter, M. 2018. Comparative efficacy of Bentonite clay, activated charcoal and *Trichosporon mycotoxinivorans* in regulating the feed-to-tissue transfer of mycotoxins. Journal of the Science of Food and Agriculture 98: 884–890. doi:10.1002/jsfa.8533.

Bhavaniramya, S., Vishnupriya, S., Al-Aboody, M.S., Vijayakumar, R. and Baskaran, D. 2019. Role of essential oils in food safety: Antimicrobial and antioxidant applications. Grain & Oil Science and Technology 2: 49–55. doi:10.1016/j.gaost.2019.03.001.

Bleve, G., Grieco, F., Cozzi, G., Logrieco, A. and Visconti, A. 2006. Isolation of epiphytic yeasts with potential for biocontrol of *Aspergillus carbonarius* and *A. niger* on grape. International Journal of Food Microbiology 108: 204–209. doi:10.1016/j.ijfoodmicro.2005.12.004.

Bzducha-Wróbel, A., Bryła, M., Gientka, I., Błażejak, S. and Janowicz, M. 2019. *Candida utilis* ATCC 9950 cell walls and β(1,3)/(1,6)-glucan preparations produced using agro-waste as a mycotoxins trap. Toxins 11: 192. doi:10.3390/toxins11040192.

Cabañes, F.J., Bragulat, M.R. and Castellá, G. 2013. Characterization of nonochratoxigenic strains of *Aspergillus carbonarius* from grapes. Food Microbiology 36: 135–141. doi:10.1016/j.fm.2013.05.004.

Cabañes, F.J., Sanseverino, W., Castellá, G., Bragulat, M.R., Cigliano, R.A. and Sanchez, A. 2015. Rapid genome resequencing of an atoxigenic strain of *Aspergillus carbonarius*. Scientific Reports 5: 9086. doi:10.1038/srep09086.

Caceres, I., Colorado Vargas, R., Muñoz, E.S., Muñoz, L.N. and Hernandez Ochoa, L. 2014. Actividad Antifúngica *in vitro* de Extractos Acuosos de Especias contra *Fusarium oxysporum, Alternaria alternata, Geotrichum candidum, Trichoderma* spp., *Penicillum digitatum* y *Aspergillus niger*. Revista Mexicana de Fitopatologia 31: 105–112. Available in: http://www.scielo.org.mx/scielo.php?script=sci_arttext&pid=S0185-33092013000200003&lng=es&tlng=es.

Caceres, I., El Khoury, R., Medina, Á., Lippi, Y., Naylies, C., Atoui, A., El Khoury, A., Oswald, I.P., Bailly, J.D. and Puel, O. 2016. Deciphering the anti-aflatoxinogenic properties of eugenol using a large-scale q-PCR approach. Toxins 8: 123. doi:10.3390/toxins8050123.

Caceres, I., Snini, S.P., Puel, O. and Mathieu, F. 2018. *Streptomyces roseolus*: A promising biocontrol agent against *Aspergillus flavus*, the main aflatoxin B1 producer. Toxins 10: 442. doi:10.3390/toxins10110442.

Calado, T., Venâncio, A. and Abrunhosa, L. 2014. Irradiation for mold and mycotoxin control: A review. Comprehensive Reviews in Food Science and Food Safety 13: 1049–1061. doi:10.1111/1541-4337.12095.

Calado, T., Fernández-Cruz, M.L., Cabo Verde, S., Venâncio, A. and Abrunhosa, L. 2018. Gamma irradiation effects on ochratoxin A: Degradation, cytotoxicity and application in food. Food Chemistry 240: 463–471. doi:10.1016/j.foodchem.2017.07.136.

Calhelha, R., Andrade, J.V., Ferreira, I.C. and Estevinho, L.M. 2006. Toxicity effects of fungicide residues on the wine-producing process. Food Microbiology 23: 393–398. doi:10.1016/j.fm.2005.04.008.

Camiletti, B.X., Moral, J., Asensio, C.M., Torrico, A.K., Lucini, E.I., Giménez-Pecci, M.D.L.P. and Michailides, T.J. 2018. Characterization of Argentinian endemic *Aspergillus flavus* isolates and their potential use as biocontrol agents for mycotoxins in maize. Phytopathology 108: 818–828. doi:10.1094/PHYTO-07-17-0255-R.

Carvalho, F.P. 2017. Pesticides, environment and food safety. Food and Energy Security 6: 48–60. doi:10.1002/fes3.108.

Casas-Junco, P.P., Solís-Pacheco, J.R., Ragazzo-Sánchez, J.A., Aguilar-Uscanga, B.R., Bautista-Rosales, P.U. and Calderón-Santoyo, M. 2019. Cold plasma treatment as an alternative for ochratoxin A detoxification and inhibition of mycotoxigenic fungi in roasted coffee. Toxins 11: 337. doi:10.3390/toxins11060337.

Castellá, G., Bragulat, M.R., Puig, L., Sanseverino, W. and Cabañes, F.J. 2018. Genomic diversity in ochratoxigenic and non ochratoxigenic strains of *Aspergillus carbonarius*. Scientific Reports 8: 5439. doi:10.1038/s41598-018-23802-8.

Castellá, G., Bragulat, M.R., Cigliano, R.A. and Cabañes, F.J. 2020. Transcriptome analysis of non-ochratoxigenic *Aspergillus carbonarius* strains and interactions between some black *Aspergilli* species. International Journal of Food Microbiology 317: 108498. doi:10.1016/j.ijfoodmicro.2019.108498.

Castellari, M., Versari, A., Fabiani, A., Parpinello, G.P. and Galassi, S. 2001. Removal of ochratoxin A in red wines by means of adsorption treatments with commercial fining agents. Journal of Agricultural and Food Chemistry 49: 3917–3921. doi:10.1021/jf010137o.

Chalwe, H.M., Ngulube, M., Njoroge, S.M.C., Mweetwa, A.M. and Lungu, O.I. 2019. The role of manual sorting of raw peanuts to minimize exposure to aflatoxin-contaminated peanuts. Journal of Postharvest Technology 7: 80–86.

Chen, W., Li, C., Zhang, B., Zhou, Z., Shen, Y., Liao, X., Yang, J., Wang, Y., Li, X., Li, Y. and Shen, X.L. 2018. Advances in biodetoxification of ochratoxin A – A review of the past five decades. Frontiers in Microbiology 9: 1386. doi:10.3389/fmicb.2018.01386.

Chlebicz, A. and Śliżewska, K. 2020. *In vitro* detoxification of aflatoxin B1, deoxynivalenol, fumonisins, T-2 toxin and zearalenone by probiotic bacteria from genus *Lactobacillus* and *Saccharomyces cerevisiae* yeast. Probiotics and Antimicrobial Proteins 12: 289–301. doi:10.1007/s12602-018-9512-x.

Cho, S.M., Jeong, S.E., Lee, K.R., Sudhani, H.P.K., Kim, M., Hong, S.Y. and Chung, S.H. 2016. Biodegradation of ochratoxin A by *Aspergillus tubingensis* isolated from meju. Journal of Microbiology and Biotechnology 26: 1687–1695. doi:10.4014/jmb.1606.06016.

Choque, E., Klopp, C., Valiere, S., Raynal, J. and Mathieu, F. 2018. Whole-genome sequencing of *Aspergillus tubingensis* G131 and overview of its secondary metabolism potential. BMC Genomics 19: 200. doi:10.1186/s12864-018-4574-4.

Cobos, C.J., Tengey, T.K., Balasubramanian, V.K. and Williams, L.D. 2018. Employing peanut seed coat cell wall mediated resistance against *Aspergillus flavus* infection and aflatoxin contamination. Preprints 2018080292. doi:10.20944/preprints201808.0292.v1.

Codex Alimentarius. 2003. Code of practice for the prevention and reduction of mycotoxin contamination in cereals, including annexes on Ochratoxin A, zearalenone, fumonisins and trichothecenes. CAC/RCP, 51-2003.

Cotty, P.J. 1988. Aflatoxin and sclerotial production by *Aspergillus flavus*: Influence of pH. Physiology and Biochemistry 78: 1250–1253.

Cotty, P.J. 1989. Virulence and cultural characteristics of two *Aspergillus flavus* strains pathogenic on cotton. Postharvest Pathology and Mycotoxins 79: 808–814.

Cotty, P.J. and Bhatnagar, D. 1994. Variability among atoxigenic *Aspergillus flavus* strains in ability to prevent aflatoxin contamination and production of aflatoxin biosynthetic pathway enzymes. Applied and Environmental Microbiology 60: 2248–2251.

Cozzi, G., Paciolla, C., Haidukowski, M., Leonardis, S.D.E., Mulè, G. and Logrieco, A.F. 2013. Increase of fumonisin B 2 and ochratoxin A production by black *Aspergillus* species and oxidative stress in grape berries damaged by powdery mildew. Journal of Food Protection 76: 2031–2036. doi:10.4315/0362-028X.JFP-13-149

Dambolachepa, H.B., Muthomi, J.W., Mutitu, E.W. and Njoroge, S.M. 2019. Effects of postharvest handling practices on quality of groundnuts and aflatoxin contamination. Novel Research in Microbiology Journal 3: 396–414. doi:10.21608/nrmj.2019.37214.

De Bellis, P., Tristezza, M., Haidukowski, M., Fanelli, F., Sisto, A., Mulè, G. and Grieco, F. 2015. Biodegradation of ochratoxin A by bacterial strains isolated from vineyard soils. Toxins 7: 5079–5093. doi:10.3390/toxins7124864.

De Pilli, T. and Alessandrino, O. 2020. Effects of different cooking technologies on biopolymers modifications of cereal-based foods: Impact on nutritional and quality characteristics review. Critical Reviews in Food Science and Nutrition 60: 556–565. doi:10.1080/10408398.2018.1544884.

De Souza, M.L., Passamani, F.R.F., Ávila, C.L. da S., Batista, L.R., Schwan, R.F. and Silva, C.F. 2017. Use of wild yeasts as a biocontrol agent against toxigenic fungi and OTA production. Acta Scientiarum – Agronomy 39: 349–358. doi:10.4025/actasciagron.v39i3.32659.

Diao, E., Shen, X., Zhang, Z., Ji, N., Ma, W. and Dong, H. 2015. Safety evaluation of aflatoxin B1 in peanut oil after ultraviolet irradiation detoxification in a photodegradation reactor. International Journal of Food Science and Technology 50: 41–47. doi:10.1111/ijfs.12648.

Dimakopoulou, M., Tjamos, S.E., Antoniou, P.P., Pietri, A., Battilani, P., Avramidis, N., Markakis, E.A. and Tjamos, E.C. 2008. Phyllosphere grapevine yeast *Aureobasidium pullulans* reduces *Aspergillus carbonarius* (sour rot) incidence in wine-producing vineyards in Greece. Biological Control 46: 158–165. doi:10.1016/j.biocontrol.2008.04.015.

Domijan, A.M., Pleadin, J., Mihaljević, B., Vahčić, N., Frece, J. and Markov, K. 2015. Reduction of ochratoxin A in dry-cured meat products using gamma-irradiation. Food Additives and Contaminants – Part A: Chemistry, Analysis, Control, Exposure and Risk Assessment 32: 1185–1191. doi:10.1080/19440049.2015.1049219.

Domijan, A.M., Marjanović Čermak, A.M., Vulić, A., Tartaro Bujak, I., Pavičić, I., Pleadin, J., Markov, K. and Mihaljević, B. 2019. Cytotoxicity of gamma irradiated aflatoxin B1 and ochratoxin A. Journal of Environmental Science and Health – Part B: Pesticides, Food Contaminants and Agricultural Wastes 54: 155–162. doi:10.1080/03601234.2018.1536578.

Dorner, J.W. 2004. Combined effects of biological control formulations, cultivars and fungicides on preharvest colonization and aflatoxin contamination of peanuts by *Aspergillus* species. Peanut Science 31: 79–86. doi:10.3146/pnut.31.2.0004.

Dragacci, S., Favrot, M., Fremy, J., Massimi, C., Prigent, P., Debongnie, P., Pussemier, L., Boudra, H., Morgavi, D., Oswald, I.P. and others. 2009. Review of mycotoxin-detoxifying agents used as feed additives: Mode of action, efficacy and feed/food safety. EFSA Scientific Report FEEDAP 2009-01. doi.org/10.2903/sp.efsa.2009.EN-22.

Ehrlich, K. 2008. Genetic diversity in *Aspergillus flavus* and its implications for agriculture. pp. 233–247. *In*: J. Varga and R.A. Samson (eds.). *Aspergillus* in the Genomics Era. doi: 10.3920/978-90-8686-635-9.

Ehrlich, K.C. 2014. Non-aflatoxigenic *Aspergillus flavus* to prevent aflatoxin contamination in crops: Advantages and limitations. Frontiers in Microbiology 5: 50. doi:10.3389/fmicb.2014.00050.

Ehrlich, K.C. and Cotty, P.J. 2004. An isolate of *Aspergillus flavus* used to reduce aflatoxin contamination in cottonseed has a defective polyketide synthase gene. Applied Microbiology and Biotechnology 65: 473–478. doi:10.1007/s00253-004-1670-y.

El-Nekeety, A.A., Abdel-Wahhab, K.G., Abdel-Aziem, S.H., Mannaa, F.A., Hassan, N.S. and Abdel-Wahhab, M.A. 2017. Papaya fruits extracts enhance the antioxidant capacity and modulate the genotoxicity and oxidative stress in the kidney of rats fed ochratoxin A-contaminated diet. Journal of Applied Pharmaceutical Science 7: 111–121. doi:10.7324/JAPS.2017.70718.

El Khoury, R., Atoui, A., Verheecke, C., Maroun, R., El Khoury, A. and Mathieu, F. 2016. Essential oils modulate gene expression and ochratoxin A production in *Aspergillus carbonarius*. Toxins 8: 242. doi:10.3390/toxins8080242.

El Khoury, R., Caceres, I., Puel, O., Bailly, S., Atoui, A., Oswald, I.P., El Khoury, A. and Bailly, J.D. 2017. Identification of the anti-aflatoxinogenic activity of *Micromeria graeca* and elucidation of its molecular mechanism in *Aspergillus flavus*. Toxins 9: 87. doi:10.3390/toxins9030087.

El Khoury, R., Choque, E., El Khoury, A., Snini, S.P., Cairns, R., Riantsiferana, C. and Mathieu, F. 2018. OTA prevention and detoxification by *Actino bacteria*l strains and activated carbon fibers: Preliminary results. Toxins 10: 137. doi:10.3390/toxins10040137.

Elmnasser, N., Guillou, S., Leroi, F., Orange, N., Bakhrouf, A. and Federighi, M. 2007. Pulsed-light system as a novel food decontamination technology: A review. Canadian Journal of Microbiology 53: 813–821. doi:10.1139/W07-042.

Espejo, F.J. and Armada, S. 2009. Effect of activated carbon on ochratoxin A reduction in "Pedro Ximenez" sweet wine made from off-vine dried grapes. European Food Research and Technology 229: 255–262. doi:10.1007/s00217-009-1055-7.

European Commission (EC). 2009. Commission Regulation 386/2009/EC of 12 May 2009 amending Regulation (EC) No 1831/2003 of the European Parliament and of the Council as regards the establishment of a new functional group of feed additives. Official Journal of the European Union.

Evelyn and Silva, F.V.M. 2015. Inactivation of *Byssochlamys nivea* as cospores in strawberry puree by high pressure, power ultrasound and thermal processing. International Journal of Food Microbiology 214: 129–136. doi:10.1016/j.ijfoodmicro.2015.07.031.

FAO. 2001. Manual on the Application of the HACCP System in Mycotoxin Prevention and Control. FAO Food and Nutrition Paper No. 73. Rome, Italy.

FAO. 2002. Guidelines for good agricultural practices. http://www.fao.org/prods/gap/docs/pdf/guidelines_for_agricultural_practices.pdf : Visited in march 2020.

Farbo, M.G., Urgeghe, P.P., Fiori, S., Marceddu, S., Jaoua, S. and Migheli, Q. 2016. Adsorption of ochratoxin A from grape juice by yeast cells immobilised in calcium alginate beads. International Journal of Food Microbiology 217: 29–34. doi:10.1016/j.ijfoodmicro.2015.10.012.

Fasoyin, O.E., Yang, K., Qiu, M., Wang, B., Wang, S. and Wang, S. 2019. Regulation of morphology, aflatoxin production and virulence of *Aspergillus flavus* by the major nitrogen regulatory gene *areA*. Toxins 11: 718. doi:10.3390/toxins11120718.

Féat, A., Federle, W., Kamperman, M. and Van der Gucht, J. 2019. Coatings preventing insect adhesion: An overview. Progress in Organic Coatings 134: 349–359. doi:10.1016/j.porgcoat.2019.05.013.

Fernandes, J.-M., Calado, T., Guimarães, A., Rodrigues, M.A.M. and Abrunosa, L. 2019. *In vitro* adsorption of aflatoxin B1, ochratoxin A and zearalenone by micronized grape stems and olive pomace in buffer solutions. Mycotoxin Research 35: 243–252. doi: 10.1007/s12550-019-00349-9.

Ferrer, A. and Cabral, R. 1991. Toxic epidemics caused by alimentary exposure to pesticides: A review. Food Additives & Contaminants 8: 755–775.

Fiori, S., Urgeghe, P.P., Hammami, W., Razzu, S., Jaoua, S. and Migheli, Q. 2014. Biocontrol activity of four non- and low-fermenting yeast strains against *Aspergillus carbonarius* and their ability to remove ochratoxin A from grape juice. International Journal of Food Microbiology 189: 45–50. doi:10.1016/j.ijfoodmicro.2014.07.020.

Freita-Silva, O., de Oliveira, P.S. and Freire Júnior, M. 2015. Potential of electron beams to control mycotoxigenic fungi in food. Food Engineering Reviews 7: 160–170. doi:10.1371/journal.pone.0023496.

Frisvad, J.C., Larsen, T.O., Thrane, U., Meijer, M., Varga, J., Samson, R.A. and Nielsen, K.F. 2011. Fumonisin and ochratoxin production in industrial *Aspergillus niger* strains. PLoS ONE 6: 2–7. doi:10.1371/journal.pone.0023496.

Gallo, A., Masoero, F., Bertuzzi, T., Piva, G. and Pietri, A. 2010. Effect of the inclusion of adsorbents on aflatoxin B1 quantification in animal feedstuffs. Food Additives and Contaminants 27: 54–63. doi:10.1080/02652030903207219.

García-Díaz, M., Patiño, B., Vázquez, C. and Gil-Serna, J. 2019. A novel niosome-encapsulated essential oil formulation to prevent *Aspergillus flavus* growth and aflatoxin contamination of maize grains during storage. Toxins 11: 646. doi:10.3390/toxins11110646.

Garcia, D., Girardi, N.S., Passone, M.A., Nesci, A. and Etcheverry, M. 2016. Evaluation of food grade antioxidant formulation for sustained antifungal, antiaflatoxigenic and insecticidal activities on peanut conditioned at different water activities. Journal of Stored Products Research 65: 6–12. doi:10.1016/j.jspr.2015.11.002.

Gauthier, T., Duarte-Hospital, C., Vignard, J., Boutet-Robinet, E., Sulyok, M., Snini, S.P., Alassane-Kpembi, I., Lippi, Y., Puel, S., Oswald, I.P. and Puel, O. 2020. Versicolorin A, a precursor in aflatoxins biosynthesis, is a food contaminant toxic for human intestinal cells. Environment International 137: 105568. doi:10.1016/j.envint.2020.105568.

Ghorbanian, M., Razzaghi-Abyaneh, M., Allameh, A., Shams-Ghahfarokhi, M. and Qorbani, M. 2007. Study on the effect of neem (*Azadirachta indica A. juss*) leaf extract on the growth of *Aspergillus parasiticus* and production of aflatoxin by it at different incubation times. Mycoses 51: 35–39. doi:10.1111/j.1439-0507.2007.01440.x.

Gil-Serna, J., Vázquez, C., González-Jaén, M. and Patiño, B. 2018. Wine contamination with ochratoxins: A review. Beverages 4: 6. doi:10.3390/beverages4010006.

Gómez, J.V., Tarazona, A., Mateo, F., Jiménez, M. and Mateo, E.M. 2019. Potential impact of engineered silver nanoparticles in the control of aflatoxins, ochratoxin A and the main aflatoxigenic and ochratoxigenic species affecting foods. Food Control 101: 58–68. doi:10.1016/j.foodcont.2019.02.019.

González-Jartín, J.M., Castro, L. De, Alfonso, A., Piñeiro, Y., Yáñez, S., González, M., Vargas, Z., Sainz, M.J., Vieytes, M.R., Rivas, J. and Botana, L.M. 2019. Detoxification agents based on magnetic nanostructured particles as a novel strategy for mycotoxin mitigation in food. Food Chemistry Journal 294: 60–66. doi:10.1016/j.foodchem.2019.05.013.

Greco, D., D'Ascanio, V., Santovito, E., Logrieco, A.F. and Avantaggiato, G. 2019. Comparative efficacy of agricultural by-products in sequestering mycotoxins. Journal of the Science of Food and Agriculture 99: 1623–1634. doi:10.1002/jsfa.9343.

Grenier, B., Bracarense, A.P.F.L., Leslie, J.F. and Oswald, I.P. 2014. Physical and Chemical Methods for Mycotoxin Decontamination in Maize. pp. 116–127. *In*: J.F. Leslie and A.F. Logrieco (eds.). Mycotoxin Reduction in Grain Chains. doi: 10.1002/9781118832790.ch9.

Gupta, P. 2018. Chapter 45 – Toxicity of Fungicides. pp. 569–580. *In*: R.C. Gupta (ed.). Veterinary Toxicology (Third Edition) Basic and Clinical Principles. doi:10.1016/B978-0-12-811410-0.00045-3.

Haidukowski, M., Casamassima, E., Cimmarusti, M.T., Branà, M.T., Longobardi, F., Acquafredda, P., Logrieco, A. and Altomare, C. 2019. Aflatoxin B1-adsorbing capability of *Pleurotus eryngii* mycelium: Efficiency and modeling of the process. Frontiers in Microbiology 10: 1386. doi:10.3389/fmicb.2019.01386.

Halford, N.G. 2019. Legislation governing genetically modified and genome-edited crops in Europe: The need for change. Journal of the Science of Food and Agriculture 99: 8–12. doi:10.1002/jsfa.9227.

Haque, M.A., Wang, Y., Shen, Z., Li, X., Saleemi, M.K. and He, C. 2020. Mycotoxin contamination and control strategy in human, domestic animal and poultry: A review. Microbial Pathogenesis 142: 104095. doi:10.1016/j.micpath.2020.104095.

Harkai, P., Szabó, I., Cserháti, M., Krifaton, C., Risa, A., Radó, J., Balázs, A., Berta, K. and Kriszt, B. 2016. Biodegradation of aflatoxin-B1 and zearalenone by *Streptomyces* sp. collection. International Biodeterioration and Biodegradation 108: 48–56. doi:10.1016/j.ibiod.2015.12.007.

Hendrix, M.C., Obed, I.L., Alice, M.M., Elijah, P., Jones, Y., Samuel, M.C.N., Rick, L.B. and David, J. 2019. Predicting aflatoxin content in peanuts using ambient temperature, soil temperature and soil moisture content during pod development. African Journal of Plant Science 13: 59–69. doi:10.5897/ajps2018.1742.

Horn, B.W., Moore, G.G. and Carbone, I. 2009a. Sexual reproduction in *Aspergillus flavus*. Mycologia 101: 423–429. doi:10.3852/09-011.

Horn, B.W., Ramirez-Prado, J.H. and Carbone, I. 2009b. The sexual state of *Aspergillus parasiticus*. Mycologia 101: 275–280. doi:10.3852/08-205.

Horn, B.W., Gell, R.M., Singh, R., Sorensen, R.B. and Carbone, I. 2016. Sexual reproduction in *Aspergillus flavus* sclerotia: Acquisition of novel alleles from soil populations and uniparental mitochondrial inheritance. PLoS ONE 11: 1–22. doi:10.1371/journal.pone.0146169.

Hu, Y., Zhang, J., Kong, W., Zhao, G. and Yang, M. 2017. Mechanisms of antifungal and anti-aflatoxigenic properties of essential oil derived from turmeric (*Curcuma longa* L.) on *Aspergillus flavus*. Food Chemistry 220: 1–8. doi:10.1016/j.foodchem.2016.09.179.

Hulikunte Mallikarjunaiah, N., Jayapala, N., Puttaswamy, H. and Siddapura Ramachandrappa, N. 2017. Characterization of non-aflatoxigenic strains of *Aspergillus flavus* as potential biocontrol agent for the management of aflatoxin contamination in groundnut. Microbial Pathogenesis 102: 21–28. doi:10.1016/j.micpath.2016.11.007.

IARC. 1993a. Aflatoxins. Some naturally occurring substances: Food items and constituents, heterocyclic aromatic amines and mycotoxins. IARC Monograph on the Evaluation of Carcinogenic Risks to Humans 56: 245–395.

IARC. 1993b. Ochratoxin A. Some naturally occurring substances: Food items and constituents, heterocyclic aromatic amines and mycotoxins. IARC Monographs on the Evaluation of Carcinogenic Risk of Chemicals to Humans 56: 489–521.

IARC. 2012. Chapter 9: Practical approaches to control mycotoxins. pp. 131–146. *In*: F.W. John I. Pitt, Christopher P. Wild, Robert A. Baan, Wentzel C.A. Gerlderblom, J. David Miller, Ronald T. Riley (eds.). IARC Scientific Publications No. 158. Lyon, France.

Isman, M. and Machial, C. 2006. Pesticides based on plant essential oils: From traditional practice to commercialization. Advances in Phytomedicine 3: 29–44.

Isman, M.B. 2000. Plant essential oils for pest and disease management. Crop Protection 19: 603–608. doi:10.1016/S0261-2194(00)00079-X.

Jahanshiri, Z., Shams-Ghahfarokhi, M., Allameh, A. and Razzaghi-Abyaneh, M. 2012. Effect of curcumin on *Aspergillus parasiticus* growth and expression of major genes involved in the early and late stages of aflatoxin biosynthesis. Iranian Journal of Public Health 41: 72–79.

Jahanshiri, Z., Shams-Ghahfarokhi, M., Allameh, A. and Razzaghi-Abyaneh, M. 2015. Inhibitory effect of eugenol on aflatoxin B1 production in *Aspergillus parasiticus* by downregulating the expression of major genes in the toxin biosynthetic pathway. World Journal of Microbiology and Biotechnology 31: 1071–1078. doi:10.1007/s11274-015-1857-7.

Jalili, M., Jinap, S. and Noranizan, M.A. 2012. Aflatoxins and ochratoxin A reduction in black and white pepper by gamma radiation. Radiation Physics and Chemistry 81: 1786–1788. doi:10.1016/j.radphyschem.2012.06.001.

Jayashree, T. and Subramanyam, C. 1999. Antiaflatoxinogenic activity of eugenol is due to inhibition of lipid peroxidation. Letters in Applied Microbiology 28: 179–183. doi:10.1046/j.1365-2672.1999.00512.x.

Ji, J. and Xie, W. 2020. Detoxification of aflatoxin B1 by magnetic graphene composite adsorbents from contaminated oils. Journal of Hazardous Materials 381: 120915. doi:10.1016/j.jhazmat.2019.120915.

Jindal, N., Mahipal, S.K. and Mahajan, N.K. 1994. Toxicity of aflatoxin B1 in broiler chicks and its reduction by activated charcoal. Research in Veterinary Science 56: 37–40.

Juarez-Segovia, K.G., Diaz-Darcia, M.D., Mendez-Lopez, M.S., Pina-Canseco, A.D., Perez, S. and Sanchez-Medina, M.A. 2019. Effect of garlic extracts (*Allium sativum*) on the development in vitro of *Aspergillus parasiticus* and *Aspergillus niger*. Polibotanica 47: 99–111. doi:10.18387/polibotanica.47.8.

Kabak, B., Dobson, A.D.W. and Var, I. 2006. Strategies to prevent mycotoxin contamination of food and animal feed: A review. Critical Reviews in Food Science and Nutrition 46: 593–619. doi:10.1080/10408390500436185.

Kedia, A., Prakash, B., Mishra, P.K. and Dubey, N.K. 2014. Antifungal and antiaflatoxigenic properties of *Cuminum cyminum* (L.) seed essential oil and its efficacy as a preservative in stored commodities. International Journal of Food Microbiology 168–169: 1–7. doi:10.1016/j.ijfoodmicro.2013.10.008.

Kihal, A., Rodriguez-Prado, M., Godoy, C., Cristofol, C. and Calsamiglia, S. 2020. *In vitro* assessment of the capacity of certain mycotoxin binders to adsorb some amino acids and water-soluble vitamins. Journal of Dairy Science 103: 3125–3132. doi:10.3168/jds.2019-17561.

Kollia, E., Proestos, C., Zoumpoulakis, P. and Markaki, P. 2017. Inhibitory effect of *Cynara cardunculus* L. extract on aflatoxin B1 production by *Aspergillus parasiticus* in sesame (*Sesamum indicum* L.). International Journal of Food Properties 20: 1270–1279. doi:10.1080/10942912.2016.1206928.

Krulj, J., Markov, S., Bočarov-Stančić, A., Pezo, L., Kojić, J., Ćurčić, N., Janić-Hajnal, E. and Bodroža-Solarov, M. 2019. The effect of storage temperature and water activity on aflatoxin B1 accumulation in hull-less and hulled spelt grains.

Journal of the Science of Food and Agriculture 99: 3703–3710. doi:10.1002/jsfa.9601.

Kumar, A., Shukla, R., Singh, P. and Dubey, N.K. 2010. Chemical composition, antifungal and antiaflatoxigenic activities of *Ocimum sanctum* L. essential oil and its safety assessment as plant based antimicrobial. Food and Chemical Toxicology 48: 539–543. doi: 10.1016/j.fct.2009.11.028.

Lappa, I.K., Simini, E., Nychas, G.J.E. and Panagou, E.Z. 2017. *In vitro* evaluation of essential oils against *Aspergillus carbonarius* isolates and their effects on ochratoxin A related gene expression in synthetic grape medium. Food Control 73: 71–80. doi:10.1016/j.foodcont.2016.08.016.

Lee, S.E., Park, B.S., Bayman, P., Baker, J.L., Choi, W.S. and Campbell, B.C. 2007. Suppression of ochratoxin biosynthesis by naturally occurring alkaloids. Food Additives and Contaminants 24: 391–397. doi:10.1080/02652030601053147.

Lee, H.J., Dahal, S., Perez, E.G., Kowalski, R.J., Ganjyal, G.M. and Ryu, D. 2017. Reduction of ochratoxin A in oat flakes by twin-screw extrusion processing. Journal of Food Protection 80: 1628–1634. doi:doi.org/10.4315/0362-028X.JFP-16-559.

Lee, L., Chan, K., Stach, J., Wellington, E.M.H. and Goh, B.-H. 2018. Editorial: The search for biological active agent(s) from *Actinobacteria*. Frontiers in Microbiology 9: 824. doi:10.3389/fmicb.2018.00824.

Levasseur-Garcia, C. 2018. Updated overview of infrared spectroscopy methods for detecting mycotoxins on cereals (corn, wheat and barley). Toxins 10: 38. doi:10.3390/toxins10010038.

Lewis, M.H., Carbone, I., Luis, J.M., Payne, G.A., Bowen, K.L., Hagan, A.K., Kemerait, R., Heiniger, R. and Ojiambo, P.S. 2019. Biocontrol strains differentially shift the genetic structure of indigenous soil populations of *Aspergillus flavus*. Frontiers in Microbiology 10: 1738. doi:10.3389/fmicb.2019.01738.

Liang, D., Xing, F., Selvaraj, J.N., Liu, X., Wang, L., Hua, H., Zhou, L., Zhao, Y., Wang, Y. and Liu, Y. 2015. Inhibitory effect of cinnamaldehyde, citral and eugenol on aflatoxin biosynthetic gene expression and aflatoxin B1 biosynthesis in *Aspergillus flavus*. Journal of Food Science 80: M2917–M2924. doi:10.1111/1750–3841.13144.

Liu, R., Jin, Q., Huang, J., Liu, Y., Wang, X., Mao, W. and Wang, S. 2011. Photodegradation of aflatoxin B1 in peanut oil. European Food Research and Technology 232: 843–849.

Liu, R., Wang, R., Lu, J., Chang, M., Jin, Q., Du, Z., Wang, S., Li, Q. and Wang, X. 2016. Degradation of AFB1 in aqueous medium by electron beam irradiation: Kinetics, pathway and toxicology. Food Control 66: 151–157. doi:10.1016/j.foodcont.2016.02.002.

Liu, N., Wang, J., Deng, Q., Gu, K. and Wang, J. 2018a. Detoxification of aflatoxin B1 by lactic acid bacteria and hydrated sodium calcium aluminosilicate in broiler chickens. Livestock Science 208: 28–32. doi:10.1016/j.livsci.2017.12.005.

Liu, R., Deng, Z. and Liu, T. 2018b. *Streptomyces* species: Ideal chassis for natural product discovery and overproduction. Metabolic Engineering 50: 74–84. doi:10.1016/j.ymben.2018.05.015.

Liu, J.J., Cai, Z., Liao, Y., Zhao, L., Moulin, J. and Hartmann, C. 2019. Validation of a laser based in-line aflatoxin sorting technology in Spanish type raw peanut in factory-scale production conditions. Journal of Food Safety 39: 1–11. doi:10.1111/jfs.12611.

Liu, Y., Li, F.M., Bai, F. and Bian, K. 2019a. Effects of pulsed ultrasound at 20 kHz on the sonochemical degradation of mycotoxins. World Mycotoxin Journal 12: 1–10. doi:10.3920/WMJ2018.2431

Liu, Y., Li, M., Liu, Y. and Bian, K. 2019b. Structures of reaction products and degradation pathways of aflatoxin B1 by ultrasound treatment. Toxins 11: 526. doi:10.3390/toxins11090526.

Liu, Y., Teng, K., Wang, T., Dong, E., Zhang, M., Tao, Y. and Zhong, J. 2020. Antimicrobial *Bacillus velezensis* HC6: Production of three kinds of lipopeptides and biocontrol potential in maize. Journal of Applied Microbiology 128: 242–254. doi:10.1111/jam.14459.

Lucas, J.A., Hawkins, N.J. and Fraaije, B.A. 2015. The evolution of fungicide resistance. Advances in Applied Microbiology 90: 29–92. doi:10.1016/bs.aambs.2014.09.001.

Luo, X., Qi, L., Liu, Y., Wang, R., Yang, D., Li, K., Wang, L., Li, Y., Zhang, Y. and Zhengxing, C. 2017. Effects of electron beam irradiation on zearalenone and ochratoxin A in naturally contaminated corn and corn quality parameters. Toxins 9: 84. doi:10.3390/toxins9030084.

Lyu, A., Yang, L., Wu, M., Zhang, J. and Li, G. 2020. High efficacy of the volatile organic compounds of *Streptomyces yanglinensis* 3-10 in suppression of *Aspergillus* contamination on peanut kernels. Frontiers in Microbiology 11: 142. doi:10.3389/fmicb.2020.00142.

Maatouk, I., Mehrez, A., Ben Amara, A., Chayma, R., Abid, S., Jerbi, T. and Landoulsi, A. 2019. Effects of Gamma Irradiation on ochratoxin A stability and cytotoxicity in methanolic solutions and potential application in Tunisian Millet Samples. Journal of Food Protection 82: 1433–1439. doi: 10.4315/0362-028X.JFP-18-557

Magan, N. and Aldred, D. 2007. Post-harvest control strategies: Minimizing mycotoxins in the food chain. International Journal of Food Microbiology 119: 131–139. doi:10.1016/j.ijfoodmicro.2007.07.034.

Mahmoud, A.L.E. 1999. Inhibition of growth and aflatoxin biosynthesis of *Aspergillus flavus* by extracts of some Egyptian plants. Letters in Applied Microbiology 29: 334–336. doi:10.1046/j.1472-765X.1999.00636.x.

Mannaa, M. and Kim, K.D. 2018. Effect of temperature and relative humidity on growth of *Aspergillus* and *Penicillium* spp. and biocontrol activity of *Pseudomonas protegens* AS15 against aflatoxigenic *Aspergillus flavus* in stored rice grains. Mycobiology 46: 287–295. doi:10.1080/12298093.2018.1505247.

Mansfield, B.E., Oltean, H.N., Oliver, B.G., Hoot, S.J., Leyde, S.E., Hedstrom, L. and White, T.C. 2010. Azole drugs are imported by facilitated diffusion in *Candida albicans* and other pathogenic fungi. PLoS Pathogens 6: e1001126. doi:10.1371/journal.ppat.1001126.

Mao, J., He, B., Zhang, L., Li, P., Zhang, Q., Ding, X. and Zhang, W. 2016. A structure identification and toxicity assessment of the degradation products of aflatoxin B1 in peanut oil under UV irradiation. Toxins 8: 332. doi:10.3390/toxins8110332.

Maor, U., Sadhasivam, S., Zakin, V., Prusky, D. and Sionov, E. 2017. The effect of ambient pH modulation on ochratoxin A accumulation by *Aspergillus carbonarius*. World Mycotoxin Journal 10: 339–348. doi:10.3390/toxins8110332.

Mateo, E.M., Gómez, J.V., Gimeno-Adelantado, J.V., Romera, D., Mateo-Castro, R. and Jiménez, M. 2017. Assessment of azole fungicides as a tool to control growth of *Aspergillus flavus* and aflatoxin B1 and B2 production in maize. Food Additives and Contaminants – Part A: Chemistry, Analysis, Control, Exposure and Risk Assessment 34: 1039–1051. doi:10.1080/19440049.2017.1310400.

Matumba, L., Van Poucke, C., Njumbe Ediage, E., Jacobs, B. and De Saeger, S. 2015. Effectiveness of hand sorting, flotation/washing, dehulling and combinations thereof on the decontamination of mycotoxin-contaminated white maize. Food Additives and Contaminants – Part A: Chemistry, Analysis, Control, Exposure and Risk Assessment 32: 960–969. doi:10.1080/19440049.2015.1029535.

Mauro, A., Garcia-Cela, E., Pietri, A., Cotty, P.J. and Battilani, P. 2018. Biological control products for aflatoxin prevention in Italy: Commercial field evaluation of atoxigenic *Aspergillus flavus* active ingredients. Toxins 10: 30. doi:10.3390/toxins10010030.

Medina, Á., Rodriguez, A. and Magan, N. 2015. Climate change and mycotoxigenic fungi: Impacts on mycotoxin production. Current Opinion in Food Science 5: 99–104. doi:10.1016/j.cofs.2015.11.002

Méndez-Albores, A., Martıinez-Bustos, F., Gaytan-Martınez, M. and Moreno-Martınez, E. 2008. Effect of lactic and citric acid on the stability of B-aflatoxins in extrusion-cooked sorghum. Letters in Applied Microbiology 47: 1–7. doi:10.1111/j.1472-765X.2008.02376.x.

Moore, G.G., Singh, R., Horn, B.W. and Carbone, I. 2009. Recombination and lineage-specific gene loss in the aflatoxin gene cluster of *Aspergillus flavus*. Molecular Ecology 18: 4870–4887. doi:10.1111/j.1365-294X.2009.04414.x.

Morcia, C., Tumino, G., Ghizzoni, R., Bara, A., Salhi, N. and Terzi, V. 2017. *In vitro* evaluation of sub-lethal concentrations of plant-derived antifungal compounds on Fusaria growth and mycotoxin production. Molecules 22: 1271 doi:10.3390/molecules22081271.

Moreau, M., Lescure, G., Agoulon, A., Svinareff, P., Orange, N. and Feuilloley, M. 2013. Application of the pulsed light technology to mycotoxin degradation and inactivation. Journal of Applied Toxicology 33: 357–363. doi:10.1002/jat.1749.

Mossini, S.A.G., Arrotéia, C.C. and Kemmelmeier, C. 2009. Effect of neem leaf extract and neem oil on *Penicillium* growth, sporulation, morphology and ochratoxin A production. Toxins 1: 3–13. doi:10.3390/toxins1010003.

Muga, F.C., Marenya, M.O. and Workneh, T.S. 2019. Effect of temperature, relative humidity and moisture on aflatoxin contamination of stored maize kernels. Bulgarian Journal of Agricultural Science 25: 271–277.

Muhammad, I., Sun, X., Wang, H., Li, W., Wang, X., Cheng, P., Li, S., Zhang, X. and Hamid, S. 2017. Curcumin successfully inhibited the computationally identified CYP2A6 enzyme-mediated bioactivation of aflatoxin B1 in Arbor Acres broiler. Frontiers in Microbiology 8: 462925. doi:10.3389/fphar.2017.00143.

Nally, M.C., Pesce, V.M., Maturano, Y.P., Rodriguez Assaf, L.A., Toro, M.E., Castellanos de Figueroa, L.I. and Vazquez, F. 2015. Antifungal modes of action of *Saccharomyces* and other biocontrol yeasts against fungi isolated from sour and grey rots. International Journal of Food Microbiology 204: 91–100. doi:10.1016/j.ijfoodmicro.2015.03.024.

Navarro García, V.M., Gonzalez, A., Fuentes, M., Aviles, M., Rios, M.Y., Zepeda, G. and Rojas, M.G. 2003. Antifungal activities of nine traditional Mexican medicinal plants. Journal of Ethnopharmacology 87: 85–88. doi:10.1016/S0378-8741(03)00114-4.

Nazarizadeh, H., Mohammad Hosseini, S. and Pourreza, J. 2019. Effect of plant extracts derived from thyme and chamomile on the growth performance, gut morphology and immune system of broilers fed aflatoxin B1 and ochratoxin A contaminated diets. Italian Journal of Animal Science 18: 1073–1081. doi:10.1080/1828051X.2019.1615851.

Nguyen, P.A., Strub, C., Fontana, A. and Schorr-Galindo, S. 2017. Crop molds and mycotoxins: Alternative management using biocontrol. Biological Control 104: 10–27. doi:10.1016/j.biocontrol.2016.10.004.

Ojiambo, P.S., Battilani, P., Cary, J.W., Blum, B.H. and Carbone, I. 2018. Cultural and genetic approaches to manage aflatoxin contamination: Recent insights provide opportunities for improved control. Phytopathology 108: 1024–1037. doi:10.1094/PHYTO-04-18-0134-RVW.

Olarte, R.A., Horn, B.W., Dorner, J.W., Monacell, J.T., Singh, R., Stone, E.A. and Carbone, I. 2012. Effect of sexual recombination on population diversity in aflatoxin production by *Aspergillus flavus* and evidence for cryptic heterokaryosis. Molecular Ecology 21: 1453–1476. doi:10.1111/j.1365-294X.2011.05398.x.

Olivares-Marín, M., Del Prete, V., Garcia-Moruno, E., Fernández-González, C., Macías-García, A. and Gómez-Serrano, V. 2009. The development of an activated carbon from cherry stones and its use in the removal of ochratoxin A from red wine. Food Control 20: 298–303. doi:10.1016/j.foodcont.2008.05.008.

Oliveira, R.C., Carvajal-Moreno, M., Mercado-Ruaro, P., Rojo-Callejas, F. and Correa, B. 2020. Essential oils trigger an antifungal and anti-aflatoxigenic effect on *Aspergillus flavus* via the induction of apoptosis-like cell death and gene regulation. Food Control 110: 107038. doi:10.1016/j.foodcont.2019.107038

Ozturkoglu-budak, S. 2017. A model for implementation of HACCP system for prevention and control of mycotoxins during the production of red dried chili pepper. Food Science and Technology 37: 24–29. doi:10.1590/1678-457x.30316

Pankaj, S.K., Shi, H. and Keener, K.M. 2018a. A review of novel physical and chemical decontamination technologies for aflatoxin in food. Trends in Food Science and Technology 71: 73–83. doi:10.1016/j.tifs.2017.11.007.

Pankaj, S.K., Wan, Z. and Keener, K.M. 2018b. Effects of cold plasma on food quality: A review. Foods 7: 4. doi:10.3390/foods7010004.

Pappas, A.C., Tsiplakou, E., Tsitsigiannis, D.I., Georgiadou, M., Iliadi, M.K., Sotirakoglou, K. and Zervas, G. 2016. The role of bentonite binders in single or concomitant mycotoxin contamination of chicken diets. British Poultry Science 57: 551–558. doi:10.1080/00071668.2016.1187712.

Paranagama, P.A., Abeysekera, K.H.T., Abeywickrama, K. and Nugaliyadde, L. 2003. Fungicidal and anti-aflatoxigenic effects of the essential oil of *Cymbopogon citratus* (DC.) Stapf. (lemongrass) against *Aspergillus flavus* Link. isolated from stored rice. Letters in Applied Microbiology 37: 86–90. doi:10.1046/j.1472-765X.2003.01351.x.

Passone, M.A., Girardi, N.S. and Etcheverry, M. 2013. Antifungal and antiaflatoxigenic activity by vapor contact of three essential oils and effects of environmental factors on their efficacy. LWT-Food Science and Technology 53: 434–444. doi:10.1016/j.lwt.2013.03.012.

Patras, A., Julakanti, S., Yannam, S., Rishipal R. Bansode, Burns, M. and Vergne, M.J. 2017. Effect of UV irradiation on aflatoxin reduction: A cytotoxicity evaluation study using human hepatoma cell line. Mycotoxin Research 33: 343–350. doi: 10.1007/s12550-017-0291-0.

Pelletier, M.J. and Reizner, J.R. 1992. Comparison of fluorescence sorting and color sorting for the removal. Peanut 19: 15–20.

Peng, Z., Chen, L., Zhu, Y., Huang, Y., Hu, X., Wu, Q., Nüssler, A.K., Liu, L. and Yang, W. 2018. Current major degradation methods for aflatoxins: A review. Trends in Food Science and Technology 80: 155–166. doi:10.1016/j.tifs.2018.08.009.

Pfliegler, W.P., Pusztahelyi, T. and Pócsi, I. 2015. Mycotoxins – Prevention and decontamination by yeasts. Journal of Basic Microbiology 55: 805–818. doi:10.1002/jobm.201400833.

Phokane, S., Flett, B.C., Ncube, E., Rheeder, J.P. and Rose, L.J. 2019. Agricultural practices and their potential role in mycotoxin contamination of maize and groundnut subsistence farming. South African Journal of Science 115: 2–7. doi: 10.17159/sajs.2019/6221

Ponsone, M.L., Chiotta, M.L., Combina, M., Dalcero, A. and Chulze, S. 2011. Biocontrol as a strategy to reduce the impact of ochratoxin A and *Aspergillus* section *Nigri* in grapes. International Journal of Food Microbiology 151: 70–77.

Ponsone, M.L., Kuhn, Y.G., Schmidt-Heydt, M., Geisen, R. and Chulze, S.N. 2013. Effect of *Kluyveromyces thermotolerans* on polyketide synthase gene expression and ochratoxin accumulation by *Penicillium* and *Aspergillus*. World Mycotoxin Journal 6: 291–297. doi:10.3920/WMJ2012.1532.

Ponsone, M.L., Nally, M.C., Chiotta, M.L., Combina, M., Köhl, J. and Chulze, S.N. 2016. Evaluation of the effectiveness of potential biocontrol yeasts against black sur rot and ochratoxin A occurring under greenhouse and field grape production conditions. Biological Control 103: 78–85. doi:10.1016/j.biocontrol.2016.07.012.

Prettl, Z., Dési, E., Lepossa, A., Kriszt, B., Kukolya, J. and Nagy, E. 2017. Biological degradation of aflatoxin B1 by a *Rhodococcus pyridinivorans* strain in by-product of bioethanol. Animal Feed Science and Technology 224: 104–114. doi:10.1016/j.anifeedsci.2016.12.011.

Price, C.L., Parker, J.E., Warrilow, A.G., Kelly, D.E. and Kelly, S.L. 2015. Azole fungicides – Understanding resistance mechanisms in agricultural fungal pathogens. Pest Management Science 71: 1054–1058. doi:10.1002/ps.4029.

Pruter, L.S., Brewer, M.J., Weaver, M.A., Murray, S.C., Isakeit, T.S. and Bernal, J.S. 2019. Association of insect-derived ear injury with yield and aflatoxin of maize hybrids varying in Bt transgenes. Enviromental Entomology 48: 1401–1411. doi: 10.1093/ee/nvz112.

Quintela, S., Villarán, M.C., López de Armentia, I. and Elejalde, E. 2013. Ochratoxin A removal in wine: A review. Food Control 30: 439–445. doi:10.1016/j.foodcont.2012.08.014.

Raghavender, C., Reddy, B. and Shobharani, G. 2007. Aflatoxin contamination of pearl millet during field and storage conditions with reference to stage of grain maturation and insect damage. Mycotoxin Research 23: 199–209. doi:10.1007/BF02946048.

Rajendran, R.M., Umesh, B. and Chirakkal, H. 2019. Assessment of H-β zeolite as an ochratoxin binder for poultry. Poultry Science 99: 76–88. doi:10.3382/ps/pez535.

Raters, M. and Matissek, R. 2008. Thermal stability of aflatoxin B1 and ochratoxin A. Mycotoxin Res. 24: 130–134. doi:10.1007/BF03032339.

Rejeb, R., Antonissen, G., De Boevre, M., Detavernier, C., Van De Velde, M., De Saeger, S., Ducatelle, R., Ayed, M.H. and Ghorbal, A. 2019. Calcination enhances the aflatoxin and zearalenone binding efficiency of a Tunisian clay. Toxins 11: 602. doi:10.3390/toxins11100602.

Ren, Z., Luo, J. and Wan, Y. 2019. Enzyme-like metal-organic frameworks in polymeric membranes for efficient removal of aflatoxin B1. ACS Applied Materials and Interfaces 11: 30542–30550. doi:10.1021/acsami.9b08011.

Ren, X., Zhang, Q., Zhang, W., Mao, J. and Li, P. 2020. Control of aflatoxigenic molds by antagonistic microorganisms: Inhibitory behaviors, bioactive compounds, related mechanisms and influencing factors. Toxins (Basel) 12: 24. doi:10.3390/toxins12010024.

Risa, A., Krifaton, C., Kukolya, J., Kriszt, B., Cserháti, M. and Táncsics, A. 2018. Aflatoxin B1 and zearalenone-detoxifying profile of rhodococcus type strains. Current Microbiology 75: 907–917. doi:10.1007/s00284-018-1465-5.

Rousk, J., Brookes, P.C. and Bååth, E. 2009. Contrasting soil pH effects on fungal and bacterial growth suggest functional redundancy in carbon mineralization. Applied and Environmental Microbiology 75: 1589–1596. doi:10.1128/AEM.02775-08.

Sarlak, Z., Rouhi, M., Mohammadi, R., Khaksar, R., Mortazavian, A.M., Sohrabvandi, S. and Garavand, F. 2017. Probiotic biological strategies to decontaminate aflatoxin M1 in a traditional Iranian fermented milk drink (Doogh). Food Control 71: 152–159. doi:10.1016/j.foodcont.2016.06.037.

Schmidt, M., Zannini, E., Lynch, K.M. and Arendt, E.K. 2019. Novel approaches for chemical and microbiological shelf life extension of cereal crops. Critical Reviews in Food Science and Nutrition 59: 3395–3419. doi:10.1080/10408398.2018.1491526.

Scott, P.M. 1984. Effects of Food Processing on Mycotoxins. Journal of Food Protection 47: 489–499.

Shakeel, Q., Lyu, A., Zhang, J., Wu, M., Li, G., Hsiang, T. and Yang, L. 2018. Biocontrol of *Aspergillus flavus* on Peanut Kernels using *Streptomyces yanglinensis* 3-10. Frontiers in Microbiology 9: 1049. doi:10.3389/fmicb.2018.01049.

Sharma, N. and Tripathi, A. 2008. Effects of *Citrus sinensis* (L.) Osbeck epicarp essential oil on growth and morphogenesis of *Aspergillus niger* (L.) Van Tieghem. Microbiological Research 163: 337–344. doi:10.1016/j.micres.2006.06.009.

Siahmoshteh, F., Siciliano, I., Banani, H., Hamidi-Esfahani, Z., Razzaghi-Abyaneh, M., Gullino, M.L. and Spadaro, D. 2017. Efficacy of *Bacillus subtilis* and *Bacillus amyloliquefaciens* in the control of *Aspergillus parasiticus* growth and aflatoxins production on pistachio. International Journal of Food Microbiology 254: 47–53. doi:10.1016/j.ijfoodmicro.2017.05.011.

Siahmoshteh, F., Hamidi-Esfahani, Z., Spadaro, D., Shams-Ghahfarokhi, M. and Razzaghi-Abyaneh, M. 2018. Unraveling the mode of antifungal action of *Bacillus subtilis* and *Bacillus amyloliquefaciens* as potential biocontrol agents against aflatoxigenic *Aspergillus parasiticus*. Food Control 89: 300–307. doi:10.1016/j.foodcont.2017.11.010.

Sibakwe, C.B., Kasambara-Donga, T., Njoroge, S.M.C., Msuku, W.A.B., Mhango, W.G., Brandenburg, R.L. and Jordan, D.L. 2017. The role of drought stress on aflatoxin contamination in groundnuts (*Arachis hypogea* L.) and *Aspergillus flavus* population in the soil. Modern Agricultural Science and Technology 3: 22–29. doi:10.15341/mast(2375-9402)/03.03.2017/005.

Sidhu, O.P., Chandra, H. and Behl, H.M. 2009. Occurrence of aflatoxins in mahua (*Madhuca indica* Gmel) seeds: Synergistic effect of plant extracts on inhibition of *Aspergillus flavus* growth and aflatoxin production. Food and Chemical Toxicology 47: 774–777. doi:10.1016/j.fct.2009.01.001.

Šimović, M., Delaš, F., Gradvol, V., Kocevski, D. and Pavlović, H. 2014. Antifungal effect of eugenol and carvacrol against foodborne pathogens *Aspergillus carbonarius* and *Penicillium* roqueforti in improving safety of fresh-cut watermelon. Journal of Intercultural Ethnopharmacology 3: 91–96. doi:10.5455/jice.20140503090524.

Singh, A., Dwivedy, A.K., Singh, V.K., Upadhyay, N., Chaudhari, A.K., Das, S. and Dubey, N.K. 2019. Essential oils based formulations as safe preservatives for stored plant masticatories against fungal and mycotoxin contamination: A review. Biocatalysis and Agricultural Biotechnology 17: 313–317. doi:10.1016/j.bcab.2018.12.003.

Sinha, B.K. 2018. Studies on the inhibitory effect of some plant extracts on mycoflora and aflatoxin production. Acta Scientific Microbiology 1: 23–25.

Snigdha, M., Hariprasad, P. and Venkateswaran, G. 2015. Transport via xylem and accumulation of aflatoxin in seeds of groundnut plant. Chemosphere 119: 524–529. doi:10.1016/j.chemosphere.2014.07.033.

Sonker, N., Pandey, A.K. and Singh, P. 2015. Efficiency of *Artemisia nilagirica* (Clarke) Pamp. essential oil as a mycotoxicant against postharvest mycobiota of table grapes. Journal of the Science of Food and Agriculture 95: 1932–1939. doi:10.1002/jsfa.6901.

Spasic, J., Mandic, M., Radivojevic, J., Jeremic, S., Vasiljevic, B., Nikodinovic-Runic, J. and Djokic, L. 2018. Biocatalytic potential of *Streptomyces* spp. isolates from rhizosphere of plants and mycorrhizosphere of fungi. International Union of Biochemistry and Molecular Biology 65: 822–833. doi:10.1002/bab.1664.

Stasiewicz, M.J., Falade, T.D.O., Mutuma, M., Mutiga, S.K., Harvey, J.J.W., Fox, G., Pearson, T.C., Muthomi, J.W. and Nelson, R.J. 2017. Multi-spectral kernel sorting to reduce aflatoxins and fumonisins in Kenyan maize. Food Control 78: 203–214. doi:10.1016/j.foodcont.2017.02.038.

Suárez-Quiroz, M.L., González-Rios, O., Barel, M., Guyot, B., Schorr-Galindo, S. and Guiraud, J.P. 2004. Effect of chemical and environmental factors on *Aspergillus ochraceus* growth and toxigenesis in green coffee. Food Microbiology 21: 629–634. doi:10.1016/j.fm.2004.03.005.

Sun, X., Sun, C., Zhang, X., Zhang, H., Ji, J., Liu, Y. and Tang, L. 2016. Aflatoxin B1 decontamination by UV-mutated live and immobilized *Aspergillus niger*. Food Control 61: 235–242. doi:10.1016/j.foodcont.2016.08.025.

Sun, X., Niu, Y., Ma, T., Xu, P., Huang, W. and Zhan, J. 2017. Determination, content analysis and removal efficiency of fining agents on ochratoxin A in Chinese wines. Food Control 73: 382–392. doi:10.1016/j.foodcont.2016.08.025.

Tang, X., Shao, Y.L., Tang, Y.J. and Zhou, W.W. 2018. Antifungal activity of essential oil compounds (geraniol and citral) and inhibitory mechanisms on grain pathogens (*Aspergillus flavus* and *Aspergillus ochraceus*). Molecules 23: 1–18. doi:10.3390/molecules23092108.

Tano, Z.J. 2011. Ecological effects of pesticides. pp. 533–540. *In*: M. Stoytcheva (ed.). Pesticides in the modern world – Risk and Benefits. doi:10.1508/cytologia.34.533.

Tao, Y., Xie, S., Xu, F., Liu, A., Wang, Y., Chen, D., Pan, Y., Huang, L., Peng, D., Wang, X. and Yuan, Z. 2018. Ochratoxin A: Toxicity, oxidative stress and metabolism. Food and Chemical Toxicology 112: 320–331. doi:10.1016/j.fct.2018.01.002.

Temba, B.A., Sultanbawa, Y., Kriticos, D.J., Fox, G.P., Harvey, J.J.W. and Fletcher, M.T. 2016. Tools for defusing a major global food and feed safety risk: Nonbiological postharvest procedures to decontaminate mycotoxins in foods and feeds. Journal of Agricultural and Food Chemistry 64: 8959–8972. doi:10.1021/acs.jafc.6b03777.

Ten Bosch, L., Pfohl, K., Avramidis, G., Wieneke, S., Viöl, W. and Karlovsky, P. 2017. Plasma-based degradation of mycotoxins produced. Toxins 9: 97. doi:10.3390/toxins9030097.

Theumer, M.G., Henneb, Y., Khoury, L., Snini, S.P., Tadrist, S., Canlet, C., Puel, O., Oswald, I.P. and Audebert, M. 2018. Genotoxicity of aflatoxins and their precursors in human cells. Toxicology Letters 287: 100–107. doi:10.1016/j.toxlet.2018.02.007.

Thippeswamy, S., Mohana, D.C., Abhishek, R.U. and Manjunath, K. 2014. Inhibitory activity of plant extracts on aflatoxin B1 biosynthesis by *Aspergillus flavus*. Journal of Agricultural Science and Technology 16: 1123–1132.

Thokchom, B., Pandit, A.B., Qiu, P., Park, B., Choi, J. and Khim, J. 2015. Ultrasonics sonochemistry: A review on sonoelectrochemical technology as an upcoming alternative for pollutant degradation. Ultrasonics Sonochemistry Journal 27: 210–234. doi:10.1016/j.ultsonch.2015.05.015.

Tibola, C.S., Fernandes, J.M.C. and Guarienti, E.M. 2016. Effect of cleaning, sorting and milling processes in wheat mycotoxin content. Food Control 60: 174–179. doi:10.1016/j.foodcont.2015.07.031.

Torlak, E. 2018. Use of gaseous ozone for reduction of ochratoxin A and fungal populations on sultanas. Australian Journal of Grape and Wine Research 25: 25–29. doi:10.1111/ajgw.12362.

Torres, A.M., Barros, G.G., Palacios, S.A., Chulze, S.N. and Battilani, P. 2014. Review on pre- and post-harvest management of peanuts to minimize aflatoxin contamination. Food Research International 62: 11–19. doi:10.1016/j.foodres.2014.02.023.

Tripathi, S. and Mishra, H.N. 2010. Enzymatic coupled with UV degradation of aflatoxin B1 in red chili powder. Journal of Food Quality 33: 186–203. doi:10.1111/j.1745-4557.2010.00334.x.

Tryfinopoulou, P., Fengou, L. and Panagou, E.Z. 2019. Influence of *Saccharomyces cerevisiae* and *Rhotodorula mucilaginosa* on the growth and ochratoxin A production of *Aspergillus carbonarius*. LWT 105: 66–78. doi:10.1016/j.lwt.2019.01.050.

Tryfinopoulou, P., Chourdaki, A., Nychas, G.J.E. and Panagou, E.Z. 2020. Competitive yeast action against *Aspergillus carbonarius* growth and ochratoxin A production. International Journal of Food Microbiology 317: 108460. doi:10.1016/j.ijfoodmicro.2019.108460.

Turek, C. and Stintzing, F.C. 2013. Stability of essential oils: A review. Comprehensive Reviews in Food Science and Food Safety 12: 40–53. doi:10.1111/1541-4337.12006.

Var, I., Kabak, B. and Erginkaya, Z. 2008. Reduction in ochratoxin A levels in white wine, following treatment with activated carbon and sodium bentonite. Food Control 19: 592–598. doi:10.1016/j.foodcont.2007.06.013.

Verheecke, C., Liboz, T., Anson, P., Diaz, R. and Mathieu, F. 2015. Reduction of aflatoxin production by *Aspergillus flavus* and *Aspergillus parasiticus* in interaction with *Streptomyces*. Microbiology 161: 967–972. doi:10.1099/mic.0.000070.

Verheecke, C., Liboz, T. and Mathieu, F. 2016. Microbial degradation of aflatoxin B1: Current status and future advances. International Journal of Food Microbiology 237: 1–9. doi:10.1016/j.ijfoodmicro.2016.07.028.

Vila-Donat, P., Marín, S., Sanchis, V. and Ramos, A.J. 2018. A review of the mycotoxin adsorbing agents, with an emphasis on their multi-binding capacity, for animal feed decontamination. Food and Chemical Toxicology 114: 246–259. doi:10.1016/j.fct.2018.02.044.

Vita, D.S., Rosa, P. and Giuseppe, A. 2014. Effect of gamma irradiation on aflatoxins and ochratoxin A reduction in almond samples. Journal of Food Research 3: 113–118. doi:10.5539/jfr.v3n4p113.

Wang, B., Mahoney, N.E., Pan, Z., Khir, R., Wu, B., Ma, H. and Zhao, L. 2016. Effectiveness of pulsed light treatment for degradation and detoxification of aflatoxin B1 and B2 in rough rice and rice bran. Food Control 59: 461–467. doi:10.1016/j.foodcont.2015.06.030.

Wang, B., Han, X., Bai, Y., Lin, Z., Qiu, M., Nie, X., Wang, S., Zhang, F., Zhuang, Z., Yuan, J. and Wang, S. 2017. Effects of nitrogen metabolism on growth and aflatoxin biosynthesis in *Aspergillus flavus*. Journal of Hazardous Materials 324: 691–700. doi:10.1016/j.jhazmat.2016.11.043.

Wang, G., Lian, C., Xi, Y., Sun, Z. and Zheng, S. 2018. Evaluation of nonionic surfactant modified montmorillonite as mycotoxins adsorbent for aflatoxin B1 and zearalenone. Journal of Colloid and Interface Science 518: 48–56. doi:10.1016/j.jcis.2018.02.020.

Wang, X., Bai, Y., Huang, H., Tu, T., Wang, Yuan, Wang, Yaru, Luo, H., Yao, B. and Su, X. 2019a. Degradation of aflatoxin B1 and zearalenone by bacterial and fungal laccases in presence of structurally defined chemicals and complex natural mediators. Toxins 11: 609. doi:10.3390/toxins11100609.

Wang, X., Qin, X., Hao, Z., Luo, H., Yao, B. and Su, X. 2019b. Degradation of four major mycotoxins by eight manganese peroxidases in presence of a dicarboxylic acid. Toxins 11: 566. doi:10.3390/toxins11100566.

Weaver, M.A. and Abbas, H.K. 2019. Field displacement of aflatoxigenic *Aspergillus flavus* strains through repeated biological control applications. Frontiers in Microbiology 10: 1788. doi:10.3389/fmicb.2019.01788.

Wielogorska, E., Ahmed, Y., Meneely, J., Graham, W.G., Elliott, C.T. and Gilmore, B.F. 2019. A holistic study to understand the detoxification of mycotoxins in maize and impact on its molecular integrity using cold atmospheric plasma treatment. Food Chemistry 301: 125281. doi:10.1016/j.foodchem.2019.125281.

Williams, S.B., Baributsa, D. and Woloshuk, C. 2014. Assessing Purdue Improved Crop Storage (PICS) bags to mitigate fungal growth and aflatoxin contamination. Journal of Stored Products Research 59: 190–196. doi:10.1016/j.jspr.2014.08.003.

Williams, S.B., Murdock, L.L. and Baributsa, D. 2017. Storage of maize in Purdue Improved Crop Storage (PICS) bags. PLoS ONE 12: 1–12. doi:10.1371/journal.pone.0168624.

Xing, F., Wang, L., Liu, X., Selvaraj, J.N., Wang, Y., Zhao, Y. and Liu, Y. 2017. Aflatoxin B1 inhibition in *Aspergillus flavus* by *Aspergillus niger* through down-regulating expression of major biosynthetic genes and AFB1 degradation by atoxigenic *A. flavus*. International Journal of Food Microbiology 256: 1–10. doi:10.1016/j.ijfoodmicro.2017.05.013.

Xiong, K., Wang, X.L., Zhi, H.W., Sun, B.G. and Li, X.T. 2017. Identification and safety evaluation of a product from the biodegradation of ochratoxin A by an *Aspergillus* strain. Journal of the Science of Food and Agriculture 97: 434–443. doi:10.1002/jsfa.7742.

Xu, Y., Doel, A., Watson, S., Routledge, M.N., Elliott, C.T., Moore, S.E. and Gong, Y. 2017. Study of an educational hand sorting intervention for reducing aflatoxin B1 in groundnuts in rural Gambia. Journal of Food Protection 80: 44–49. doi:10.4315/0362-028X.JFP-16-152.

Yang, M., Lu, L., Pang, J., Hu, Y., Guo, Q., Li, Z., Wu, S. and Liu, H. 2019. Biocontrol activity of volatile organic compounds from *Streptomyces alboflavus* TD-1 against *Aspergillus flavus* growth and aflatoxin production. Journal of Microbiology 57: 396–404. doi: 10.1007/s12275-019-8517-9.

Yang, X., Zhang, Q., Chen, Z.Y., Liu, H. and Li, P. 2017. Investigation of *Pseudomonas fluorescens* strain 3JW1 on preventing and reducing aflatoxin contaminations in peanuts. PLoS ONE 12: 1–11. doi:10.1371/journal.pone.0178810.

Yu, B., Jiang, H., Pandey, M.K., Huang, L., Huai, D., Zhou, X., Kang, Y., Varshney, R.K., Sudini, H.K., Ren, X., Luo, H., Liu, N., Chen, W., Guo, J., Li, W., Ding, Y., Jiang, Y., Lei, Y. and Liao, B. 2020. Identification of two novel peanut genotypes resistant to aflatoxin production and their SNP markers associated with resistance. Toxins 12: 156. doi:10.3390/toxins12030156.

Yunes, N.B.S., Oliveira, R.C., Reis, T.A., Baquião, A.C., Rocha, L.O. and Correa, B. 2019. Effect of temperature on growth, gene expression and aflatoxin production by *Aspergillus nomius* isolated from Brazil nuts. Mycotoxin Research December: 1–8. doi:10.1007/s12550-019-00380-w.

Zavala-Franco, A., Hernández-Patlán, D., Solís-Cruz, B., López-Arellano, R., Tellez-Isaias, G., Vázquez-Durán, A. and Méndez-Albores, A. 2018. Assessing the aflatoxin B1 adsorption capacity between biosorbents using an *in vitro* multicompartmental model simulating the dynamic conditions in the gastrointestinal tract of poultry. Toxins 10: 484. doi:10.3390/toxins10110484.

Zhai, S.S., Ruan, D., Zhu, Y.W., Li, M.C., Ye, H., Wang, W.C. and Yang, L. 2020. Protective effect of curcumin on ochratoxin A-induced liver oxidative injury in duck is mediated by modulating lipid metabolism and the intestinal microbiota. Poultry Science 99: 1124–1134. doi:10.1016/j.tifs.2016.03.012.

Zhang, H., Apaliya, M.T., Mahunu, G.K., Chen, L. and Li, W. 2016. Control of ochratoxin A-producing fungi in grape berry by microbial antagonists: A review. Trends in Food Science and Technology 51: 88–97. doi:10.1016/j.tifs.2016.03.012.

Zhang, Z.S., Xie, Q.F. and Che, L.M. 2018. Effects of gamma irradiation on aflatoxin B1 levels in soybean and on the properties of soybean and soybean oil. Applied Radiation and Isotopes 139: 224–230. doi:10.1016/j.apradiso.2018.05.003.

Zheng, H., Wei, S., Xu, Y. and Fan, M. 2015. Reduction of aflatoxin B1 in peanut meal by extrusion cooking. LWT-Food Science and Technology 64: 515–519. doi:10.1016/j.lwt.2015.06.045.

Zhu, C., Shi, J., Jiang, C. and Liu, Y. 2015. Inhibition of the growth and ochratoxin A production by *Aspergillus carbonarius* and *Aspergillus ochraceus in vitro* and *in vivo* through antagonistic yeasts. Food Control 50: 125–132. doi:10.1016/j.foodcont.2014.08.042.

Zivoli, R., Gambacorta, L., Piemontese, L. and Solfrizzo, M. 2016. Reduction of aflatoxins in apricot kernels by electronic and manual color sorting. Toxins 8: 26. doi:10.3390/toxins8010026.

Zubrod, J., Bundschuh, M., Arts, G., Brühl, C., Imfeld, G., Knäbel, A., Payraudeau, S., Rasmussen, J., Rohr, J., Scharmüller, A., Smalling, K., Sthele, S., Schulz, R. and Schäfer, R.B. 2019. Fungicides: An overlooked pesticide class? Environmental Science & Technology 53: 3347–3365. doi:10.1021/acs.est.8b04392

Zucchi, T.D., De Moraes, L.A.B. and De Melo, I.S. 2008. *Streptomyces* sp. ASBV-1 reduces aflatoxin accumulation by *Aspergillus parasiticus* in peanut grains. Journal of Applied Microbiology 105: 2153–2160. doi:10.1111/j.1365-2672.2008.03940.x.

CHAPTER

4

Microbial Characterization of Organic Amendments and Their Potential for Biocontrol of Phytopathogenic and Mycotoxigenic Fungi in Amended Soils

Caroline Strub*, Phuong-Anh Nguyen, Sabine Schorr-Galindo and Angélique Fontana
Qualisud, Univ Montpellier, CIRAD, Montpellier SupAgro, Univ d'Avignon, Univ. de La Réunion, Montpellier, France

1. Introduction

Fungal pathogens are an international concern regarding their impact on numerous crops with diseases such as the *Fusarium* wilt, grey molds, blue molds or downy mildews that cause colossal yield losses both at pre- and post-harvest stages all over the world. The fungal diseases are usually associated with the infection and accumulation of mycotoxins in plant products and derivate foods. These fungal metabolites cause public health problems, thus phytopathogenic and mycotoxigenic fungi become the subject of many in-depth researches aiming their efficient control. Frequently applied approaches have been based on the use of fungicides with negative effects on health and environment (Ishii 2006, Goodson et al. 2015). Alternative approaches of biological control that are non-chemical and eco-friendly have been studied such as the crop rotations or the application of organic phyto-protective amendments using living organisms. The use of organic amendments is considered as a promising strategy because of the multiple biochemical and biological mechanisms and the absence of limitations unlike chemical fungicides (Scotti et al. 2020). As previously reported, these amendments with variable natural compositions on one hand contain indispensable nutritive elements to enhance the development of plants (Ninh et al. 2015, Song et al. 2015, Nguyen et al. 2017); on the other hand,

*Corresponding author: caroline.strub@umontpellier.fr

the amendments contribute substantially to reduce the pathogen viability, their diffusion in soils and their mycotoxin productions (Bonanomi et al. 2010, Larney and Angers 2012). As such, they control the diseases caused by soil-borne pathogens and the fatal risks linked to mycotoxins. Soils are possibly the most complex natural environment of microbes including bacterial and fungal species. One gram of soil can contain up to 10 billion microorganisms in which about 10^8 are prokaryotic cells that belong to thousands different species (Torsvik et al. 1990, Portillo et al. 2013, Regan et al. 2014). Historically, these soil microbes were reported to be an important source of natural bioactive products (Handelsman et al. 1998, Daniel 2004). Similarly, studies have demonstrated that the phyto-protective properties of amendments are partly due to the presence of interesting microorganisms with antimicrobial/antifungal activities (Pugliese et al. 2009, Suárez-Estrella et al. 2013, Cloutier et al. 2020) named PGPM (Plant Growth-Promoting Microorganisms). They are known for their phyto-beneficial and phyto-protective effects including antifungal potential (Navarro et al. 2019). Among the microbial genera most frequently cited for their biocontrol activities in amendments are *Bacillus, Pseudomonas, Streptomyces* and *Trichoderma* employing both antibiosis and induction of host resistance (Krauss et al. 2006, Vida et al. 2017, Kaushal and Wani 2017, Inderbitzin et al. 2018, Nguyen et al. 2018, Stark et al. 2018). The PGPM are reported to be able to promote the growth of plants and protect the plants from diseases and pathogens (Verma et al. 2019). They are involved in diverse direct mechanisms that support the plant growth in term of phytohormone productions, siderophore productions, N_2 fixation, phosphate solubilization etc. On the other hand, the PGPM could be used as biocontrol agents by suppressing the pathogenic fungi by antagonism or by reinforcing the resistance system in plants. The main modes of biocontrol activity employed by PGPM are the competition for nutrients or habitat, the antifungal metabolite production and induced systemic resistance in plants. Sometimes the effectiveness can be synergized when different PGPM interact with a same plant host (Alizadeh et al. 2013, Song et al. 2015). PGPM consist of a large number of bacterial (also called PGPR – Plant Growth-Promoting Rhizobacteria) and fungal species belonging *inter alia* to genera *Azospirillum, Alcaligenes, Acinetobacter, Bacillus, Burkholderia, Enterobacter, Flavobacterium, Pseudomonas, Xanthomonas, Rhizobium, Serratia, Streptomyces, Penicillium,* and *Trichoderma*.

Among the strategies for investigating biodiversity, metabarcoding is now emerging as an effective method for studying the different microbial communities and is recognized for its reliability for soil biodiversity monitoring (Taberlet et al. 2012, Ji et al. 2013, Schmidt et al. 2013, Deiner et al. 2017).

In this chapter, we describe the results of the meta barcoding analysis of three organic amendments, and corresponding amended soils. The purpose was to use the organic amendments in large scale on cereal crops such as maize, wheat or barley. Emphasis will relate to several microbial families of interest as biocontrol agents that could act in the protection of soil and plants against the phytopathogenic and/or mycotoxigenic fungi.

2. Description of the Organic Amendments and Their Origin

The organic amendments used were three industrial products, named A, B and C. They were provided by Frayssinet company (Rouairoux, France). Amendment A is a compost derived from vegetable matter (50% w/w) composed of cakes and fruit pulps (coffee, olive, cocoa, bark), sheep manure and raw wool. Its composting process took place during a thermophilic phase at 65°C lasting three months during which the compost was mixed by inversion followed by a five months mesophilic maturation phase. Amendment B underwent the same phase of thermophilic maturation as amendment A but its mesophilic phase was shorter (1 to 2 months). It is composed of vegetable matter and composted manure. Plant matter represented only 20% of the amendment while composted manure was the major component. The composted base of amendment A accounted for 75% (w/w) of composted base of amendment C. To this base were added 25% (w/w) of concentrated organic supplements (powder of feathers and bones, hydrolysates of meats and beetroot wine). Amendments B and C shared the same composting process.

The soil employed as reference soil was collected on a loamy zone and was prepared as described by Bonanomi et al. (2010) and Nguyen et al. (2018). The mix of reference soil and of the amendments were prepared as described by Nguyen et al. (2018). Briefly, the soil preparation was carried out as follows: 30 g of reference soil, 0.35 g of amendments (approximately corresponding to 1.15% w/w) and 5.1 mL of water, corresponding to 2/3 of the water holding capacity (WHC) of the reference soil (Kaisermann et al. 2015), were mixed in a pot. The WHC was determined according to protocol described previously (Grace et al. 2006). Knowing that an incubation of soils at 28°C for 28 days simulates a 3-month field crop, pots were placed in a climatic chamber at 28°C without lid. Samples were taken after 3, 14 and 28 days of incubation.

3. Microbial Diversity Indices of the Different Amendments and of the Soils

Many indices allow us to provide an estimation of the biodiversity status of natural samples and their impact factors (Kim et al. 2017). In this study, the evolution of microbial community was assessed by richness, including number of observed OTUs (Operational Taxonomic Units) and Chao1 indices (Kim et al. 2017), and diversity (Shannon index) (Peet 1974). The number of observed OTUs in bacterial libraries of all samples were greater than that in fungal libraries (Table 1). Among three amendments, the amendment C showed highest richness in term of OTU number (415 and 251 for bacterial and fungal communities respectively). The three organic amendments had a lower richness of both bacterial and fungal communities than the amended soils and reference soil. However, the richness in soils amended with these organic products increased during the incubation and was generally higher than that in reference soils. Moreover, the number of observed OTUs was higher in amended soils compared to the reference soil after three days. The opposite was observed after 28 days both for bacterial and fungal communities. This result concerning the number of observed OTU could correspond to the flush of the microbial growth which could be explained by a rapid mineralization of the organic matter by the microbial communities of the soil (Pathak and Rao 1998, Thuriès et al. 2002, Pansu et al. 2003, Zhang and Marschner 2018). The Chao1 index also represents the microbial richness of a sample. It varies overall in the same way as the number of observed OTUs. The Chao1 in amendments ranged from 378 to 495 for bacteria and from 125 to 380 for fungi. The highest Chao1 index was observed in amendment C.

Similarly, to the richness indicators, the Shannon indices of samples fluctuated in the same way. For both bacterial and fungal communities, Shannon index values of amendments were lower than those of amended soils and reference soil. According to the Shannon indices, amendment C was characterized by the highest diversity for the bacteria and for the fungi among the three amendments. The index values in our study were found greater than those observed for bacteria and fungi in biochar-amended soils in study of Hu et al. (2014). In amended soils, values of the Shannon index were systematically higher for bacteria than for fungi, suggesting that the bacterial communities were more diverse than fungal population and that bacteria were more adaptive than fungi to the change of soil environment induced by adding amendments. Furthermore, the addition of amendment had a perennial positive effect (0, 3 and 28 days) on soil bacterial diversity but not on fungal diversity, since for the bacterial community, the indexes of amended soils were higher than those of the reference soil. The opposite trend was observed for molds. This was in agreement with the results on amended soil with biochar (Hu et al. 2014). Although amendment C was the richest and most diverse amendment in terms of bacterial and fungal communities, it did not confer the most diverse and rich microbial community to the amended soil. The endemic microbiota of the amendment did not explain alone those of the amended soil. This confirmed the importance of the organic matter brought to the soil by the amendment. Soils complemented with amendment A or B had the richest and most diverse bacterial community. This bacterial diversity was able to effectively control *in vitro* the development of the mycotoxigenic mold *Fusarium verticillioides*. Indeed, Nguyen et al. (2018) observed that the inhibition of *Fusarium verticillioides* growth was higher with soils amended with these amendments than with the reference soil. The metabolites produced by the microbiota of the amended soils also exhibited an important inhibition rate on the fumonisin production, up to 68.7% and 92.5% for fumonisins B1 and B2 respectively. Some *Streptomyces* strains isolated from these amendments were able to strongly reduce both fungal growth and fumonisin production. Thus, fungal growth inhibition reached more than 60% in the presence of extracellular metabolites of several *Streptomyces* strains. Moreover, the fumonisin B1 and fumonisin B2 amounts decreased by up to 87.5% and 98.2% respectively, depending on strains and cultivation time.

Table 1. Diversity in genomic libraries from different amendments, amended soils and reference soils during incubation time

Bacterial samples	*Richness*		*Diversity*	*Fungal samples*	*Richness*		*Diversity*
	Observed OTUs	*Chao1*	*Shannon*		*Observed OTUs*	*Chao1*	*Shannon*
A	186	438	2.71	A	118	113	1.95
B	225	378	3.1	B	135	232	2.88
C	415	495	3.33	C	251	392	3.69
A-3	1475	2375	6.65	A-3	298	476	3.65
B-3	1054	2124	6.48	B-3	378	773	3.43
C-3	1930	3609	6.86	C-3	343	664	3.92
Ref-3	1085	1815	6.58	Ref-3	265	436	4.02
A-28	3920	4763	7.52	A-28	741	1060	4.03
B-28	3722	4812	7.5	B-28	626	873	4.28
C-28	1608	2483	6.45	C-28	753	918	4.51
Ref-28	3232	4002	7.33	Ref-28	981	1185	4.78

A: amendment A. A-3: amended soil with amendment A at 3rd day of incubation. A-28: amended soil with amendment A at 28th day of incubation. B: amendment B. B-3: amended soil with amendment B at 3rd day of incubation. B-28: amended soil with amendment B at 28th day of incubation. C: amendment C. C-3: amended soil with amendment C at 3rd day of incubation. C-28: amended soil with amendment C at 28th day of incubation. Ref-3: reference soil at 3rd day of incubation. Ref-28: reference soil at 28th day of incubation.

Principal Coordinate Analyses (PCoA) of microbial communities of different amendments, amended soils and reference soil were carried out. PCoA of bacterial pyrosequencing data (Fig. 1A), based on weighted Unifrac metric, which is widely used in microbial ecology for comparing biological communities by accounting for abundance of observed organisms (Lozupone et al. 2011), on three principal coordinates that accounted for over 92% of the variability showed a clear discrimination between the three amendments. The results demonstrated also that the raw amendments were clearly different from the amended soils. The amended soils and the reference soils were found as well separated. A similar discrimination was observed on PCoA plot of ITS reads from samples (Fig. 1B).

4. Bacterial Communities in Different Amendments and Amended Soils

4.1 Bacterial Communities in Amendments

The bacterial compositions of the three amendments were quite different, especially comparing A and B to C (Fig. 2). Only families with relative abundances greater than 1% are represented. The bacterial diversity was

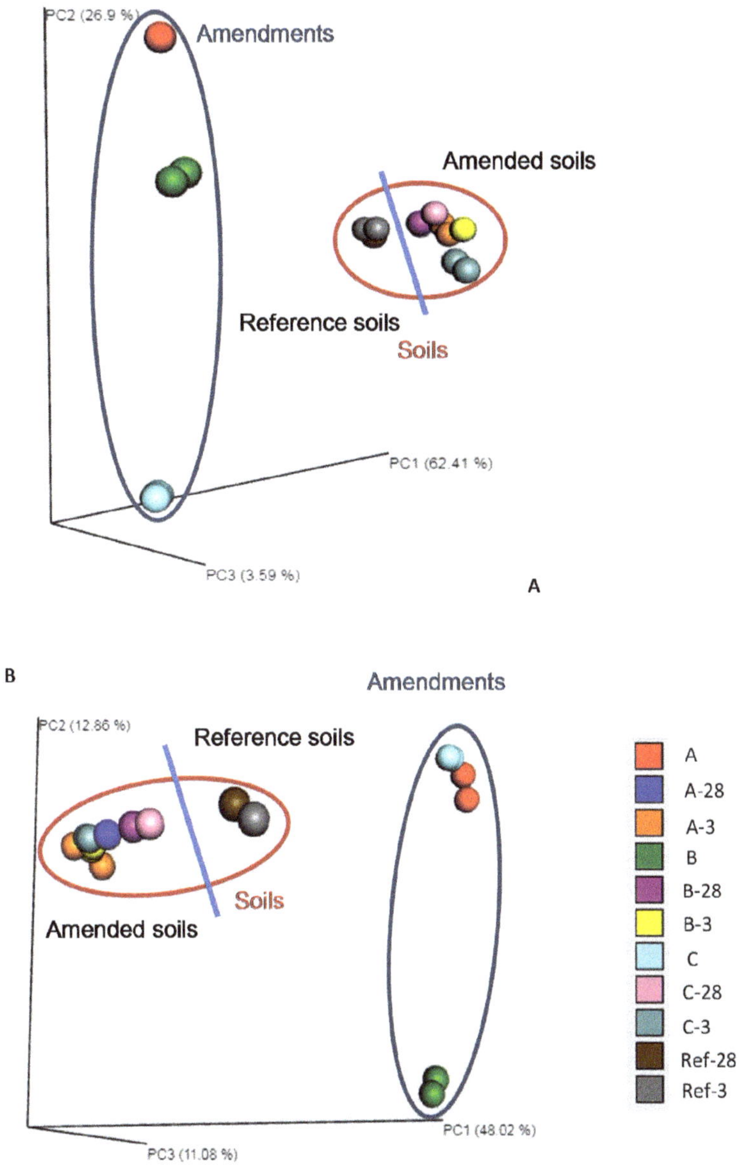

Figure 1. Principal Coordinate Analyses (PCoA) plot of microbial communities in amended soils and reference soil. A: PCoA plot of bacterial communities. B: PCoA plot of fungal communities. A-3: amended soil with amendment A at 3rd day of incubation. A-28: amended soil with amendment A at 28th day of incubation. B-3: amended soil with amendment B at 3rd day of incubation. B-28: amended soil with amendment B at 28th day of incubation. C-3: amended soil with amendment cat 3rd day of incubation. C-28: amended soil with amendment C at 28th day of incubation. Ref-3: reference soil at 3rd day of incubation. Ref-28: reference soil at 28th day of incubation

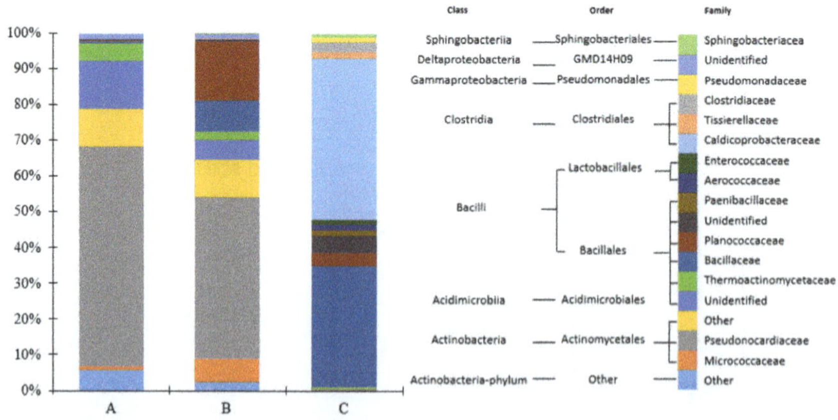

Figure 2. Relative abundance (%) of the sequences corresponding to bacterial family identified in amendments. A: amendment A. B: amendment B. C: amendment C.

distributed within only 9, 12 and 13 identified clusters in the amendments A, B and C respectively.

The *Pseudonocardiaceae* family is the most abundant in bacterial communities of both amendments A and B with levels of 60.1% and 42.5% of the relative abundance. This family belonging to the order *Actinomycetales* has also been found predominant in soils treated with organic amendments. This family was reported to include various genera that exhibit antimicrobial activities (Maleka et al. 2015, Rao et al. 2015, Jurado et al. 2019) especially by production of bioactive compounds. Strains of this family were detected in amendments and have been expected to contribute to antimicrobial activities. Besides, a considerable proportion of sequences belonging to the order *Actinomycetales* was reported in the amendments A and B. The genus *Streptomyces* which is included in this order was largely applied in many phytoprotection commercial products such as Mycostop®, Actinovate®, Bla S®, Valinum®, Polyoxin®, etc. and achieved inhibitory effects against phytopathogens including *Fusarium, Rhizotonia, Pyricularia, Botritys,* and *Phytium* (Palazzini et al. 2017).

However, these clusters were found scarcely in amendment C with only less than 1% for the *Pseudonocardiaceae* family meanwhile *Caldicoprobacteraceae* and *Bacillaceae* families predominated in the DNA libraries from amendment C by representing up to 41.5% and 31.1% of the total identified bacterial sequences, respectively. The *Caldicoprobacteraceae* family, which belongs to the order *Clostridiales* and was found predominant in amendment C was not known for antimicrobial activity but includes thermophilic bacteria that are found in herbivore feces. To the contrary, the *Bacillaceae* was found in many amendments (Ntougias et al. 2013, Tortosa et al. 2017). The members of these two families are reported to form endospores which allow them to survive in extreme environments such as under high-temperature during composting stages (Mandic-Mulec et al. 2016). Numerous genera belonging to *Bacillaceae* family are able to produce antimicrobial compounds against a wide range of foodborne bacterial and fungal pathogens (Pandin et al. 2019). Another cluster of the order *Bacillales* was also found in amendment C with a proportion of 4.2%. Various strains of *Bacillus* genus that belong to these families showed antifungal in general and particularly against *Fusarium* species (Blacutt et al. 2016, Martínez-Álvarez et al. 2016, Kulimushi et al. 2017, Cucu et al. 2020). Another interesting family, the *Pseudomonadaceae*, was slightly present in the amendment C (1.2%). The genus *Pseudomonas* which belongs to this family was recognized to produce antifungal metabolites and to protect the plants from *Fusarium* spp. (Selvaraj et al. 2014, Yasmin et al. 2014, Quecine et al. 2016, Scotti et al. 2020). Obviously, the bacterial compositions of the different amendments appeared to be greatly impacted by the specific compositions and composting conditions of each amendment.

4.2 Bacterial Communities in Soils and Amended Soils

Contrary to the amendments, the bacterial compositions of the reference soils and amended soils were found to be much more diversified (Fig. 3). Many different bacterial families have been identified but accounting for low abundance in the total communities. In 3-day-old incubated reference soil, the predominant families were *Sphingomonadaceae* (6%), *Hyphomicrobiaceae* (5.8%), *Cytophagaceae* (5.7%) and *Rhodospirillaceae* (4.7%). The bacterial composition of 28-day-old reference soils was as diverse as that of the 3-day-old soil, but different in the contribution levels of the different bacterial clusters. The participation of the predominant families in 3-day-old soil decreased after 28 days of incubation. Conversely, several families which were less present in the 3-day-old reference soil were found more abundant after incubation and became the predominant families in 28-day-old soils. For example, the families of *Chitinophaceae* and *Sinobacteraceae* accounted for 6% and 5.3% respectively of total bacteria after 28 days compared to 4.1% and 3.5% at 3rd day, respectively. For families of interest for biocontrol, like the *Streptomycetaceae* and *Bacillaceae*, their level in reference soil of both 3rd day and 28th day of incubation was much less than 1% and 1.5% respectively.

Similarly to reference soil, the bacterial families in amended soils were diversified and each one represented less than 10% of total bacterial population (Fig. 3). Since a low percentage (1.15%) of amendments was added into the reference soil, the most abundant families were found to be the same of that in the 3 and 28-day-old incubated reference soils. Furthermore, the predominant family of amendments A and B, *Pseudonorcadiaceae* was found at low abundance in amended soils. Conversely, amended soil with amendment C exhibited a higher level for the *Bacillaceae* family (3.6% at 3rd day and 1.5% at 28th day) than in the reference soil. This suggested that the strains of the *Bacillaceae* family from the amendment C established well in the amended soils or the organic matter that it contains promotes the bacilli present in the soil. In the same way, the family *Streptomycetaceae* which was insignificant in reference soil was detected in all amended soils with a proportion ranging

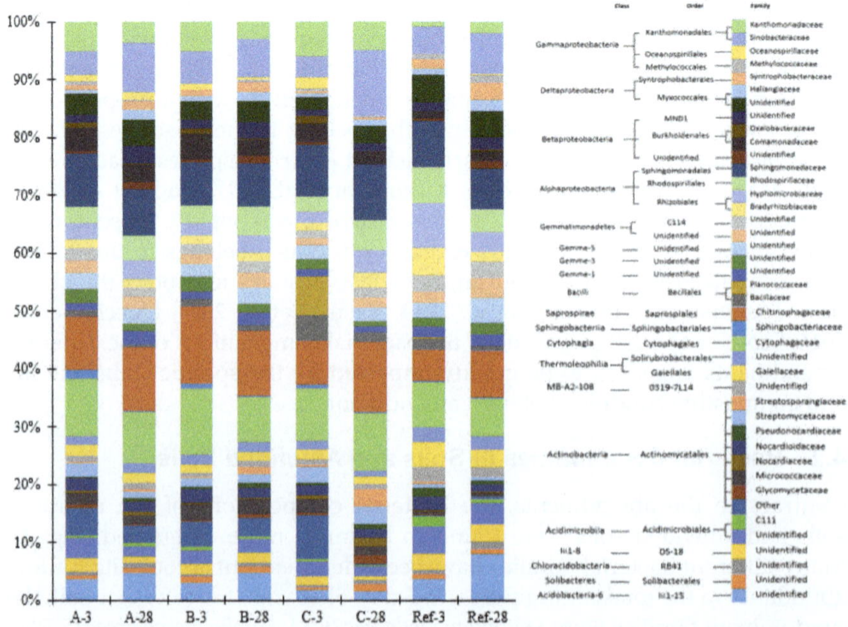

Figure 3. Relative abundance (%) of the sequences corresponding to bacterial family identified in amended soils and reference soils during incubation time. A-3: amended soil with amendment A at 3rd day of incubation. A-28: amended soil with amendment A at 28th day of incubation. B-3: amended soil with amendment B at 3rd day of incubation. B-28: amended soil with amendment B at 28th day of incubation. C-3: amended soil with amendment C at 3rd day of incubation. C-28: amended soil with amendment C at 28th day of incubation. Ref-3: reference soil at 3rd day of incubation. Ref-28: reference soil at 28th day of incubation.

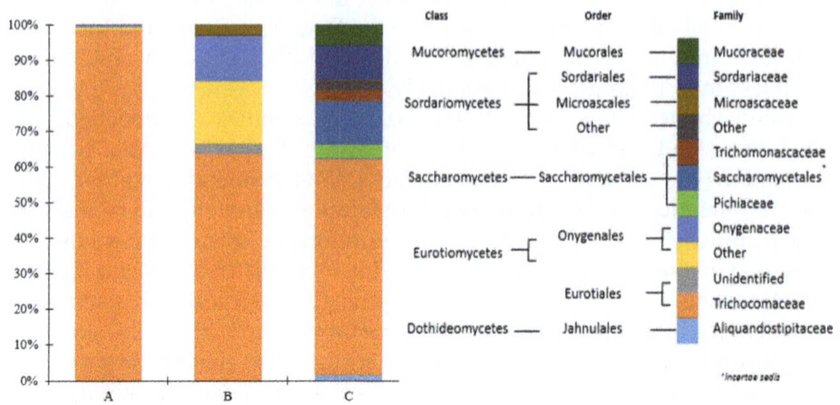

Figure 4. Relative abundance (%) of the sequences corresponding to fungal family identified in amendments. A: amendment A. B: amendment B. C: amendment C.

from 0.8% to 2.5%. Moreover, the family level remained almost constant between 3 and 28 day of incubation in amended soils with amendments B and C. These strains probably came from the amendments and spread out in the amended soils.

Some OTUs were found in both the amended soil and the amendment used. In parallel, they were absent from all other soils. This exclusive occurrence is an argument in favor of the possible transfer and persistence of microorganisms from the amendment to the soil.

5. Fungal Communities in Different Amendments and Amended Soils

5.1 Fungal Communities in Amendments

A total of 12 clusters of fungal communities were identified in the three amendments (Fig. 4). The fungal microflora of the amendment A was almost only constituted with the *Trichocomaceae* family that accounted for 98.2% of the total families. This family also represents the majority of fungal communities in amendments B and C where it represented 62% and 55.3% of total population, respectively. The *Trichocomaceae* family includes the major anamorph genera *Aspergillus* and *Penicillium*. Although some species of these genera are opportunistic pathogens (Tóth et al. 2012, Rodrigues et al. 2013, Ruadrew et al. 2013, Ok et al. 2014, Kange et al. 2015), many other strains are exploited in both industry and medicine for the production of enzymes, antibiotics and bioactive compounds and could be found in different substrates. *Penicillium* spp. and *Aspergillus* spp. were reported to enhance the growth and induce self-defense of plants and produce numerous antifungal compounds against phytopathogenic fungi (Yang et al. 2008, Hegedűs et al. 2011, Awaad et al. 2012, Patil et al. 2013, Sreevidya et al. 2015, Pandey and Yarzábal 2019). The fungal communities of the amendment C were found to be the most diversified. Besides *Trichocomaceae* family that predominated, an important presence of strains from the class *Sordariomycetes* was noticed. A wide range of genera that were reported to be PGPM could be included in this class. For example, the *Trichoderma* genus which is well-described for PGP and antifungal effects on a broad spectrum of plants (Cucu et al. 2020). The species *T. harzianum, T. ghanense* and *T. hamatum* help reducing the input of chemical pesticides in agricultures and induce phytohormone production in melon plants to fight against *F. oxysporum* infection (Martínez-Medina et al. 2014). The *Trichoderma* species have been widely applied for the production of various antifungal compounds such as chitinase, peptaibol and as commercial biocontrol agents like Canna®, Trichosan®, Vitalin®, TrichoMax® (Degenkolb et al. 2015, Abdelrahman et al. 2016, Küçük 2017). Another genus belonging to the class *Sordariomycetes* that was reported PGPM is *Gliocladium* (*Clonostachys*). Several commercial products based on *Gliocladium* spp. (PreStop® and Primastorp® (Kimera Agro, Japan), SoilGard® (Certis, USA) and Gliomix® (Verdera Oy, Finland)) employed for soil treatments as antiparasites, anti-*Rhizotonia*, and anti-*Phytium*. The *Clonostachys rosea* biocontrol agent strain showed effective control of FUM1 gene expression that regulates the fumonisin production of *F. verticillioides* on maize cobs. *Clonostachys rosea* strains have also exhibited biocontrol effect on various *Fusarium* spp. including *F. graminearum, F. verticillioides* and *F. avenaceum* on wheat stalks (Samsudin et al. 2017, Sun et al. 2020).

5.2 Fungal Communities in Soils and Amended Soils

The fungal community of the reference soil was more diversified as compared to the fungal communities of the amendments (Fig. 5). At day 3 of incubation period, the main clusters identified in reference soils were the *Hypocreales* (3.4%), *Chaetomiaceae* (14.5%), *Plectosphaerellaceae* (11.3%) and *Nectriaceae* (7.9%). These clusters remained the major contributors of the 28-day-old reference soil communities. During the incubation, no significant variation was found for the *Chaetomiaceae* family compared to total communities. Meanwhile, the *Plectosphaerellaceae* and *Hypocreales* cluster levels decreased and the *Nectriaceae* family level increased up to 20% of the total fungi. Within the *Nectriaceae* family, species of *Fusarium* might be found, in which numerous are soil-borne phytopathogenic and/or mycotoxigenic such as *F. graminearum, F. verticillioides, F. oxysporum, F. solani*, etc. (Saremi and Saremi 2013, Karim et al. 2016).

Fungal families involved in the communities of the amended soils with the amendments A, B and C are almost the same found in reference soils notably those belonging to the *Sordariomycetes* class. Among them, the *Nectriaceae* family level was different in the three amended soils during the incubation period. The amendment A induced a decrease of the relative abundance of the *Nectriaceae* family in this amended soil. Furthermore, this observation suggests that the abundance in the amended soil of the harmful species of *Fusarium* has also decreased. The presence of antifungal biocontrol agents in amendment A was expected to impact this decrease. The amendment A was expected to decrease pathogenic *Fusarium* species existing in reference soil. The barcoding study conducted on the amended soil microbiota indicates that amendment A should be the most effective in terms of *Fusarium* biocontrol. This observation agrees with the results observed *in vitro* (Nguyen et al. 2018). Conversely, the amendment B and C increased the level of this family as in the reference soil. Despite the pathogenicity of several species, the genus *Fusarium* can include PGPM and antifungal species. For example, *Fusarium* spp. was reported to promote the germination in spinach, increase the shoot length and root length of cucumber (Islam et al. 2014). Several non-pathogenic strains of *Fusarium* can serve as antifungal agents towards pathogenic *Fusarium* and other fungal pathogens (Ishimoto et al. 2004, Zhang et al. 2015, Alberts et al. 2016, Sharma 2019). A non-pathogenic *F. oxysporum* strain was also applied in the production of anti-fusariose commercial preparations such as Biofox C® (SIAPA, Italy), Fusaclean® (Natural Plant Protection, France). Besides, many other families described as PGPMs were identified as belonging to the *Sordariomycetes* class (Jahagirdar et al. 2019).

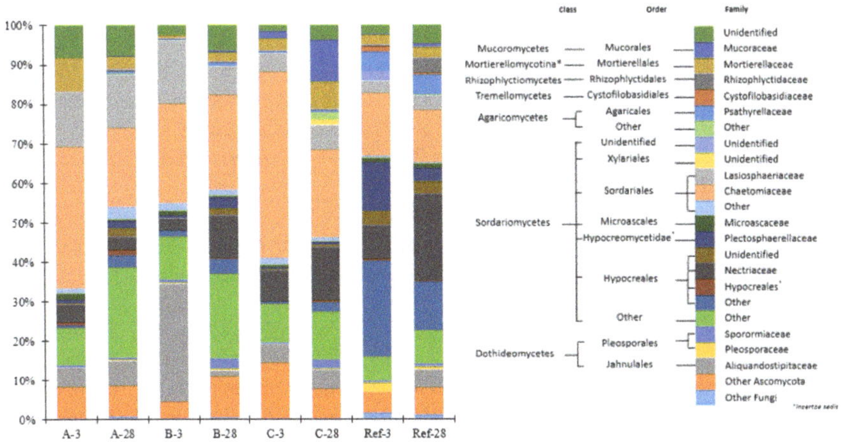

Figure 5. Relative abundance (%) of the sequences corresponding to fungal family identified in amended soils and reference soils during incubation time. A-3: amended soil with amendment A at 3rd day of incubation. A-28: amended soil with amendment A at 28th day of incubation. B-3: amended soil with amendment B at 3rd day of incubation. B-28: amended soil with amendment B at 28th day of incubation. C-3: amended soil with amendment C at 3rd day of incubation. C-28: amended soil with amendment C at 28th day of incubation. Ref-3: reference soil at 3rd day of incubation. Ref-28: reference soil at 28th day of incubation.

6. Conclusion and Perspectives

Metabarcoding analysis highlighted interesting differences in microbial communities of different organic amendments. Their bacterial and fungal composition appeared to be greatly impacted by the raw materials and the composting conditions used to produce the amendment. In addition, the composition of the soil microbiota seems to be impacted by a supplementation of only 1.15% (w/w) of organic amendments and moreover these modifications tend to persist over time. In general, the organic amendments contained various PGPM species (including bacterial and fungal). Interestingly, PGPM content of the soil increased by adding the amendment. Metabarcoding analysis showed that amendment A was expected to decrease pathogenic *Fusarium* species existing in reference soil. Real time PCR with specific primers targeting *phytopathogenic fusaria* in green house or field experimentations could support this hypothesis. In addition, this work tends to demonstrate the possibility of the transfer and the persistence of certain microorganisms from the amendment to the soil such as *Streptomyces* genus with amendments A and B and *Bacillus* genus for amendment C. Thus, a new generation of amendments could be developed. Their microbial composition could be tuned in order to promote phytoprotective microbiota by adjusting their technical production route.

References

Abdelrahman, M., Abdel-Motaa, F., El-Sayed, M., Jogaiah, S., Shigyo, M., Ito, S.I. and Tran, L.S.P. 2016. Dissection of *Trichoderma longibrachiatum*-induced defense in onion (*Allium cepa* L.) against *Fusarium oxysporum* f. sp. *cepa* by target metabolite profiling. Plant Science 246: 128–138. https://doi.org/10.1016/j.plantsci.2016.02.008

Alberts, J.F., van Zyl, W.H. and Gelderblom, W.C.A. 2016. Biologically based methods for control of fumonisin-producing *Fusarium* species and reduction of the fumonisins. Frontiers in Microbiology 7: 1–33. https://doi.org/10.3389/fmicb.2016.00548

Alizadeh, H., Behboudi, K., Ahmadzadeh., M., Javan-Nikkhah, M., Zamioudis, C., Pieterse, C.M.J. and Bakker, P.A.H.M. 2013. Induced systemic resistance in cucumber and *Arabidopsis thaliana* by the combination of *Trichoderma harzianum* Tr6 and *Pseudomonas* sp. Ps14. Biological Control 65: 14–23.https://doi.org/10.1016/j.biocontrol.2013.01.009

Awaad, A.S., Nabilah, A.J.A. and Zain, M.E. 2012. New antifungal compounds from *Aspergillus terreus* isolated from desert soil. Phytotherapy Research 26: 1872–1877. https://doi.org/10.1002/ptr.4668

Bernal-Vicente, A., Ros, M., Tittarelli, F., Intrigliolo, F. and Pascual, J. 2008. Citrus compost and its water extract for cultivation of melon plants in greenhouse nurseries: Evaluation of nutriactive and biocontrol effects. Bioresource Technology 99: 8722–8728. https://doi.org/10.1016/j.biortech.2008.04.019

Blacutt, A., Mitchell, T., Bacon, C. and Gold, S. 2016. *Bacillus mojavensis* RRC101 lipopeptides provoke physiological and metabolic changes during antagonism against *Fusarium verticillioides*. Molecular Plant-Microbe Interactions 29: 713–723. https://doi.org/10.1094/MPMI-05-16-0093-R

Bonanomi, G., Antignani, V., Capodilupo, M. and Scala, F. 2010. Identifying the characteristics of organic soil amendments that suppress soilborne plant diseases. Soil Biology and Biochemistry 42: 136–144. https://doi.org/10.1016/j.soilbio.2009.10.012

Cloutier, A., Tran, S. and Avis, T.J. 2020. Suppressive effect of compost bacteria against grey mould and *Rhizopus* rot on strawberry fruit. Biocontrol Science and Technology 30: 143–159. https://doi.org/10.1080/09583157.2019.1695745

Cucu, M.A., Gilardi, G., Pugliese, M., Gullino, M.L. and Garibaldi, A. 2020. An assessment of the modulation of the population dynamics of pathogenic *Fusarium oxysporum* f. sp. *lycopersici* in the tomato rhizosphere by means of the application of *Bacillus subtilis* QST 713, *Trichoderma* sp. TW2 and two composts. Biological Control 142: 104–158. https://doi.org/10.1016/j.biocontrol.2019.104158

Daniel, R. 2004. The soil metagenome – A rich resource for the discovery of novel natural products. Current Opinion in Biotechnology 15: 199–204. https://doi.org/10.1016/j.copbio.2004.04.005

Degenkolb, T., Fog Nielsen, K., Dieckmann, R., Branco-Rocha, F., Chaverri, P., Samuels, G.J., Thrane, U., Von Döhren, H., Vilcinskas, A. and Brückner, H. 2015. Peptaibol, secondary-metabolite, and hydrophobin pattern of commercial biocontrol agents formulated with species of the *Trichoderma harzianum* complex. Chemistry & Biodiversity 12: 662–684. https://doi.org/10.1002/cbdv.201400300

Deiner, K., Bik, H.M., Mächler, E., Seymour, M., Lacoursière-Roussel, A., Altermatt, F., Creer, S., Bista, I., Lodge, D.M., de Vere, N., Pfrender, M.E. and Bernatchez, L. 2017. Environmental DNA metabarcoding: Transforming how we survey animal and plant communities. Molecular Ecology 26: 5872–5895. https://doi.org/10.1111/mec.14350

Goodson, W.H., Lowe, L., Carpenter, D.O., Gilbertson, M., Manaf, A., Lopez de Cerain Salsamendi, A., ... and Charles, A.K. 2015. Assessing the carcinogenic potential of low-dose exposures to chemical mixtures in the environment: The challenge ahead. Carcinogenesis 36: S254–S296. https://doi.org/10.1093/carcin/bgv039

Grace, C., Hart, M. and Brookes, P.C. 2006. Laboratory manual of the soil microbial biomass group. Rothamsted Research 65. https://doi.org/10.13140/RG.2.1.3911.2407

Handelsman, J., Rondon, M.R., Brady, S.F., Clardy, J. and Goodman, R.M. 1998. Molecular biological access to the chemistry of unknown soil microbes: A new frontier for natural products. Chemistry & Biology 5: R245–R249. https://doi.org/10.1016/S1074-5521(98)90108-9

Hegedűs, N., Leiter, É., Kovács, B., Tomori, V., Kwon, N.J., Emri, T., Marx, F., Batta, G., Csernoch, L., Haas, H., Yu, J.H. and Yu, J.H. 2011. The small molecular mass antifungal protein of *Penicillium chrysogenum* – A mechanism of action oriented review. Journal of Basic Microbiology 51: 561–571. https://doi.org/10.1002/jobm.201100041

Hu, L., Cao, L. and Zhang, R. 2014. Bacterial and fungal taxon changes in soil microbial community composition induced by short-term biochar amendment in red oxidized loam soil. World Journal of Microbiology and Biotechnology 30: 1085–1092. https://doi.org/10.1007/s11274-013-1528-5

Inderbitzin, P., Ward, J., Barbella, A., Solares, N., Izyumin, D., Burman, P., Chellemi, D.O. and Subbarao, K.V. 2018. Soil microbiomes associated with *Verticillium* wilt-suppressive broccoli and chitin amendments are enriched with potential biocontrol agents. Phytopathology 108: 31–43. https://apsjournals.apsnet.org/doi/pdf/10.1094/PHYTO-07-17-0242-R

Ishii, H. 2006. Impact of fungicide resistance in plant pathogens on crop disease control and agricultural environment. Japan Agricultural Research Quarterly 40: 205–211. https://doi.org/10.6090/jarq.40.205

Ishimoto, H., Fukushi, Y. and Tahara, S. 2004. Non-pathogenic *Fusarium* strains protect the seedlings of *Lepidium sativum* from *Pythium ultimum*. Soil Biology and Biochemistry 36: 409–414. https://doi.org/10.1016/j.soilbio.2003.10.016

Islam, S., Akanda, A.M., Prova, A., Sultana, F. and Hossain, M.M. 2014. Growth promotion effect of *Fusarium* spp. PPF1 from bermudagrass (*Cynodon dactylon*) rhizosphere on Indian spinach (*Basella alba*) seedlings are linked to root colonisation. Archives of Phytopathology and Plant Protection 47: 2319–2331. https://doi.org/10.1080/03235408.2013.876745

Jahagirdar, S., Kambrekar, D.N., Navi, S.S. and Kunta, M. 2019. Plant growth-promoting fungi: Diversity and classification. pp. 25–34. *In*: S. Jogaiah and M. Abdelrahman (eds.). Bioactive Molecules in Plant Defense. Springer, Cham. https://doi.org/10.1007/978-3-030-27165-7_2

Ji, Y., Ashton, L., Pedley, S.M., Edwards, D.P., Tang, Y., Nakamura, A., Kitching, R., Dolman, P.M., Woodcock, P., Edwards, F.A., Larsen, T.H., Hsu, W.W., Benedick, S., Hamer, K.C., Wilcove, D.S., Bruce, C., Wang, X., Levi, T., Lott, M., Emerson B.C. and Yu, D.W. 2013. Reliable, verifiable and efficient monitoring of biodiversity via metabarcoding. Ecology Letters 16: 1245–1257. https://onlinelibrary.wiley.com/doi/pdf/10.1111/ele.12162

Jurado, M.M., Suárez-Estrella, F., López, M.J., López-González, J.A. and Moreno, J. 2019. Bioprospecting from plant waste composting: Actinobacteria against phytopathogens producing damping-off. Biotechnology Reports 23: e00354. https://doi.org/10.1016/j.btre.2019.e00354

Kaisermann, A., Maron, P., Beaumelle, L. and Lata, J. 2015. Fungal communities are more sensitive indicators to non-extreme soil moisture variations than bacterial communities. Applied Soil Ecology 86: 158–164. https://doi.org/10.1016/j.apsoil.2014.10.009

Kange, A.M., Cheruiyot, E.K., Ogendo, J.O. and Arama, P.F. 2015. Effect of sorghum (*Sorghum bicolor* L. Moench) grain conditions on occurrence of mycotoxin-producing fungi. Agriculture & Food Security 4: 1. https://doi.org/10.1186/s40066-015-0034-4

Karim, N.F.A., Mohd, M., Nor, N.M.I.M. and Zakaria, L. 2016. Saprophytic and potentially pathogenic *Fusarium* species from peat soil in Perak and Pahang. Tropical Life Sciences Research 27: 1. https://www.ncbi.nlm.nih.gov/pmc/articles/PMC4807956/pdf/tlsr-27-1-1.pdf

Kaushal, M. and Wani, S.P. 2017. Efficacy of biological soil amendments and biocontrol agents for sustainable rice and maize production. pp. 2792–2798. *In*: V. Meena, P. Mishra, J. Bisht, A. Pattanayak (eds.). Agriculturally Important Microbes for Sustainable Agriculture. Springer, Singapore. https://doi.org/10.1007/978-981-10-5589-8_13

Kim, B.R., Shin, J., Guevarra, R., Lee, J.H., Kim, D.W., Seol, K.H., Kim, H.B. and Isaacson, R.E. 2017. Deciphering diversity indices for a better understanding of microbial communities. Journal of Microbiology and Biotechnology 27: 2089–2093. https://doi.org/10.4014/jmb.1709.09027

Krauss, U., ten Hoopen, G.M., Hidalgo, E., Martínez, A., Stirrup, T., Arroyo, C., Garcia, J. and Palacios, M. 2006. The effect of cane molasses amendment on biocontrol of frosty pod rot (*Moniliophthora roreri*) and black pod (*Phytophthora* spp.) of cocoa (*Theobroma cacao*) in Panama. Biological Control 39: 232–239. https://doi.org/10.1016/j.biocontrol.2006.06.005

Küçük, Ç. 2017. *In vitro* antagonistic activity against *Fusarium* species of local *Trichoderma* spp. isolates. Journal of Biological and Environmental Sciences 11: 67–74. https://www.researchgate.net/publication/321905548_In_vitro_Antagonistic_Activity_against_Fusarium_Species_of_Local_Trichoderma_spp_Isolates

Kulimushi, P.Z., Basime, G.C., Nachigera, G.M., Thonart, P. and Ongena, M. 2018. Efficacy of *Bacillus amyloliquefaciens* as biocontrol agent to fight fungal diseases of maize under tropical climates: From lab to field assays in south Kivu. Environmental Science and Pollution Research 25: 29808–29821. https://doi.org/10.1007/s11356-017-9314-9

Larney, F.J. and Angers, D.A. (2012). The role of organic amendments in soil reclamation: A review. Canadian Journal of Soil Science 92: 19–38. https://www.nrcresearchpress.com/doi/full/10.4141/cjss2010-064#.XpsZJ5k6_IU

Lozupone, C., Lladser, M.E., Knights, D., Stombaugh, J. and Knight, R. 2011. UniFrac: An effective distance metric for microbial community comparison. The ISME Journal 5: 169–172. https://doi.org/10.1038/ismej.2010.133

Malek, N.A., Zainuddin, Z., Chowdhury, A.J.K. and Abidin, Z.A.Z. 2015. Diversity and antimicrobial activity of mangrove soil actinomycetes isolated from Tanjung Lumpur, Kuantan. Jurnal Teknologi 77: 37–43. https://doi.org/10.11113/jt.v77.6734

Mandic-Mulec, I., Stefanic, P. and van Elsas, J.D. 2016. Ecology of *Bacillaceae*. pp. 59–85. *In*: A. Driks and P. Eichenberger (eds.). The Bacterial Spore: From Molecules to Systems. ASM Press, Washington, DC. https://doi.org/10.1128/9781555819323.ch3

Martínez-Álvarez, J.C., Castro-Martínez, C., Sánchez-Peña, P., Gutiérrez-Dorado, R. and Maldonado-Mendoza, I.E. 2016. Development of a powder formulation based on *Bacilluscereus* sensu lato strain B25 spores for biological control of *Fusarium verticillioides* in maize plants. World Journal of Microbiology and Biotechnology 32: 75. https://doi.org/10.1007/s11274-015-2000-5

Martínez-Medina, A., Del Mar Alguacil, M., Pascual, J.A. and Van Wees, S.C.M. 2014. Phytohormone profiles induced by *Trichoderma* isolates correspond with their biocontrol and plant growth-promoting activity on melon plants. Journal of Chemical Ecology 40: 804–815. https://doi.org/10.1007/s10886-014-0478-1

Navarro, M.O., Barazetti, A., Niekawa, E.T., Dealis, M.L., Matos, J.M.S., Liuti, G., Modolon, F., Oliveira, I.M., Andreata, M., Torres Cely, M.V. and Andrade, G. 2019. Microbial biological control of diseases and pests by PGPR and PGPF. pp. 75–122. *In*: D. Singh, V. Gupta, R. Prabha (eds.). Microbial Interventions in Agriculture and Environment. Springer, Singapore. https://doi.org/10.1007/978-981-13-8383-0_3

Nguyen, P.A., Strub, C., Fontana, A. and Schorr-Galindo S. 2017. Crop molds and mycotoxins: Alternative management using biocontrol. Biological Control 104: 10–27. https://doi.org/10.1016/j.biocontrol.2016.10.004

Nguyen, P.A., Strub, C., Durand, N., Alter, P., Fontana, A and Schorr-Galindo, S. 2018. Biocontrol of *Fusarium verticillioides* using organic amendments and their actinomycete isolates. Biological Control 118: 55–66. https://doi.org/10.1016/j.biocontrol.2017.12.006

Ninh, H.T., Grandy, A.S., Wickings, K., Snapp, S.S., Kirk, W. and Hao, J. 2015. Organic amendment effects on potato productivity and quality are related to soil microbial activity. Plant and Soil 386: 223–236. https://doi.org/10.1007/s11104-014-2223-5

Ntougias, S., Bourtzis, K. and Tsiamis, G. 2013. The microbiology of olive mill wastes. BioMed Research International 2013: Article ID 784591. https://doi.org/10.1155/2013/784591

Ok, H.E., Kim, D.M., Kim, D., Chung, S.H., Chung, M.S., Park, K.H. and Chun, H.S. 2014. Mycobiota and natural occurrence of aflatoxin, deoxynivalenol, nivalenol and zearalenone in rice freshly harvested in South Korea. Food Control 37: 284–291. https://doi.org/10.1016/j.foodcont.2013.09.020

Palazzini, J.M., Yerkovich, N., Alberione, E., Chiotta, M. and Chulze S.N. 2017. An integrated dual strategy to control *Fusarium graminearum* sensu stricto by the biocontrol agent *Streptomyces* sp. RC 87B under field conditions. Plant Gene 9: 13–18. https://doi.org/10.1016/j.plgene.2017.07.002

Pandey, A. and Yarzábal, L.A. 2019. Bioprospecting cold-adapted plant growth promoting microorganisms from mountain environments. Applied Microbiology and Biotechnology 103: 643–657. https://doi.org/10.1007/s00253-018-9515-2

Pandin, C., Darsonval, M., Mayeur, C., Le Coq, D., Aymerich, S. and Briandet, R. 2019. Biofilm formation and synthesis of antimicrobial compounds by the biocontrol agent *Bacillus velezensis* QST713 in an *Agaricus bisporus* compost micromodel. Applied and Environmental Microbiology 85: e00327–19. https://doi.org/10.1128/AEM.00327-19

Pansu, M., Thuriès, L., Larré-Larrouy, M.C. and Bottner, P. 2003. Predicting N transformations from organic inputs in soil in relation to incubation time and biochemical composition. Soil Biology and Biochemistry 35: 353–363. https://doi.org/10.1016/S0038-0717(02)00285-7

Pathak, H. and Rao, D. 1998. Carbon and nitrogen mineralization from added organic matter in saline and alkali soils. Soil Biology and Biochemistry 30: 695–702. https://doi.org/10.1016/S0038-0717(97)00208-3

Patil, N.S., Waghmare, S.R. and Jadhav, J.P. 2013. Purification and characterization of an extracellular antifungal chitinase from *Penicillium ochrochloron* MTCC 517 and its application in protoplast formation. Process Biochemistry 48: 176–183. https://doi.org/10.1016/j.procbio.2012.11.017

Peet, R.K. 1974. The measurement of species diversity. Annual Review of Ecology and Systematics 5: 285–307. https://www.annualreviews.org/doi/pdf/10.1146/annurev.es.05.110174.001441

Portillo, M.C., Leff, J.W., Lauber, C.L. and Fierer, N. 2013. Cell size distributions of soil bacterial and archaeal taxa. Applied and Environmental Microbiology 79: 7610–7617. https://aem.asm.org/content/aem/79/24/7610.full.pdf

Pugliese, M., Gullino, M. and Garibaldi, A. 2009. Efficacy of microorganisms selected from compost to control soil-borne pathogens. Communications in Agricultural and Applied Biological Sciences 75: 665-669. https://www.researchgate.net/publication/51090612_Efficacy_of_microorganisms_selected_from_compost_to_control_soil-borne_pathogens

Qiu, M., Zhang, R., Xue, C., Zhang, S., Li, S., Zhang, N. and Shen, Q. 2012. Application of bio-organic fertilizer can control *Fusarium* wilt of cucumber plants by regulating microbial community of rhizosphere soil. Biology and Fertility of Soils 48: 807–816. https://doi.org/10.1007/s00374-012-0675-4

Quecine, M.C., Kidarsa, T.A., Goebel, N.C., Shaffer, B.T., Henkels, M.D., Zabriskie, T.M. and Loper, J.E. 2016. An interspecies signaling system mediated by fusaric acid has parallel effects on antifungal metabolite production by *Pseudomonas protegens* strain Pf-5 and antibiosis of *Fusarium* spp. Applied and Environmental Microbiology 82: 1372–1382. https://aem.asm.org/content/aem/82/5/1372.full.pdf

Rao, H.C.Y., Rakshith, D. and Satish, S. 2015. Antimicrobial properties of endophytic actinomycetes isolated from *Combretum latifolium* Blume: A medicinal shrub from Western Ghats of India. Frontiers in Biology 10: 528–536. https://doi.org/10.1007/s11515-015-1377-8

Regan, K.M., Nunan, N., Boeddinghaus, R.S., Baumgartner, V., Berner, D., Boch, S., Oelmann, Y., Overmann, J., Prati, D., Schloter, M., Schmitt, B., Sorkau, E., Steffens, M., Kandeler, E. and Marhan, S. 2014. Seasonal controls on grassland microbial biogeography: Are they governed by plants, abiotic properties or both? Soil Biology and Biochemistry 71: 21–30. https://doi.org/10.1016/j.soilbio.2013.12.024

Rodrigues, P., Venâncio, A. and Lima, N. 2013. Incidence and diversity of the fungal genera *Aspergillus* and *Penicillium* in Portuguese almonds and chestnuts. European Journal of Plant Pathology 137: 197–209. https://doi.org/10.1007/s10658-013-0233-4

Ruadrew, S., Craft, J. and Aidoo, K. 2013. Occurrence of toxigenic *Aspergillus* spp. and aflatoxins in selected food commodities of Asian origin sourced in the West of Scotland. Food and Chemical Toxicology 55: 653–658. https://doi.org/10.1016/j.fct.2013.02.001

Samsudin, N.I.P., Rodriguez, A., Medina, A. and Magan, N. 2017. Efficacy of fungal and bacterial antagonists for controlling growth, FUM1 gene expression and fumonisin B1 production by *Fusarium verticillioides* on maize cobs of different ripening stages. International Journal of Food Microbiology 246: 72–79. https://doi.org/10.1016/j.ijfoodmicro.2017.02.004

Saremi, H. and Saremi, H. 2013. Isolation of the most common *Fusarium* species and the effect of soil solarisation on main pathogenic species in different climatic zones of Iran. European Journal of Plant Pathology 137: 585–596. https://doi.org/10.1007/s10658-013-0272-x

Schmidt, P.A., Bálint, M., Greshake, B., Bandow, C., Römbke, J. and Schmitt, I. 2013. Illumina metabarcoding of a soil fungal community. Soil Biology and Biochemistry 65: 128–132. https://doi.org/10.1016/j.soilbio.2013.05.014

Scotti, R., Mitchell, A.L., Pane, C., Finn, R.D. and Zaccardelli, M. 2020. Microbiota characterization of agricultural green waste-based suppressive composts using omics and classic approaches. Agriculture 10: 61–78. https://doi.org/10.3390/agriculture10030061

Selvaraj, S., Ganeshamoorthi, P., Anand, T., Raguchander, T., Seenivasan, N. and Samiyappan, R. 2014. Evaluation of a liquid formulation of *Pseudomonas fluorescens* against *Fusarium oxysporum* f. sp. *cubense* and *Helicotylenchus multicinctus* in banana plantation. BioControl 59: 345–355. https://doi.org/10.1007/s10526-014-9569-8

Sharma, A. 2019. Fungi as biological control agents. pp. 395–411. *In*: B. Giri, R. Prasad, Q.-S. Wu, A. Varma (eds.). Biofertilizers for Sustainable Agriculture and Environment. Springer, Cham. https://doi.org/10.1007/978-3-030-18933-4_18

Song, X., Liu, M., Wu, D., Griffiths, B.S., Jiao, J., Li, H. and Hu, F. 2015. Interaction matters: Synergy between vermicompost and PGPR agents improves soil quality, crop quality and crop yield in the field. Applied Soil Ecology 89: 25–34. https://doi.org/10.1016/j.apsoil.2015.01.005

Sreevidya, M., Gopalakrishnan, S., Melø, T.M., Simic, N., Bruheim, P., Sharma, M., Srinivas, V. and Alekhya, G. 2015. Biological control of *Botrytis cinerea* and plant growth promotion potential by *Penicillium citrinum* in chickpea (*Cicer arietinum* L.). Biocontrol Science and Technology 25: 739–755. https://doi.org/10.1080/09583157.2015.1010483

Stark, C.H., Hill, R.A., Cummings, N.J. and Li, J.H. 2018. Amendment with biocontrol strains increases *Trichoderma* numbers in mature kiwifruit (*Actinidia chinensis*) orchard soils for up to six months after application. Archives of Phytopathology and Plant Protection 51: 54–69. https://doi.org/10.1080/03235408.2018.1438818

Suárez-Estrella, F., Arcos-Nievas, M.A., López, M.J., Vargas-García, M.C. and Moreno, J. 2013. Biological control of plant pathogens by microorganisms isolated from agro-industrial composts. Biological Control 67: 509–515. https://doi.org/10.1016/j.biocontrol.2013.10.008

Sun, Z.B., Li, SD., Ren, Q., Xu, J.L., Lu, X. and Sun, M.H. 2020. Biology and applications of *Clonostachys rosea*. Journal of Applied Microbiology. https://doi.org/10.1111/jam.14625

Taberlet, P., Coissac, E., Pompanon, F., Brochmann, C. and Willerslev, E. 2012. Towards next-generation biodiversity assessment using DNA metabarcoding. Molecular Ecology 21: 2045–2050. https://onlinelibrary.wiley.com/doi/epdf/10.1111/j.1365-294X.2012.05470.x

Thuriès, L., Pansu, M., Larré-Larrouy, M.C. and Feller, C. 2002. Biochemical composition and mineralization kinetics of organic inputs in a sandy soil. Soil Biology and Biochemistry 34: 239–250. https://doi.org/10.1016/S0038-0717(01)00178-X

Tiquia, S., Richard, T. and Honeyman, M. 2000. Effect of windrow turning and seasonal temperatures on composting of hog manure from hoop structures. Environmental Technology 21: 1037–1046. https://doi.org/10.1080/09593332108618048

Torsvik, V., Goksøyr, J. and Daae, F.L. 1990. High diversity in DNA of soil bacteria. Applied and Environmental Microbiology 56: 782–787. https://aem.asm.org/content/aem/56/3/782.full.pdf

Tortosa, G., Castellano-Hinojosa, A., Correa-Galeote, D. and Bedmar, E.J. 2017. Evolution of bacterial diversity during two-phase olive mill waste ("alperujo") composting by 16S rRNA gene pyrosequencing. Bioresource Technology 224: 101–111. https://doi.org/10.1016/j.biortech.2016.11.098

Tóth, B., Baranyi, N., Berki, A., Török, O., Kótai, É., Mesterházy, Á. and Varga, J. 2012. Occurrence of *Aspergillus flavus* on cereals in Hungary. Review on Agriculture and Rural Development 1: 446–451. http://www.analecta.hu/index.php/rard/article/download/13248/13104

Verma, P.P., Shelake, R.M., Das, S., Sharma, P. and Kim, J.Y. 2019. Plant growth-promoting rhizobacteria (PGPR) and fungi (PGPF): Potential biological control agents of diseases and pests. pp. 2281–2311. *In:* D. Singh, V. Gupta, R. Prabha (eds.). Microbial Interventions in Agriculture and Environment. Springer, Singapore. https://doi.org/10.1007/978-981-13-8391-5_11

Vida, C., Cazorla, F.M. and de Vicente, A. 2017. Characterization of biocontrol bacterial strains isolated from a suppressiveness-induced soil after amendment with composted almond shells. Research in Microbiology 168: 583–593. https://doi.org/10.1016/j.resmic.2017.03.007

Yang, L., Xie, J., Jiang, D., Fu, Y., Li, G. and Lin, F. 2008. Antifungal substances produced by *Penicillium oxalicum* strain PY-1—potential antibiotics against plant pathogenic fungi. World Journal of Microbiology and Biotechnology 24: 909–915. https://doi.org/10.1007/s11274-007-9626-x

Yasmin, S., Hafeez, F.Y. and Rasul, G. 2014. Evaluation of *Pseudomonas aeruginosa* Z5 for biocontrol of cotton seedling disease caused by *Fusarium oxysporum*. Biocontrol Science and Technology 24: 1227–1242. https://doi.org/10.1080/09583157.2014.932754

Zhang, Q., Yang, L., Zhang, J., Wu, M., Chen, W., Jiang, D. and Li, G. 2015. Production of anti-fungal volatiles by non-pathogenic *Fusarium oxysporum* and its efficacy in suppression of *Verticillium* wilt of cotton. Plant and Soil 392: 101–114. https://doi.org/10.1007/s11104-015-2448-y

Zhang, Y. and Marschner, P. 2018. Respiration, microbial biomass and nutrient availability are influenced by previous and current soil water content in plant residue amended soil. Journal of Soil Science and Plant Nutrition 18: 173–187. http://dx.doi.org/10.4067/S0718-95162018005000703

CHAPTER

5

Advances and Criticisms on the Use of Mycotoxin Detoxifying Agents

Giuseppina Avantaggiato*, Donato Greco, Vito D'Ascanio and Antonio F. Logrieco

Institute of Sciences of Food Production (CNR-ISPA), National Research Council (CNR), Via Amendola 122/O, 70126 Bari (Italy)

1. Introduction

Mycotoxins are secondary metabolites produced by a wide array of diverse fungal species, mostly belonging to the three genera *Aspergillus*, *Penicillium* and *Fusarium* (Avantaggiato and Visconti 2009, Jarda et al. 2011, Lerda 2011, Miller et al. 2014, JECFA 2017, Logrieco et al. 2018, Vila-Donat et al. 2018). They can be produced within the growing crop and during storage. Mycotoxin-contaminated feeds impair farm operations as well as feed production in various ways: mycotoxins are invisible, odourless and cannot be detected by smell or taste, but can reduce performance in animal production significantly. All species of livestock are affected by mycotoxins (Avantaggiato and Visconti 2009). Monogastrics (swine, horses) are the most sensitive, followed by poultry and ruminants. In general, young stock and animals under environmental, nutritional and production stresses are most sensitive. Direct effects of mycotoxins include acute diseases, where severe conditions of altered health may exist prior to death as a result of exposure to the toxin. These conditions are more likely following exposure to high levels of a mycotoxin. Other, more insidious or occult conditions (e.g. growth retardation, reproduction troubles, impaired immunity, decreased disease resistance, decreased milk or egg production) or more chronic disease manifestations (e.g. tumor formation) may result from prolonged exposure to small quantities of a toxin. Mycotoxins display a diversity of chemical structures, accounting for their different biological effects (Laan et al. 2006, Liew and Mohd-Redzwan 2018, Bryden 2012). Depending on their structure, these toxins can be carcinogenic, teratogenic, mutagenic, immunosuppressive, tremorgenic, hemorrhagic,

*Corresponding author: giuseppina.avantaggiato@ispa.cnr.it

hepatotoxic, nephrotoxic, and neurotoxic. Diagnosis of mycotoxicoses in animals is difficult as they may be similar to diseases with other causations (Logrieco et al. 2018). This is even more difficult in cases where more than one mycotoxin is involved because the toxins can produce additive, and sometimes synergistic, effects in animals (Ruiz 2011, Grenier Oswald 2011). In addition, the possible presence of toxic residues in animal products such as milk, meat and eggs may have some detrimental effects on human health (Bryden 2012). Considerable economic losses are attributed to reduced crop yields and grain quality following fungal contamination, to the downgrading of cereals from human food grade to animal feed, and to decreased animal performance and increased incidence of disease in livestock consuming mycotoxin contaminated grain (Miller et al. 2014, FAO 2013, Sassi et al. 2018). Due to the modern methods and to a growing interest in this field of research more than 300 different mycotoxins have been differentiated to date (Paterson and Lima 2010). Mycotoxins commonly found in animal feed include aflatoxins (EFSA 2004a), ochratoxin A (EFSA 2004b), zearalenone (EFSA 2004c), trichothecenes (e.g. deoxynivalenol) (EFSA 2004d), and fumonisins (EFSA 2005), which differ in their toxic effects and their prevalence across regions. Aflatoxins cause liver damage and impaired immune function. Trichothecenes reduce feed intake and weight gain and, at higher concentrations, cause emesis and complete feed refusal. Fumonisin B_1 is a carcinogen that is associated with equine leucoencephalomalacia, porcine pulmonary oedema, and spiking mortality in poultry. Zearalenone is an oestrogen that causes reproductive problems. Ochratoxin A is a nephrotoxin. Safe levels of mycotoxins in feed, below which there are no effects on animal health or production, are not well established. Regulatory officials worldwide are very concerned about the presence of mycotoxins in food and animal feed. In the EU, two mycotoxins (aflatoxin B_1 and rye ergot) are regulated (EC Regulation 2002) under strict limits, while deoxynivalenol, zearalenone, ochratoxin A, T-2/HT-2, and fumonisins are subject to recommend guidance values (EC Recommendation 2006). Use of feeding stuffs with levels above the maximum permitted are not allowed in the EU, neither is the mixing of contaminated feed with non-contaminated feed in order to reduce the concentration of mycotoxins. Controlling mould growth and mycotoxin production is very important to feed manufacturers and livestock producer (Colovic et al. 2019, Logrieco et al. 2018, Pinotti et al. 2018, Avantaggiato and Visconti 2009). Although desirable, the prevention of the mycotoxin contamination of grain in the field is currently impossible. Destruction of contaminated products or diversion to non-animal uses are not always practical, and could seriously compromise the feed supply. Control of mould growth in feeds can be accomplished by keeping moisture low, feed fresh, equipment clean and using mould inhibitors. Mycotoxin decontamination refers to methods by which mycotoxins are removed or neutralised from the contaminated feed while detoxification refers to methods by which the toxic properties of mycotoxins are removed. Feed processing may involve physical and/or chemical decontamination and can destroy or redistribute mycotoxins

(Colovic et al 2019, Vila-Donat et al. 2018, Karlovsky et al. 2016, Kolosova and Stroka 2011). Physical procedures such as sorting, thermal inactivation, irradiation or extraction of contaminated products have been attempted with different levels of success. Chemical procedures such as treatment with acid/base solutions or other chemicals, ammoniation, ozonation, and reaction with food grade additives such as sodium bisulfite have been shown to be effective in degrading and detoxifying aflatoxin contaminated feedstuffs. Biological methods primarily involving toxin degradation by microorganisms are receiving increasing interest and have shown promising results. Dietary supplementation with large neutral aminoacids, antioxidants, and omega-3 polyunsaturated fatty acids as well as the inclusion of mycotoxin-sequestering agents can ameliorate the harmful effects of mycotoxins in contaminated feeds. Guidelines for evaluating mycotoxin detoxification and decontamination procedures have been established by the Food and Agriculture Organization (FAO) and the U.S. Food and Drug Administration (FDA) (Park et al. 1988). The process should be able to (1) inactivate, destroy, or remove the mycotoxin; (2) avoid the formation of toxic substances, metabolites, or by-products in the feed; (3) retain nutrient value and feed acceptability of the product or commodity; (4) avoid significant alterations of the product's technological properties; and if possible, (5) destroy fungal spores. In addition to these criteria, the process(es) should be readily available, easily utilized, and inexpensive. Although a variety of decontamination/detoxification methods show potential for commercial application, large-scale, practical, and costs-effective methods for a complete mycotoxin decontamination are currently not available (Colovic et al. 2019). Moreover, no single decontamination method that is equally effective against the variety of naturally occurring mycotoxins has been developed. Therefore, the use of feed additives that reduce the exposure of the animals to these mycotoxins may be regarded as a way to improve animal welfare (Boudergue et al. 2009). These additives are defined as substances (e.g. clay mineral, enzyme, micro-organism, yeast cell wall) that are mixed into feed and then adsorb or denature (biologically detoxify) mycotoxins in the digestive tract of animals.

There is a vast scientific literature covering mycotoxin detoxification in animal feed and a large number of substances have been proposed as physical or biological adsorbents or microbiological/enzymatic transformation agents (Colovic et al. 2019, Vila-Donat et al. 2018, Karlovsky et al. 2016, Kolosova and Stroka 2011, Avantaggiato and Visconti 2009, Avantaggiato et al. 2003, 2005, 2007, Kabak et al. 2006, Huwig et al. 2001). Particularly exhaustive is a 2009 report commissioned by the European Food Safety Authority (EFSA) to review mycotoxin-detoxifying agents used as feed additives, their mode of action, efficacy and feed/food safety (Boudergue et al. 2009).

This chapter will present an overview on advances and criticisms regarding the use of mycotoxin-detoxifying agents, focusing on adsorbent materials for mycotoxin removal from feedstocks.

2. EU Approach for Evaluation and Authorization of Mycotoxin-Detoxifying Agents

Efficacy/safety assessment and authorisation of additives for mycotoxin reduction in feeds differs across the world. So far,most countries where these feed additives are used on a regular basis lacks on regulations regarding their use and/or evaluation. In EU, mycotoxin-detoxifying agents were regulated in 2009. This regard, the European Regulation (EC) No 1831/2003 of 22 September 2003 on additives for use in animal nutrition was amended, and a new functional group was added in the category of technological feed additives. This group was defined by the Commission Regulation (EC) No 386/2009 of 12 May 2009 as *'substances for reduction of the contamination of feed by mycotoxins: substances that can suppress or reduce the absorption, promote the excretion of mycotoxins or modify their mode of action'* (EC 2009). It should be pointed out that the use of such products does not mean that the animal feed exceeding the established maximum limits may be used. Their use should rather improve the quality of the feed, which is lawfully on the market, providing additional guarantee for the protection of animal and public health. These additives may not be applied to camouflage non-compliant consignments as compliant following the addition of an additive.

According to the EC Regulation 386/2009, to register a feed additive in the EU, an application must be submitted to the EC, a technical dossier to the EFSA, and reference samples of the feed additive must be provided to European Union Reference Laboratory for evaluation. Thereof, EFSA evaluates the data submitted for authorisation, and following a positive opinion on safety and efficacy of the additive, the EC assigns the final authorisation for the product which can be placed on the market.

To date, three feed additives have received authorisation by the EC to be used as substances for reduction of the contamination of feed by mycotoxins:

- Bentonite was authorised as aflatoxin binder for all animal species in 2013 (EC 2013). This was based on several, published *in vitro* and *in vivo* studies, including a study on dairy cows showing a reduction of aflatoxin M_1 transfer to milk by 40% (EFSA 2011).
- The product BBSH 797 is a bio-transforming agent for deoxynivalenol detoxification. It was authorised in pigs, in 2013, after revising *in vivo* studies proving significant reduction of deoxynivalenol concentration in pigs' serum of animals fed contaminated feeds, due to its bio-transformation into a less toxic metabolite (de-epoxy-deoxynivalenol) (EFSA 2013). In 2017, authorisation for all avian species was also obtained (EC 2017a).
- FUMzyme is a fumonisin esterase, which was authorised in 2014 after a thorough review of *in vivo* evidences of efficacy in pigs (EFSA 2014). In 2017, it obtained an authorization also for avian species (EC 2017b).

2.1 EFSA's Approach for the Assessment of Mycotoxin-detoxifying Agents

In July 2010,EFSA through its Panel on Additives and Products or Substances used in Animal Feed (FEEDAP) issued a statement where it detailed the additional information that would be required to perform an assessment of the safety and efficacy of this new group of additives (EFSA 2010). This statement lists only the requirements which are not common to the rest of technological additives. In 2012, EFSA published several guidelines on its website for the bringing to market of several feed additives and safety measures (EFSA 2012). This guidance document follows the structure and definitions of Regulation (EC) No 1831/2003 and it is intended to assist the applicant in the preparation and the presentation of its application, as foreseen in Article 7.6 of Regulation (EC) No 1831/2003.

Characterisation of the additive and conditions of use: The target mycotoxin(s) against which the additive is active should be clearly specified. Evidence should be provided to show that the use of the additive does not interfere with the analytical determination of mycotoxins in feed.

Safety of the additive: The additive must be safe for the animals receiving it, the consumers of food derived from these animals, the users of the product and also for the environment. The safety of a feed additive is assessed by a series of tolerance studies in the target species, and with a series of *in vitro* and *in vivo* toxicological studies with laboratory animals for consumers and users. No specific studies are required for mycotoxin-detoxifying agents other than for those that modify the structure of the mycotoxin. In that case, the risks associated with the metabolites/degradation products of the mycotoxin should be examined in appropriate studies (metabolic and residue studies, and toxicological studies). For mycotoxin binders, evidence should be provided to assess if, and to what extent, the supply of nutrients, micronutrients and other additives can be affected.

Efficacy of the additive: Unlike most other technological additives, mycotoxin-detoxifying agents do not have an effect in feed until after ingestion by the animal. Therefore, the FEEDAP Panel of EFSA considers that efficacy can only be fully demonstrated by *in vivo* studies. The dietary concentration of mycotoxin(s) used in such studies should not exceed official or advisory limits. *In vitro* studies have to be used to support the mode of action, but cannot substitute for *in vivo* studies. A minimum of three *in vivo* studies showing significant effects should be provided to demonstrate efficacy at the lowest recommended dose. For additives intended to be used in all animal species except fish, studies should be performed in at least three major species, poultry, monogastric mammal and ruminant. Specific studies in fish (preferably salmonids) are required. The efficacy of mycotoxin-detoxifying agents observed in laboratory animals cannot be normally taken as a basis to conclude on efficacy in target animals. The end-points selected will depend on the mycotoxin and target species, but in general, will include mycotoxin/metabolites excretion in faeces/urine, concentration in blood/plasma/serum, tissues or products (milk or eggs) or other relevant biomarkers. Zootechnical parameters, such as feed efficiency or weight gain should be reported, but by themselves do not constitute a measure of the detoxifiers efficacy.

3. Mycotoxin Adsorbing Agents

Dietary supplementation with non-nutritive mycotoxin-adsorbents is by far the most practical and most widely studied method for reducing the effects of mycotoxin exposure. An effective mycotoxin adsorbent should be a nutritionally inert adsorbent, which when incorporated into contaminated feed, should prevent or limit toxin absorption from the gastrointestinal tract of animals, and carryover of mycotoxins into animal products, such as aflatoxin M_1 in milk (Boudergue et al. 2009). Adsorbent materials should act like 'chemical sponges', thus preventing uptake and subsequent distribution of mycotoxins to the blood and target organs. Ideally, a mycotoxin adsorbent should be effective towards several mycotoxins as feedstuffs are commonly contaminated with several mycotoxins. To remain practical, mycotoxin adsorbents should be reasonably priced and should not occupy a large portion of the complete diet. Additionally, they should be free of impurities, off-flavours and odours. They should be safe without affecting the bioavailability of nutrients of veterinary drugs.

The efficacy of mycotoxin adsorbing agents depends on the chemical structure of both the adsorbent and the mycotoxin (Kolosova and Stroka 2011). The most important feature is the physical structure of the adsorbent, i.e. the total charge and charge distribution, the size of the pores, and the accessible surface area. On the other hand, the properties of the adsorbed mycotoxins, like polarity, solubility, shape and charge distribution, also play a significant role. Criteria considered important in the evaluation of potential mycotoxin binders are the stability of the sorbent-toxin bond, in order to prevent desorption of the toxin, as well as their effectiveness over a broad pH range since a product must work throughout the gastrointestinal tract. Composition of feed can also have a great influence on adsorbent efficacy.

In the last decades, various binders of different origins have been investigated for their efficacy and capacity insequestering mycotoxins (Colovic et al. 2019, Kolawole et al. 2019, Vila-Donat et al. 2018, Karlovsky et al. 2016, Kolosova and Stroka 2011). The first generation of binders, so called mineral adsorbents or inorganic adsorbents, are mainly phyllosilicates of the clay mineral group and the most significant are smectite, hydrated sodium aluminosilicates (HSCAS), and specifically bentonite or montmorillonite. Other materials used as mycotoxin binders include tectosilicates (like zeolites), activated carbon, complex indigestible carbohydrates (cellulose, polysaccharides in the cell walls of yeast and bacteria such as glucomannans, peptidoglycans, and others), and synthetic polymers. Several studies have shown that these adsorbent materials have high affinity for mycotoxins by the formation of stable linkages, which can occur in several liquid systems, including water, beer, wine, milk, and peanut oil. Although many currently available adsorbents effectively sequester aflatoxins, they appear to be ineffective at binding other non-aflatoxin mycotoxins, especially trichothecenes.

3.1 The Use of Minerals and Their Modified Forms as Mycotoxin Adsorbents

3.1.1 Raw Clay Minerals

Clay minerals, as the materials in "greening 21st century material worlds", have attracted more attention owing to their excellent adsorption performance, high chemical stability and biocompatibility advantages. Moreover, clay minerals are naturally abundant, green, non-toxic and low-cost. Aluminosilicates are the largest and most important class of clay minerals, composed of silica, alumina, and significant amounts of alkaline and alkaline earth ions (Elliott et al. 2020, D'Ascanio et al. 2019, Bergaya and Lagaly 2006, WHO 2005, Phillips et al. 1995). There are two major sub-classes in this group, phyllosilicate, and tectosilicate, with a wide range of applications, including mycotoxin adsorption. The phyllosilicate sub-class mineral clays include significant adsorbents such as the montmorillonite/smectite group, the kaolinite group and the illite (or clay-mica) group. Montmorillonite is a predominantly layered, oxygen-coordinated, phyllosilicate consisting of octahedral aluminum and tetrahedral silicon layers. The bentonite is usually impure smectite clay. The tectosilicates include important and highly studied zeolites.

Clay minerals, primarily montmorillonite, have been used in the early 1970s to reduce aflatoxin toxicity (Masimango et al. 1978, 1979). There is ample literature on this subject, mainly in the field of *in vitro* water studies (Li et al. 2018, Phillips et al. 2008, Charturvedi et al. 2002, Ramos and Hernández 1996, Phillips et al. 1995), and animal feed trials (Vila-Donat et al., 2018, Magnoli et al. 2008, Desheng et al. 2005, Nahm 1995, Marquez and Hernandez 1995, Phillips et al. 1988). The use of smectite in human nutrition was also tested for its safety, as well as the efficacy in the decrease of aflatoxin biomarkers (Afriyie-Gyawu et al. 2008a, 2008b, Phillips et al. 2008, Wang et al. 2008, Wang et al. 2005). The chemisorption of aflatoxin to smectites involves the formation of a complex by the β-keto-lactone or bilactone system of aflatoxin with uncoordinated metal ions in the mineral. In addition, aflatoxin B_1 can be adsorbed on the surface of the mineral particle and in its interlayers.

As mentioned above, in Europe, bentonites (as smectite clays) are allowed as feed additives (binders, substances for control of radionuclide contamination, and anticaking agents) (1m558*i*) for all animal species, as well as for mitigation of mycotoxin contamination for ruminants, swine, and poultry (1m558). However, bentonites containing $\geq$ 70% smectite as dioctahedral montmorillonite can be used for aflatoxin control in animal feeds (EC 2013).

Due to their functional properties, clays are often grouped into a single category. This is misleading information, as different types of clays with different features can be identified. Indeed, it should be pointed out, that most clays do not sequester mycotoxins, and not all clays that adsorb mycotoxins are equally effective in protecting animals against their toxic effects. In addition, similar clays may vary from one mine to another. It is critical that the origin (deposit) of the clay always be from the same source. Any scientific information obtained from a clay-based product is directly related to its origin and does not apply if there is a change in its geological deposit. This is the main criticism in the use of clays as binders for mycotoxins, including aflatoxins. The results of different studies show that, when compared with smectite clays (montmorillonite/bentonite), zeolitic minerals demonstrate much lower adsorption properties with respect to aflatoxins (Kolosova and Stroka 2011). However, a big difference in the effectiveness of smectite clays in adsorbing aflatoxin B_1 has been also recorded (D'Ascanio et al. 2019). Several studies indicate that the efficacy of bentonites in sequestering aflatoxin B_1can be affected by physical, chemical and mineralogical characteristics of the smectite, including clay contents, the capacities of the cation exchange (CECs), the interlayer cation hydrate radius, distributions of particle size and the specific surface area. Notwithstanding these findings, no significant correlation between any single physical or chemical property of clays and its mycotoxin binding capacity has been found. Therefore, there is still no predictive model of aflatoxin B_1 adsorption by the bentonite, as the crystal-chemical variation in the smectite group is complex. However, the findings of a recent study (D'Ascanio et al. 2019) showed a good correlation between aflatoxin adsorption parameters and the geological origin of samples. To adsorb the toxin at gastric or intestinal pH values, sedimentary bentonites performed better than samples mined from hydrothermal bentonite sites (D'Ascanio et al. 2019). In addition, the extent of aflatoxin B_1-adsorption by all samples was negatively and linearly correlated with the extent of desorption. Mineralogical and physicochemical analyses confirmed that some physical and chemical properties of bentonites correlate linearly with aflatoxin adsorption. However, these studies cannot be deemed to be conclusive since it is still hard to depict the link between properties of these mineral adsorbents and aflatoxin B_1 adsorption/desorption. Due to the complexity of interactions and factors that can affect the adsorption of aflatoxins by smectites, further research is required to describe the mechanisms of adsorption.

Concerning the use of clays as mycotoxin adsorbing agents, during the last 10 years, three legends have been created: clays inactivate solely aflatoxin; high adsorbent dosages (20 g/kg) are required to sequester aflatoxin *in vivo*; and clays deprive feeds of micronutrients or veterinary substances. Last research evidences have disproven these statements. Few clays have been shown to sequester mycotoxins other than aflatoxins, such as FB_1 (Mitchell et al. 2014) or ZEA (Ramos et al. 1996b). They are effective at levels 10-time lower than the maximum permitted use in animal feeds (20 g/kg) (Elliot et al. 2020, Li et al. 2018). Clays may adsorb micronutrients or medical substances when used at a dosage higher than 5 g/kg in feeds, but at the same time, they can be a source of some micronutrients. Micronutrients, including essential minerals or vitamins, are required in very small amounts (less than 100 mg/kg per day) by animals for proper functioning of enzymes and hormones, to maintain growth and development (Smith et al. 2018). Clays, having an anionic framework with well-defined microstructures containing chemical elements mostly alkali metal ions and trace elements, can act as a source of minerals for animals. The presence of hydrochloric acid in the stomach or bile salts in the intestine can modify the physico-chemical properties of minerals increasing their ion-exchange capacity. This process is accomplished by the release of mineral elements from the surface of the adsorbents into the chyme, increasing the concentration of minerals in the systemic circulation and subsequent accumulation in the body (MambaI et al. 2010, Mascolo et al. 2004, Park et al. 2002). However, on the other hand, the ion-exchange process of clays may lead to adsorption of minerals and nutrients from feeds, causing micronutrients deficiencies in farm animals (Elliott et al. 2020, Vila-Donat et al. 2018, Li et al. 2018, Ralla et al. 2010). Most of these studies reported symptoms of micronutrients deficiency when feeds were supplemented with high dosages of adsorbent minerals (higher than 0.5% in the feed). Thereof, nutritional balance of animal diets should take into account the proper amount of clays and trace elements to be included into the feedstuffs, to guarantee animal welfare and productivity. In addition to micronutrients, mineral adsorbents can impair intestinal absorption of veterinary substances, such as antibiotics (tilmicosin, tylosin, paromomycin, doxycycline) and coccidiostats (monensin and salinomycin) (Elliott et al. 2020, Li et al. 2018, Vila-Donat et al. 2018). As observed for micronutrients, supplementation of clays to animal diets may reduce the effectiveness of antibiotics or coccidiostats at levels higher than 0.5% in the feed. Thereof, considering that mineral mycotoxin binders can decrease or enhance the bioavailability of veterinary drugs, it may have a significant consequence on animal health, withdrawal time of the drugs and potentially public health in terms of exposure to antibiotic residues.

Due to the potential of mineral adsorbents to enter the body through different routes, such as inhalation, ingestion, and dermal penetration, their toxicity needs to be addressed. Recently, the EFSA Panel FEEDAP assessed the safety and efficacy of bentonite when used as a technological feed additive (substances for reduction of the contamination of feed by mycotoxins) for all animal species (EFSA 2017). It was reiterated that bentonites are safe for all animal species, the consumers and the environment when used at a maximum level of 20,000 mg/kg complete feed. The results of a new genotoxicity study reinforced the previous conclusion that smectites are non-genotoxic. Bentonites are not skin irritants but might be mildly irritant to the eye; based on a new study submitted, the additive is not a skin sensitizer. Owing to its silica content, the additive is a hazard by inhalation for the users. Similarly, several authors (Maisanaba et al. 2013, Li et al. 2010, Geh 2006) observed any genotoxic effect when cells were exposed to natural or unmodified mineral adsorbents. However, different results were observed in the case of modified mineral adsorbents (Maisanaba et al. 2014). Exposure of HepG2 cells to organo-modified montmorillonite significantly increased the frequency of micronuclei by 2.7-fold compared to control group using both comet and CBMN assays. Additionally, genes involved in DNA damage, metabolism, and oxidative stress were upregulated (Maisanaba et al. 2016). Similarly, HepG2 and Caco-2 cells exposed to modified montmorillonite showed a significant increase in DNA damage after 48 h of exposure (Maisanaba et al. 2013, Sharma et al. 2010).

3.1.2 Modified Clay Minerals

Modified mineral adsorbents or organoaluminosilicates have been developed by chemical modification of minerals, mainly aluminosilicates, to produce multi-mycotoxin mineral adsorbents. The main drawback in using aluminosilicates as mycotoxin adsorbents is due to their selectivity in the chemisorption of aflatoxins, with little or no beneficial effect against other mycotoxins. Raw clays exhibited unsatisfactory adsorption efficiency toward low polar and hydrophobic mycotoxins, such as zearalenone, ochratoxin A, and trichothecenes (Elliott et al. 2020, Li et al. 2018, Vila-Donat et al. 2018, Phillips et al.1995). This defect has restricted the application of raw clays as high-efficient adsorbents for mycotoxins. To overcome these limitations, numerous strategies have been developed, including purification and modification methods, such as organo-functionalization, pillarization and thermal activation (Galvano et al. 2001, Huwig et al. 2001). Indeed, the modified forms of clays, such as organo-clays, pillared montmorillonite and thermal-modified clays, exhibited far better adsorption capacity compared with the raw ones (Li et al. 2018).

Organo-clays refers to the clays modified with organic molecules, organic ions or polymers, which are generally bonded to clays by covalent bond, ion bonds, hydrogen bonds, dipoles or van der Waals forces (Papaioannou et al. 2005). *In vitro* results have verified the binding efficacy of modified montmorillonite and clinoptilolite against zearalenone and ochratoxin A (Jiang et al. 2012, Papaioannou et al. 2005, Dakovic et al. 2003). Moreover, other authors have shown that organically modified clays are more effective than natural clays towards fumonisins (Baglieri et al. 2013, Dakovic et al. 2010, Döll et al. 2005). By exchange of structural charge-balance cations with high molecular weight quaternary amines, the modified mineral adsorbents have an increased hydrophobicity, hence, an improved adsorption of non-aflatoxin mycotoxins (Li et al 2018). In organo-functionalization method, mono- or di-alkyl cationic surfactants are generally used to modify clays. The adsorption efficiency of organo-clays is positively correlated to the dosage of cationic surfactants; thus more cationic surfactants are bound to clays, more efficiently the organo-clays will sequester mycotoxins. The adsorption of cationic surfactants at the clay interface is strongly influenced by a number of factors, such as the nature of the structural groups on the clay surface, the molecular structure of the surfactant, and the temperature and concentration (Rosen and Kunjappu 2004). However, such alterations do not always reduce observed toxicities. Some cationic surfactants are toxic, which are not suitable for clay mineral modifiers used in feeds or foods. Therefore, it is also extremely significant to pay attention to synthesis conditions and security issues in practice. Enhanced toxicity of ZEA was reported in animal trials when modified clays (hexadecyltrimethyl ammonium low-pH montmorillonite and hexadecylamine low-pH montmorillonite) were tested for their mycotoxin-sequestering properties (Marroquín-Cardona et al. 2009, Lemke et al. 2001).

In order to enhance mycotoxin adsorption from animal feeds by clays, pillared-montmorillonite adsorbents have been synthesized. Pillared montmorillonite refers to the montmorillonite modified with metal cations or polyhydroxy metal cations (Barrer and Macleod 1955, Jiang 2004). The modification process includes the introduction of metal cations (or polyhydroxy metal cations) into the interlayer of montmorillonite by expansibility, adsorption or inter-layer cation exchanges. Then, thermal treatment transforms metal cations to columnar metal oxide clusters by dihydroxylation of the structural OH units or hydrogen. Finally, series of modified montmorillonites with larger specific surface area, wider pore diameter, and higher surface activity are obtained. Several *in vitro* studies showed that pillared clays are better mycotoxin adsorbents, especially for aflatoxin B_1. The metals to active clays can be Cu^{2+}, Zn^{2+}, Fe^{3+}, etc., being copper salts the most effective ones (Nones et al. 2017, Zeng et al. 2013, Daković et al. 2008). Although this technology seems promising in enhancing mycotoxin adsorption by clays, so far, literature lacks of any scientific evidence focused on removal of mycotoxins from feeds *in vivo* using pillared clays as sequestering agents. This needs to be addressed in the future. Furthermore, potential toxicity of excessive use of copper or magnesium sulphates should be taken into account.

Apart from clay pillarization, thermal treatment is another important way to activate clays for enhancing mycotoxin adsorption features. This process can produce the desorption of free water, interlayer water and loss of OH units. After thermal treatment, the adsorptive resistance of water film to organic matter is reduced, which is conducive to mycotoxin adsorption. In addition, thermal treatment may cause the delamination of montmorillonite-clays, which would increase the available space for mycotoxins between mineral layers. Considering published studies on the improved efficiency of thermal treated clays in adsorbing mycotoxins, heat activation seems effective within a certain temperature range. Indeed, too harsh treatment temperature may produce deleterious effects on mycotoxin adsorption capacity of clays (Nones et al. 2015, Mutturee et al. 2012). Therefore, it should pay more attention to the temperature control, when using thermal treatment method to activate clays in practice.

In addition to the aforementioned strategies, to prevent side effects of some modified forms of clays (such as organo-clays), natural substances have been used to activate raw clays, such as humic acids, choline, carnitine or vitamins (Velazquez and Deng 2020, Wang et al. 2017, Yao et al. 2012, Jaynes and Zartman 2011). Modification of a montmorillonite by humic acid improved by 39% the efficacy of this clay in adsorbing aflatoxin B_1, probably due to an increase of adsorption sites induced by humic acid (Yao et al. 2012). Similarly, choline- or carnitine-modified clays showed higher adsorption capacity and adsorption affinity for aflatoxin B_1 than raw clays (Wang et al. 2017, Jaynes and Zartman 2011). According to the authors (Wang et al. 2017, Jaynes and Zartman 2011), amendment of montmorillonite by choline or carnitine changes the mineral surface polarity and water occupation in the interlayer, facilitating the incorporation of aflatoxin molecules in the interlayer of the montmorillonite. Aflatoxin B_1 is slightly soluble in water based on its high K_{ow} (octanol/water partition coefficient) with an estimated solubility range of 11–33 μg/mL. Thereof, the change in polarity of choline- or carnitine-modified clays should facilitate the adsorption of aflatoxins by amended clay surfaces. Interestingly, the recent study by Velazquez and Deng (2020) showed that intercalation of organic nutrients, such as arginine, histidine, choline, lysine, and vitamin B_1, into the interlayer space of montmorillonites improved the aflatoxin B_1 adsorption by restricting the adsorption of pepsin. Indeed, proteins in gastric fluids can reduce smectite's adsorption capacity for aflatoxins. In these studies, the protective roles of raw and amended clays were identified using the adult hydra assay. As a result, amended clays at a rate of inclusion as low as 0.005% (w/v) resulted in significant protection of hydra against aflatoxin B_1, whereas at the same inclusion levels for unmodified clays no protection was observed. Additionally, choline and carnitine are non-toxic and nutrient, which is suitable for clay additives used in feeds or foods. All these studies support the conclusion that nutrient modifiers can improve aflatoxin binding under different conditions. However, further studies are warranted to determine the potential for dissociation of the modifiers (nutrients) and their potential interactions in GI tract as well as their safety and efficacy in animals and humans.

3.2 Nanoparticles for Mycotoxin Adsorption

Clay binders are the most widely used products for mycotoxin elimination to reduce their impact. Although conventional methods are constantly improving, current research trends are looking for innovative solutions. Nanotechnology approaches seem to be a promising, effective, and low-cost way to minimize the health effects of mycotoxins. The key properties in this field of research are based on the nanoscale (Horky et al. 2018). Nanomaterials are believed suitable in sequestering biological or chemical contaminants for their high surface area and high affinity to organic compounds, and also because they can be properly engineered or modified to increase selectivity to specific target pollutants. Mycotoxins show a structural diversity resulting in different chemical and physical properties. Mycotoxins can be classified as polar or nonpolar molecules; however, there are several that fall in between (Stroka and Maragos 2016; IARC, 2012). Most promising nanomaterials for mycotoxin removal are carbon nanostructures, chitosan polymeric nanoparticles, and nanoclays.

3.2.1 Activated Carbons and Carbon Nanostructures

Activated carbon is an amorphous form of carbon heated in the absence of air, and then treated by chemical or physical agents to develop a highly porous structure. The carbon source is selected from a variety of materials including the shells of nuts, wood (coal), moss, etc. This non-soluble, highly adsorbent powder has been utilised as a medical treatment for severe intoxications since 19th century. Based on literature data, activated carbon seems to be the most effective adsorbent with high affinity for different mycotoxins (including deoxynivalenol) *in vitro* (Mezes et al. 2010, Sabater-Vilar et al. 2007, Avantaggiato et al. 2005, 2004,2003, Diaz and Smith 2005, Ramos et al. 1996a, b). However, the *in vitro* efficacy of activated carbon toward some mycotoxins has not been confirmed *in vivo* (Avantaggiato et al. 2005, Solfrizzo et al. 2001a). Activated carbon is a relatively nonspecific sequestrant, and the great variability in the results of long-term exposure experiments and its potential for adsorbing important nutrients has diminished its overall practical effectiveness for routine dietary inclusion. This limitation driven the development of alternative carbon based structures for mycotoxin removal.

Actually, carbon nanostructures are considered promising successors of activated carbons due to their excellent stability, inertness, high adsorptive properties, large surface area per weight, and colloidal stability upon various pH values, which is important to preserve in the gastrointestinal tract. Graphene, graphene oxide, nanodiamonds, fullerenes, fibres, and nanotubes have a great potential to become novel adsorbents of mycotoxins. Nanocarbon structures are amphoteric and their surface could be protonated or deprotonated, which results in the binding capacity of polar or nonpolar compounds. In an *in vitro* study using nanodiamond aggregates (~40 nm), aflatoxin B_1 and ochratoxin A were adsorbed via electrostatic interactions depending on the types of functional groups on the surface of nanodiamonds (Gibson et al. 2011, Puzyr et al. 2007).

Further carbon structures showing some potential to be used as mycotoxin sequestering agents are those containing magnetic graphene (MGO). This innovative material is synthesized from iron oxide nanostructures and graphene oxide, and is relatively inexpensive and easily accessible. Oxygen functional groups on the MGO surface can interact with some *Fusarium* mycotoxins, subsequently MGO with bounded mycotoxins are removed via magnet. The approach reduced deoxynivalenol or T-2 toxin by 40, 70%, respectively (Pirouz et al. 2017). In addition, surface active maghemite nanoparticles showed chelating properties for citrinin and ochratoxin A (Magro et al. 2016). This product represents an ideal material, as its synthetic protocol is suitable for being scaled up to industrial level and is carried out in water without using of any organic solvent.

3.2.2 Chitosan and Chitosan Polymeric Nanoparticles

Several studies have demonstrated that chitosan has adsorption capacity for several contaminants including mycotoxins (Zhao et al. 2015). Chitosan is a natural cationic polysaccharide produced from chitin, which is the structural element found in the exoskeleton of crustaceans. It differs from similar polysaccharide celluloses, as it contains hydroxyl groups, acetylamine, or free amino groups which have attracted attention in many fields of applications. It is nontoxic, biodegradable, and possesses low immunogenicity. This material is easily subjected to nanoparticles via a gelatation process, which are able to encapsulate various compounds. Glutaraldehyde crosslinked chitosan showed promising adsorption ability for aflatoxin B_1 (73%), ochratoxin A (97%), zearalenone (94%), and fumonisin B_1 (99%), but it failed in adsorbing deoxynivalenol and T-2 toxin in a buffer system simulating gastrointestinal conditions (Zhao et al. 2015). However, practical application of this material in removing mycotoxins is limited due to the toxicity of glutaraldehyde.

3.2.3 Nanoclays

Nanoclay binders represent a large group of nanoparticles including minerals that are largely used for mycotoxin detoxification in food and feed, such as montmorillonite, bentonite, zeolite, or hydrated sodium (calcium) aluminosilicate. With respect to raw clays, nanocomposites show sizable surface area, higher porosity, strong cation exchange activities, and more active sites, which enable its interaction with mycotoxins. A montmorillonite nanocomposite was found effective in adsorbing aflatoxins in *in vitro* tests, and in reducing toxic effects in broilers at 3 g/kg in the diet (Shi et al. 2006).

Unlike nanocomposites, halloysite naturally occurs as a small cylinder (nanotubes) that has a wall thickness of 10–15 atomic alumosilicate sheets, an outer diameter of 50–60 nm, an inner diameter of 12–15 nm, and a length of 0.5–10 μm. Its outer surface is mostly composed of SiO_2 and the inner surface of Al_2O_3, and, hence, those surfaces are oppositely charged. Thereof, the external siloxane and internal aluminol surface of halloysite is an efficient adsorbent for both cations and anions. In addition, halloysite nanotubes can be modified by various surfactants to enhance their sorption properties and specificity. The study of Zhang et al. (2015) showed that modified halloysite nanotubes adsorbed zearalenone from simulated gastric and intestinal fluids, and mitigated toxic and estrogenic effects of zearalenone in rat animal model. Beneficial effects by halloysite nanotubes were recorded in swine exposed to zearalenone by contaminated feeds.

Taking into account the current findings on nanoparticles, it can be concluded that nanomaterials have interesting adsorption properties, which make them promising for mycotoxin removal. However, further research is highly demanded to investigate their side effects and toxicity. It should be assumed that newly synthetized nanoparticles are immediately examined for their toxicity and applicability. A key problem is that most studies on efficacy and safety assessment of nanoparticles against mycotoxins are unilaterally focused, and no data are in the literature on the correlation between nanomaterials and mycotoxins in terms of safety and toxicity. Further studies including well-designed *in vivo* trials are required to confirm effectiveness and safety of these adsorbents for animal species. The aspects of practical and economical feasibility should also be considered.

3.3 Organic Mycotoxin Binders: Microorganism Derived Materials

Due to some limitations in using inorganic materials as mycotoxin adsorbing agents, such as narrow spectrum of action, unwanted adsorption of vitamins, amino acids and minerals in feed, as well as the potential risks of complexing chemicals to the mineral adsorbents, organic binders (natural or synthetic) have been suggested. The main candidate microorganisms for mycotoxin removal include yeast (Shetty et al. 2007), lactic acid bacteria (LAB) (Gratz et al. 2005) and conidia of Aspergilli (Jard et al. 2009). Yeasts and LAB occur as part of natural microbial population in spontaneous food fermentation and as starter cultures in the food and beverage industry. The mechanism of detoxification by microbiological agents is a physical adsorption (ion-exchange) facilitated by inactivated cellular walls rather than the catabolism of mycotoxins by living microorganisms. Polysaccharide, protein, and lipid constituents of such walls provide numerous potential sites for mycotoxin adsorption through hydrogen bonds, ionic, and hydrophobic interactions.

Mycotoxin binding by some selected LAB has been described as a reversible phenomenon, strain- and dose-dependent, and did not affect the viability of LAB. Some LAB strains were effective in adsorbing aflatoxin B_1 and zearalenone in the small intestine; cell wall peptidoglycans, polysaccharides and teichoic acid were suggested as the key elements involved in adsorption process (Kabak et al. 2006). The strength of the mycotoxin-LAB interaction seems to be influenced by the peptidoglycan structure and, more precisely, by its amino acid composition (Dalié et al. 2010). Moreover, *L. rhamnosus* is considered a safe and effective chemopreventive because of its use in various dairy products including yogurt. It should be pointed out that in addition to mycotoxin removal by physical adsorption, the inhibition of their biosynthesis may be involved in the reduction of mycotoxin accumulation by LAB. Thereof, LAB cultures with high antifungal, antimycotoxigenic and mycotoxin binding potential could be of high value in limiting mycotoxin exposure (Kabak et al. 2006).

Saccharomyces cerevisiae is one of the most important microorganisms in food fermentation showing the ability to bind mycotoxins (Vila-Donat et al. 2018). Mycotoxin removal using yeast cell walls has been found higher than using whole cells. Several studies showed that mycotoxin removal by yeasts or their by-products occurs by mycotoxin adhesion to cell wall components rather than by covalent binding or by metabolism, as the dead cells do not lose adsorption ability (Shetty et al. 2007, Shetty and Jespersen 2006). The cell walls harbouring polysaccharides (glucan, mannan), proteins, and lipids exhibit numerous different and easily accessible adsorption centres as well as different binding mechanisms, e.g. hydrogen bonds, ionic, or hydrophobic interactions (Ringot et al. 2007). The study of Yiannikouris et al. (2003) demonstrated that the β-D-glucan fraction of yeast cell wall is directly involved in the adsorption process with zearalenone, and that the structural organisation of β-D-glucans modulates the binding strength. Probably, a similar mechanism is involved in the binding process of mycotoxins by LAB. It appears that the carbohydrate components are common sites for binding, with different toxins having different binding sites (Pfliegler et al. 2015).

The study of Jard et al. (2009) showed the ability of conidia of Aspergilli in removing zearalenone. This mycotoxin was successfully removed by living as well as heat-treated conidia just after inoculation. The adsorption capacity was also evaluated *in vitro* in the conditions of porcine GI tract and better adsorption in acidic stomach conditions was observed. The use of heat-treated conidia can be an alternative way of decreasing zearalenone levels in animal feed. However, *in vivo* trials are necessary before drawing a final conclusion.

Since non-viable organisms (i.e., heat- or acid-treated LAB) or their parts (i.e., yeast cell wall) retain their ability in adsorbing mycotoxins, this is an essential point which should be taken into account. In most cases, viable bacteria do not survive in the stomach due to the extremely low pH value. In addition, some LAB showing the potential to adhere to intestinal cells, thus acting as probiotics, can lose this ability when they sequester mycotoxins. Consequently, in the GI tract, the bacteria-mycotoxin complex is rapidly excreted (Dalié et al., 2010). In conclusion, strains of both *S. cerevisiae* and LAB with high mycotoxin binding ability have considerable prospect as mycotoxin-detoxifying agents. Several microorganisms are also able to degrade toxins to less-toxic or even non-toxic substances. This intensively researched field would greatly benefit from a deeper knowledge on the genetic and molecular basis of toxin degradation. Moreover, microorganisms and their biotechnologically important enzymes may exhibit sensitivity to certain mycotoxins, thereby mounting a considerable problem for the biotechnological industry. It is noted that yeasts or LAB are generally regarded as safe; however, some of them may even cause human infections. More research regarding the chemistry of mycotoxin reduction (adsorption and/or metabolization), stability of the complex (especially under the conditions of the GI tract), and safety needs to be investigated.

3.4 Organic Mycotoxin Binders: Synthetic Polymers

A few studies showed that synthetic polymers, including cholestyramine, divinylbenzene-styrene and polyvinylpyrrolidone, can sequester mycotoxins *in vitro* and *in vivo* (Avantaggiato et al. 2005, Ramos et al. 1996b). Cholestyramine (an anion exchange resin) has been shown to bind bile acids in the GI tract and reduce low density lipoproteins and cholesterol. This compound was proven to be an effective binder for ochratoxin A, fumonisins and zearalenone *in vitro* (Avantaggiato et al. 2005, 2003, Döll et al. 2004, Solfrizzo et al. 2001b, Ramos et al. 1996b). Its efficacy towards zearalenone was confirmed using a validated, dynamic gastro-intestinal model simulating the physiological conditions of the gastro-intestinal tract of pigs (TIM models) (Avantaggiato et al. 2005, 2003), and in mice (Underhill et al. 1995). In addition, cholestyramine has been found effective for fumonisins in rats using the biomarker assay (Solfrizzo et al. 2001b), while its *in vitro* efficacy in reducing deoxynivalenol levels in buffer at pH 7 was not confirmed in a test using Caco-2 cells (Cavret et al. 2010). Similarly, cholestyramine was not effective in reducing the bioaccessibility of deoxynivalenol in the dynamic gastrointestinal (TIM) model (Avantaggiato et al. 2005). New polymeric forms (cryogels) of crosslinked polyvinylpyrrolidone adsorbed zearalenone *in vitro* under isothermal conditions and simulating pH values of the GI tract (Alegakis et al. 1999).

Although synthetic materials showed promising results for mycotoxin removal in *in vitro* and *in vivo* studies, unfortunately they did not have great successes in practice, probably due to their high cost.

3.5 Organic Mycotoxin Binders: Agricultural By Products

Several studies report that undegradable dietary fibres can readily adsorb various substances (such as, steroid hormones, bile acids, cholesterol, pesticides, heavy metals) and hydrophobic carcinogens and promoters, thus preventing toxicoses resulting from xenobiotic compounds, and from the development of cancer (Sera et al. 2005, Harris et al. 1998, Ferguson and Harris 1996, Harris and Ferguson 1993). In addition, different types of biosorbents containing biomolecules such as polysaccharides, proteins, etc., with specific functional groups, have been proposed for the removal of heavy metals from wastewater (Saranya et al. 2017, Rangabhashiyam et al. 2016, 2014, Nakkeeran et al. 2016). In this context, biosorption has emerged as an alternative technique with the merits of being technically simple, eco-friendly, and highly economical. The literature contains some references to the efficacy of dietary fibres to counteract toxic effects of mycotoxins in animals (Vila-Donat et al. 2018). Diets rich in undegradable fibres have been shown to overcome the toxicity of zearalenone in rats and swine (Stangroom and Smith 1984, James and Smith 1982, Smith 1980). Alfalfa feeding in rats was found to reduce intestinal transit time, to increase faecal excretion, as well as to decrease T-2 toxicosis (Carson and Smith 1983). A significant protective effect of micronized wheat fibres against ochratoxin A toxicity was demonstrated *in vivo* for rats and piglets (Aoudia et al. 2008, 2009). The efficacy of fibres in counteracting mycotoxicoses was assumed to be due to the adsorption of mycotoxins to some components of the food plants. In addition, apple pomace (rich in fibres and pectin) was tested *in vivo* as mycotoxin adsorbent to reduce toxic effects of zearalenone and deoxynivalenol in pigs (Gutzwiller et al. 2007). Apple pomace was included up to 8% of the diet in a mycotoxin contaminated feed. The results showed that pomace alleviated the negative effect of deoxynivalenol on growth but did not counteract the hormonal effects of zearalenone. Furthermore, interesting findings were obtained from *in vitro* studies regarding humic acids, originating from natural decaying of organic plant materials. They also have shown the capacity to adsorb mycotoxins, especially aflatoxin B_1, ochratoxin A and zearalenone (Sabater-Vilar et al. 2007, Santos et al. 2011). Polymeric humic substances contain various binding sites, and have been introduced into human medicine as compounds to reduce the absorption and systemic availability of bacterial endotoxins. This latter effect would also be beneficial in the protection of animal health. Hence, these compounds deserve further *in vivo* testing. So far, humic substances have been tested *in vivo* towards aflatoxin B_1 only. The study of Van Rensburg et al. (2006) showed that a humic acid, oxihumate, with high *in vitro* adsorption capacity for aflatoxin B_1, was also able to alleviate some of the toxic effects of this mycotoxin, in growing broilers. These results were supported by the study of Ghahri et al. (2009) showing that the addition of humic acid to an aflatoxins contaminated diet protected broiler chicks against adverse effects of aflatoxin on humoral immunity.

Despite these studies, literature holds few experimental evidences on the *in vitro* ability of dietary fibres to act as multi-mycotoxin binders. However, the studies of Avantaggiato et al. (2014) and Greco et al. (2018) showed that a grape pomace from a red grape variety (pulp and skin) is able to sequester rapidly and simultaneously different mycotoxins. Aflatoxin B_1 was the most adsorbed mycotoxins followed by zearalenone, ochratoxin A, and fumonisin B_1, whereas the adsorption of deoxynivalenol was negligible. The efficacy of grape pomaces in sequestering mycotoxins was confirmed in pigs using the urinary biomarkers approach (Gambacorta et al. 2016). Grape pomace reduced significantly ($p<0.05$) urinary mycotoxin biomarker of aflatoxin B_1 (67%) and zearalenone (69%), whereas reductions statistically not significant were observed for fumonisin B_1, ochratoxin A and deoxynivalenol. Taking into account these results, authors stated that grape pomace shows the potential to be used as a broad-spectrum adsorbent material.

Evidence of the ability of food plants and by-products other than grape pomace and wheat fibres in adsorbing several mycotoxins are limited. Recently, the study of Adunphatcharaphon et al. (2020) showed that a typical agricultural by-product of South-East Asia (Durian peel) can be considered as a promising waste material for mycotoxin biosorption. Durian peel is an agricultural waste that is widely used in dyes and for organic and inorganic pollutant adsorption. In the study of Adunphatcharaphon et al. (2020), durian peel was acid-treated to enhance its mycotoxin adsorption efficacy. The acid-treated form of durian peel was assessed for simultaneous adsorption of most occurring mycotoxins. The structure of the acid treated durian peel was also characterized by SEM–EDS, FT–IR, a zetasizer, and a surface-area analyzer. The results indicated that acid treated durian peel exhibited the highest mycotoxin adsorption towards aflatoxin B_1 (98%), zearalenone (98%), and ochratoxin A (97%), followed by fumonisin B_1(86%) and deoxynivalenol (2.0%). The pH significantly affected ochratoxin A and fumonisin B_1 adsorption, whereas aflatoxin B_1 and zearalenone adsorption was not affected. Interestingly, acid treated durian peel reduced the bioaccessibility of these mycotoxins after gastrointestinal digestion using an *in vitro*, validated, static model. The acid treated durian peel showed a more porous structure, with a larger surface area and a surface charge modification. These structural changes following acid treatment may explain the higher efficacy of acid treated durian peel in adsorbing mycotoxins. This study pointed out that most biomaterials are unsuitable for adsorption in their raw form and must be pre-treated to improve their innate adsorption capacities. These pre-treatments may include physical processes (drying, autoclaving, grinding, milling, or sieving) and chemical modification with reagents. Physico-chemical modification can enhance adsorption by reducing particle size and increasing surface area as demonstrated by the work of Adunphatcharaphon et al. (2020).Similarly, the patent of Tranquil et al. (2013) showed an adsorption method using ligno-cellulosic materials modified by ambivalent proteins. The ligno-cellulosic materials originated from grain, legume and plant biomass by-products were treated by mechanical (micronization) and chemical (enzyme) treatments to increase the hemicellulose content, adsorption area, and hydrophobicity of the surface. The modified plant lingo-cellulosic components derived from grain hulls, distiller's grains, brewer's spent grains and sugar cane bagasse exhibited 50% to 60% residual adsorption of T-2 toxin and ochratoxin A. One major advantage of this plant-derived formulation is that it is likely to be better tolerated by animals, in comparison to the mineral and microbiological binders, due to inherent similarities with regular feed components.

However, further research is required to clarify the components of agricultural by-products that are involved in the biosorption of mycotoxins and to confirm their efficacy *in vivo*.

4. Conclusions and Perspectives

Frequent contamination of animal feed by mycotoxins has become an intractable problem to solve. Most of animal feeds are contaminated with more than one mycotoxin, and only a low percentage of them is contaminated above permitted/guideline levels. However, farm animals exhibit symptoms of mycotoxicosis even when exposed to feed contaminated with mycotoxins below the guidance levels, probably as a consequence of negative synergistic effects produced when different mycotoxins are simultaneously present in feed. In 2009, the use of mycotoxin adsorbing agents as technological feed additives has been officially allowed in the European Union. The growing interest to this matter is demonstrated by numerous studies and several extensive reviews. The low levels of mycotoxin found in feed and their easy management led to a widespread acceptance of these products by the farm animal industry. However, some considerations should be taken into account.

Any desirable mycotoxin adsorbent should possess the following properties: (a) binding ability against either a wide range of mycotoxins, especially mycotoxins with low hydrophobicity; (b) high adsorption capacity in order to detoxify high load of contaminating mycotoxins; (c) reduced non-specific binding to nutrients such as minerals, vitamins, and amino acids in feeds; and (d) offer similar characteristics to feed compositions (e.g. vegetable fibres) to provide better tolerance in animals. In addition, the efficacy of developed adsorbents shown in *in vitro* studies should be verified through *in vivo* animal trials. Thereof, it is essential to assess for any mycotoxin adsorbing agent its reliability and safety while considering the economic feasibility.

Looking at the data in the literature, it is clear that a gap still exists in our understanding and applicability of mycotoxin adsorbing agents as additives for mycotoxin reduction in feed, and therefore more work is necessary. Current research is mainly focused on removal of single type of mycotoxin, thus researchers should pay more attention to developing some other adsorbents that can sequester atleast two kinds of mycotoxins. Due to the chemical complexity of mycotoxins, the ability of a substance in adsorbing one mycotoxin does not imply same efficacy in sequestering other structurally different mycotoxins, and generalisations should be avoided. Mycotoxin adsorbing agents can differ in efficacy even within the same category, and their efficacy may be affected by the level of contamination as well as duration of exposure, type and dose of the binder, animal species and physiological condition of treated animals.

Current research was focused towards studying the selective properties of some adsorbents (particularly mineral based adsorbents). When these materials are applied to removal of mycotoxins from feeds in *in vivo* studies, they may adsorb nutrient meanwhile. For most substances proposed as mycotoxin adsorbents, the results are restricted to the laboratory, which has not verified by practice. *In vitro* methods are good indicator tool for identifying the potential of mycotoxin adsorbing agents and for investigating their mechanism of action as well as favourable detoxification conditions. However, overall *in vitro* tests, including the most sophisticated GI models, cannot provide the real *in vivo* conditions. Results of the previous investigations indicate that there is great variability in the efficacy of adsorbing agents *in vivo*, even though the compounds may show potential for mycotoxin binding *in vitro*. In addition, the absorption, transformation and metabolism of mycotoxins in animals and humans are in particular needs to be studied. Toxicity and side effects of mycotoxin adsorbents should be taken into account, since they may induce a risk to animals and humans.

References

Adunphatcharaphon, S., Petchkongkaew, A., Greco, D., D'Ascanio, V., Visessanguan, W. and Avantaggiato, G. 2020. The effectiveness of durian peel as a multi-mycotoxin adsorbent. Toxins 12: 108. https://doi.org/10.3390/toxins12020108.

Afriyie-Gyawu, E., Ankrah, N.A., Huebner, H., Ofosuhene, M., Kumi, J., Johnson, N., Tang, L., Xu, L., Jolly, P., Ellis, P., Ofori-Adjei, D., Williams, J.H., Wang, J.S. and Phillips, T.D. 2008a. NovaSil clay intervention in Ghanaians at high risk for aflatoxicosis, Part I: Study design and clinical outcomes. Food Addit. Contam. 25(1): 76–87. https://doi.org/10.1080/02652030701458105.

Afriyie-Gyawu, E., Wang, Z., Ankrah, N.A., Xu, L., Johnson, N.M., Tang, L., Guan, H., Huebner, H.J., Jolly, P.E., Ellis, W.O., Taylor, R., Brattin, B., Ofori-Adjei, D., Williams, J.H., Wang, J.S. and Phillips, T.D. 2008b. NovaSil clay does not affect the concentrations of vitamins A and E and nutrient minerals in serum samples from Ghanaians at high risk for aflatoxicosis. Food Addit. Contam. Part A 25(7): 872–884. https://doi.org/10.1080/02652030701854758.

Alegakis, A.K., Tsatsakis, A.M., Shtilman, M.I., Lysovenko, D.L. and Vlachonikolis, I.G. 1999. Deactivation of mycotoxins. I. An in vitro study of zearalenone adsorption on new polymeric adsorbents. Journal of Environmental Science and Health Part B 34: 633–644. https://doi.org/10.1080/03601239909373218.

Aoudia, N., Callu, P., Grosjean, F. and Larondelle, Y. 2009. Effectiveness of mycotoxin sequestration activity of micronized wheat fibres on distribution of ochratoxin A in plasma, liver and kidney of piglets fed a naturally contaminated diet. Food and Chemical Toxicology 47: 1485–1489. https://doi.org/10.1016/j.fct.2009.03.033.

Aoudia, N., Tangni, E.K. and Larondelle, Y. 2008. Distribution of ochratoxin A in plasma and tissues of rats fed a naturally contaminated diet amended with micronized wheat fibres: Effectiveness of mycotoxin sequestering activity. Food and Chemical Toxicology 46: 871–878. https://doi.org/10.1016/j.fct.2007.10.029.

Avantaggiato, G., Havenaar, R. and Visconti, A. 2003. Assessing the zearalenone-binding activity of adsorbent materials during passage through a dynamic in vitro gastrointestinal model. Food Chem. Toxicol. 41: 1283–1290. https://doi.org/10.1016/S0278-6915(03)00113-3.

Avantaggiato, G., Havenaar, R. and Visconti, A. 2004. Evaluation of the intestinal absorption of deoxynivalenol and nivalenol by an in vitro gastrointestinal model, and the binding efficacy of activated carbon and other adsorbent materials. Food Chem. Toxicol. 42: 817–824. https://doi.org/10.1016/j.fct.2004.01.004.

Avantaggiato, G., Solfrizzo, M. and Visconti, A. 2005. Recent advances on the use of adsorbent materials for detoxification of Fusarium mycotoxins. Food Addit. Contam. 22: 379–388. https://doi.org/10.1080/02652030500058312.

Avantaggiato, G., Havenaar, R. and Visconti, A. 2007. Assessment of the multi-mycotoxin binding efficacy of a carbon/aluminosilicate-based product in an in vitro gastrointestinal model. J. Agric. Food Chem. 55: 4810–4819. https://doi.org/10.1021/jf0702803.

Avantaggiato, G. and Visconti, A. 2009. Mycotoxin issues in farm animals and strategies to reduce mycotoxins in animal feeds. pp. 149–189. *In*: Garnsworthy, P.C., Wiseman, J. (eds.). Recent Advances in Animal Nutrition – 2009. Nottingham: Bonington Campus, School Biosci. ISBN 978-1-907284-65-6.

Avantaggiato, G., Greco, D., Damascelli, A., Solfrizzo, M. and Visconti, A. 2014. Assessment of multi-mycotoxin adsorption efficacy of grape pomace. J. Agric. Food Chem. 62: 497–507. https://doi.org/10.1021/jf404179h.

Baglieri, A., Reyneri, A., Gennari, M. and Negre, M. 2013. Organically modified clays as binders of fumonisins in feedstocks. J. Environ. Sci. Health 48: 776–783. https://doi.org/10.1080/03601234.2013.780941.

Barrer, R.M. and Macleod, D.M. 1955. Activation of montmorillonite by ion exchange and sorption complexes of tetra-alkyle ammonium montmorillonites. Trans. Faraday Soc. 51: 1290–1300. https://doi.org/10.1039/TF9555101290.

Bergaya, F. and Lagaly, G. 2006. Chapter 1 General Introduction: Clays, clay minerals, and clay science. pp. 1–18. *In*: Bergaya, F., Theng, B.K.G., Lagaly, G. (Eds.). Developments in Clay Science: Handbook of Clay Science, Volume 1. Elsevier: Amsterdam, The Netherlands. ISBN: 9780080441832.

Boudergue, C., Burel, C., Dragacci, S., Favrot, M.C., Fremy, J.M., Massimi, C., Prigent, P., Debongnie, P., Pussemier, L., Boudra, H., Morgavi, D., Oswald, I.,

Perez, A. and Avantaggiato, G. 2009. Review of mycotoxin-detoxifying agents used as feed additives: Mode of action, efficacy and feed/food safety. External Scientific Report Submitted to EFSA Available on: https://www.efsa.europa.eu/en/supporting/pub/en-22. https://doi.org/10.2903/sp.efsa.2009.EN-22.
Bryden, W.L. 2012. Mycotoxin contamination of the feed supply chain: Implications for animal productivity and feed security. Anim Feed Sci Tech. 173: 134–158. https://doi.org/10.1016/j.anifeedsci.2011.12.014
Carson, M.S. and Smith, T.K. 1983. Effect of feeding alfalfa and refined plant fibers on the toxicity and metabolism of T-2 toxin in rats. J. Nutr. 113(2): 304–313. https://doi.org/10.1093/jn/113.2.304.
Cavret, S., Laurent, N., Videmann, B., Mazallon, M. and Lecoeur, S. 2010. Assessment of deoxynivalenol (DON) adsorbents and characterisation of their efficacy using complementary in vitro tests. Food Additives and Contaminants Part A 27: 4353. https://doi.org/10.1080/02652030903013252
Charturvedi, V.B., Singh, K.S. and Agnihotri, A.K. 2002. In vitro aflatoxin adsorption capacity of some indigenous aflatoxin sorbent. Indian J. Anim. Sci. 72(3): 257–260.
Colovic, R., Puvaca, N., Cheli, F., Avantaggiato, G., Greco, D., Duragic, O., Kos, J. and Pinotti, L. 2019. Decontamination of mycotoxin-contaminated feedstuffs and compound feed. Toxins 11: 617. https://doi.org/10.3390/toxins11110617
D'Ascanio, V., Greco, D., Menicagli, E., Santovito, E., Catucci, L., Logrieco, A.F. and Avantaggiato, G. 2019. The role of geological origin of smectites and of their physico-chemical properties on aflatoxin adsorption. Appl. Clay Sci. 181: 105209. https://doi.org/10.1016/j.clay.2019.105209.
Dakovic, A., Kragovic, M., Rottinghaus, G.E., Sekulic, Z., Milicevic, S., Milonjic, S.K. and Zaric, S. 2010. Influence of natural zeolitic tuff and organozeolites surface charge on sorption of ionizable fumonisin B1. Colloids Surf., B 76: 272–278. https://doi.org/10.1016/j.colsurfb.2009.11.003.
Daković, A., Matijasevic, S., Rottinghaus, G.E., Ledoux, D.R., Butkeraitis, P. and Sekulic, Z. 2008. Aflatoxin B1 adsorption by natural and copper modified montmorillonite. Colloids Surf. B 66: 20–25. https://doi.org/10.1016/j.colsurfb.2008.05.008.
Dakovic, A., Tomasevic-Canovic, M., Rottinghaus, G., Dondur, V. and Masic, Z. 2003. Adsorption of ochratoxin A on octadecyldimethyl benzyl ammonium exchanged clinoptilolite-heulandite tuff. Colloids Surf., B 30: 157–165. https://doi.org/10.1016/S0927-7765(03)00067-5.
Dalié, D.K.D., Deschamps, A.M. and Richard-Forget, F. 2010. Lactic acid bacteria – potential for control of mould growth and mycotoxins: A review. Food Contr. 21: 370–380. https://doi.org/10.1016/j.foodcont.2009.07.011.
Desheng, Q., Fan, L., Yanhu, Y. and Niya, Z. 2005. Adsorption of aflatoxin B-1 on montmorillonite. Poult. Sci. 84(6): 959–961. https://doi.org/ 10.1093/ps/84.6.959.
Diaz, D.E. and Smith, T.K. 2005. Mycotoxin sequestering agents: Practical tools for the neutralisation of mycotoxins. pp. 323–339. *In*: The Mycotoxin Blue Book. Nottingham University Press, Nottingham, United Kingdom, ISBN: 9781899043521
Döll, S., Danicke, S., Valenta, H. and Flachowsky, G. 2004. In vitro studies on the evaluation of mycotoxin detoxifying agents for their efficacy on

deoxynivalenol and zearalenone. Arch. Anim. Nutr. 58: 311–324. https://doi.org/10.1080/00039420412331273268.
Döll, S., Gericke, S., Danicke, S., Raila, J., Ueberschar, K.H., Valenta, H., Schnurrbusch, U., Schweigert, F.J. and Flachowsky, G. 2005. The efficacy of a modified aluminosilicate as a detoxifying agent in Fusarium toxin contaminated maize containing diets for piglets. J. Anim. Physiol. Anim. Nutr. 89: 342–358. https://doi.org/10.1111/j.1439-0396.2005.00527.x
EFSA (European Food Safety Authority). 2004a. Opinion of the Scientific Panel on Contaminants in the Food Chain on a request from the commission related to aflatoxin B_1 as undesirable substance in animal feed. The EFSA Journal 39: 1–27. https://doi.org/10.2903/j.efsa.2004.39.
EFSA (European Food Safety Authority). 2004b. Opinion of the Scientific Panel on Contaminants in Food Chain on a request from the commission related to ochratoxin A (OTA) as undesirable substance in animal feed. The EFSA Journal 101: 1–36. https://doi.org/10.2903/j.efsa.2004.101.
EFSA (European Food Safety Authority). 2004c. Opinion of the Scientific Panel on Contaminants in the Food Chain on a request from the commission related to zearalenone as undesirable substance in animal feed. The EFSA Journal 89: 1–35. https://doi.org/10.2903/j.efsa.2004.89.
EFSA (European Food Safety Authority). 2004d. Opinion of the Scientific Panel on Contaminants in the Food Chain on a request from the commission related to deoxynivalenol (DON) as undesirable substance in animal feed. The EFSA Journal 73: 1–41. https://doi.org/10.2903/j.efsa.2004.73.
EFSA (European Food Safety Authority). 2005. Opinion of the Scientific Panel on Contaminants in Food Chain on a request from the commission related to fumonisins as undesirable substances in animal feed. The EFSA Journal 235: 1–32. https://doi.org/10.2903/j.efsa.2005.235.
EFSA Panel on Additives and Products or Substances used in Animal Feed (FEEDAP), Rychen, G., Aquilina, G., Azimonti, G., Bampidis, V., Bastos, M.L., Bories, G., Chesson, A., Cocconcelli, P.S., Flachowsky, G., Gropp, J., Kolar, B., Kouba, M., Lopez-Alonso, M., Mantovani, A., Mayo, B., Ramos, F., Saarela, M., Villa, R.E., Wallace, R.-J., Wester, P., Martelli, G., Renshaw, D., Lopez-Galvez, G. and Lopez Puente S. 2017. Scientific opinion on the safety and efficacy of bentonite as a feed additive for all animal species. EFSA Journal 2017 15(12): 5096, 13 pp. https://doi.org/10.2903/j.efsa.2017.5096.
EFSA Panel on Additives and Products or Substances used in Animal Feed (FEEDAP). 2010. Statement on the establishment of guidelines for the assessment of additives from the functional group 'substances for reduction of the contamination of feed by mycotoxins'. EFSA Journal 2010 8(7): 1693, 8 pp. https://doi.org/10.2903/j.efsa.2010.1693.
EFSA Panel on Additives and Products or Substances used in Animal Feed (FEEDAP). 2011. Scientific opinion on the efficacy of bentonite (dioctahedral montmorillonite) for all species. EFSA Journal 2011 9(6): 2276, 9 pp.https://doi.org/10.2903/j.efsa.2011.2007.
EFSA Panel on Additives and Products or Substances used in Animal Feed (FEEDAP). 2012. Guidance for the preparation of dossiers for technological additives. EFSA Journal 2012 10(1): 2528, 23 pp. https://doi.org/10.2903/j.efsa.2012.2528.

EFSA Panel on Additives and Products or Substances used in Animal Feed (FEEDAP). 2013. Scientific opinion on the safety and efficacy of micro-organism DSM 11798 when used as a technological feed additive for pigs. EFSA Journal 2013 11(5): 3203, 18 pp. https://doi.org/10.2903/j.efsa.2013.3203.

EFSA Panel on Additives and Products or Substances used in Animal Feed (FEEDAP). 2014. Scientific opinion on the safety and efficacy of fumonisin esterase (FUMzyme®) as a technological feed additive for pigs. EFSA Journal 2014 12(5): 3667, 19 pp. https://doi.org/10.2903/j.efsa.2014.3667.

Elliott, C.T., Connolly, L. and Kolawole, O. 2020. Potential adverse effects on animal health and performance caused by the addition of mineral adsorbents to feeds to reduce mycotoxin exposure. Mycotoxin Research 36: 115–126. https://doi.org/10.1007/s12550-019-00375-7.

European Commission. 2002. Commission Regulation No. 32/2002 of 7 May 2002 on undesirable substances in animal feed. *In*: Official Journal of the European Union, 2002; Vol. 30.5.2002, p L 140/10. http://data.europa.eu/eli/dir/2002/32/oj.

European Commission. 2006. Commission Regulation No. 576/2006 of 17 August 2006 on the presence of deoxynivalenol, zearalenone, ochratoxin A, T-2 and HT-2 and fumonisins in products intended for animal feeding. *In*: Official Journal of the European Union, 2006; Vol. 23.8.2006, p L 229/7. http://data.europa.eu/eli/reco/2006/576/oj.

European Commission. 2009. Commission Regulation (EC) No. 386/2009 of 12 May 2009 amending Regulation (EC) No. 1831/2003 of the European Parliament and of the Council as regards the establishment of a new functional group of feed additives. Off. J. EU. L 118/66. http://data.europa.eu/eli/reg/2009/386/oj.

European Commission. 2013. Commission Implementing Regulation (EU) No 1060/2013of 29 October 2013 concerning the authorisation of bentonite as a feed additive for all animal species. Off. J. EU. L. 289/33. http://data.europa.eu/eli/reg_impl/2013/1060/oj.

European Commission. 2017a. Commission Implementing Regulation (EU) No 2017/930 of 31 May 2017 concerning the authorisation of a preparation of a microorganism strain DSM 11798 of the *Coriobacteriaceae* family as a feed additive for all avian species and amending Implementing Regulation (EU) No 1016/2013. Off. J. EU. L. 141/6. http://data.europa.eu/eli/reg_impl/2017/930/oj.

European Commission. 2017b. Commission Implementing Regulation (EU) No 2017/913 of 29 May 2017 concerning the authorisation of a preparation of fumonisin esterase produced by *Komagataella pastoris* (DSM 26643) as a feed additive for all avian species. Off. J. EU. L. 139/33. http://data.europa.eu/eli/reg_impl/2014/1115/oj.

Ferguson, L.R. and Harris, P.J. 1996. Studies on the role of specific dietary fibres in protection against colorectal cancer. Mutat. Res. Mol. Mech. Mutagen 350(1): 173–184. http://doi.org/10.1016/0027-5107(95)00105-0.

Food and Agriculture Organization of the United Nations (FAO). 2013. FAO Rice Market Monitor, 3. http://www.fao.org/economic/est/publications/rice-publications/rice-market-monitor-rmm/en/.

Galvano, F., Piva, A., Ritieni, A. and Galvano, G. 2001. Dietary strategies to counteract the effects of mycotoxins: A review. J. Food Prot. 64: 120–131. https://doi.org/10.4315/0362-028x-64.1.120.

Gambacorta, L., Pinton, P., Avantaggiato, G., Oswald, I.P. and Solfrizzo, M. 2016. Grape pomace, an agricultural byproduct reducing mycotoxin absorption: In vivo assessment in pig using urinary biomarkers. J Agric Food Chem. 64(35): 6762–6771. https://doi.org/10.1021/acs.jafc.6b02146.

Geh, S., Shi, T., Shokouhi, B., Schins, R., Armbruster, L., Rettenmeier, A. and Dopp, E. 2006. Genotoxic potential of respirable bentonite particles with different quartz contents and chemical modifications in human lung fibroblasts. Inhal. Toxicol. 18: 405–412. https://doi.org/10.1080/08958370600563524.

Ghahri, H., Talebi, A., Chamani, M., Lotfollahian, H. and Afzali, N. 2009. Ameliorative effect of esterified glucomannan, sodium bentonite, and humic acid on humoral immunity of broilers during chronic aflatoxicosis. Turkish Journal of Veterinary and Animal Sciences 33: 419–425. https://doi.org/10.3906/vet-0805-23.

Gibson, N.M., Luo, T.J.M., Brenner, D.W. and Shenderova, O. 2011. Immobilization of mycotoxins on modified nanodiamond substrates. Biointerphases 6: 210–217. https://doi.org/10.1116/1.3672489.

Gratz, S., Mykkanen, H. and El-Nezami, H. 2005. Aflatoxin B1 binding by a mixture of Lactobacillus and Propionibacterium: In vitro versus ex vivo. Journal of Food Protection 68: 2470–2474. https://doi.org/10.4315/0362-028x-68.11.2470.

Greco, D., D'Ascanio, V., Santovito, E., Logrieco, A.F. and Avantaggiato, G. 2019. Comparative efficacy of agricultural by-products in sequestering mycotoxins: Multi-mycotoxin adsorption efficacy of agricultural by-products. J. Sci. Food Agric. 99: 1623–1634. https://doi.org/10.1002/jsfa.9343.

Grenier, B. and Oswald, I.P. 2011. Mycotoxin co-contamination of food and feed: Metaanalysis of publications describing toxicological interactions. World Mycotoxin J. 4: 285–313. https://doi.org/10.3920/wmj2011.1281.

Gutzwiller, A., Czegledi, L., Stoll, P. and Bruckner, L. 2007. Effects of Fusarium toxins on growth, humoral immune response and internal organs in weaner pigs, and the efficacy of apple pomace as an antidote. J. Anim. Physiol. Anim. Nutr. 91: 432–438. https://doi.org/10.1111/j.1439-0396.2006.00672.x.

Harris, P.J. and Ferguson, L.R. 1993. Dietary fibre: Its composition and role in protection against colorectal cancer. Mutat. Res. Mol. Mech. Mutagen. 290(1): 97–110. https://doi.org/10.1016/0027-5107(93)90037-g.

Harris, P.J., Sasidharan, V.K., Roberton, A.M., Triggs. C.M., Blakeney, A.B. and Ferguson, L.R. 1998. Adsorption of a hydrophobic mutagen to cereal brans and cereal bran dietary fibres. Mutat. Res. Genet. Toxicol. Environ. Mutagen. 412(3): 323–331. https://doi.org/10.1016/s1383-5718(98)00003-5.

Horky, P., Skalickova, S., Baholet, D. and Skladanka, J. 2018. Nanoparticles as a solution for eliminating the risk of mycotoxins. Nanomaterials 8: 727. https://doi.org/10.3390/nano8090727.

Huwig, A., Freimund, S., Kappeli, O. and Dutler, H. 2001. Mycotoxin detoxication of animal feed by different adsorbents. Toxicol. Lett. 122: 179–188. https://doi.org/10.1016/s0378-4274(01)00360-5.

IARC. 2012. Chemical and physical characteristics of the principal mycotoxins. IARC Sci. Public. 158: 31–38. ISBN 978-92-832-2214-9.

James, L.J. and Smith, T.K. 1982. Effect of dietary alfalfa on zearalenone toxicity and metabolism in rats and swine. J. Anim. Sci. Oxford University Press 55(1): 110–118. https://doi.org/10.2527/jas1982.551110x.

Jard, G., Liboz, T., Mathieu, F., Guyonvarc'h, A. and Lebrihi, A. 2009. Adsorption of zearalenone by *Aspergillus japonicus* conidia: New trends for biological decontamination in animal feed. World Mycotoxin Journal 2: 391–397. https://doi.org/10.3920/WMJ2008.1128.

Jarda, G., Liboz, T., Mathieua, F., Guyonvarc'h, A. and Lebrihi, A. 2011. Review of mycotoxin reduction in food and feed: From prevention in the field to detoxification by adsorption or transformation. Food Additives & Contaminants 28(11): 1590–1609. https://doi.org/10.1080/19440049.2011.595377.

Jaynes, W.F. and Zartman, R.E. 2011. Aflatoxin toxicity reduction in feed by enhanced binding to surface-modified clay additives. Toxins 3: 551–565. https://doi.org/10.3390/toxins3060551.

JECFA. 2017. Evaluation of certain contaminants in food: Eighty-third Report of the Joint FAO/WHO Expert Committee on Food Additives; WHO Technical Report Series No. 1002; WHO Press: Geneva, Switzerland, 2017. ISBN 978-92-4-121-002-7.

Jiang, S.Z., Yang, Z.B., Yang, W.R., Wang, S.J., Liu, F.X., Johnston, L.A., Chi, F. and Wang, Y. 2012. Effect of purified zearalenone with or without modified montmorillonite on nutrient availability, genital organs and serum hormones in post-weaning piglets. Livest. Sci. 144: 110–118. https://doi.org/10.1016/j.livsci.2011.11.004.

Jiang, Y. 2004. Preparation of Al-pillared bentonite by microwave irradiation method and its properties. Non-metallic Mines 27: 19–21.

Kabak, B., Dobson, A.D.W. and Var, I. 2006. Strategies to prevent mycotoxin contamination of food and animal feed: A review. Crit. Rev. Food Sci. Nutr. 46: 593–619. https://doi.org/10.1080/10408390500436185.

Karlovsky, P., Suman, M., Berthiller, F., De Meester, J., Eisenbrand, G., Perrin, I., Oswald, I.P., Speijers, G., Chiodini, A. and Recker, T. 2016. Impact of food processing and detoxification treatments on mycotoxin contamination. Mycotoxin Res. 32: 179–205. https://doi.org/10.1007/s12550-016-0257-7.

Kolawole, O., Meneely, J., Greer, B., Chevallier, O., Jones, D.S., Connolly, L. and Elliott, C. 2019. Comparative in vitro assessment of a range of commercial feed additives with multiple mycotoxin binding claims. Toxins 11: 659. https://doi.org/10.3390/toxins11110659.

Kolosova, A. and Stroka, J. 2011. Substances for reduction of the contamination of feed by mycotoxins: A review. World Mycotoxin J. 4: 225–256. https://doi.org/10.3920/WMJ2011.1288.

Laan, T., Bull, S., Pirie, R. and Fink-Gremmels, J. 2006. The role of alveolar macrophages in the pathogenesis of recurrent airway obstruction in horses. Journal of Veterinary Internal Medicine 20: 167–174. https://doi.org/10.1111/j.1939-1676.2006.tb02837.x.

Lemke, S.L., Mayura, K., Reeves, W.R., Wang, N., Fickey, C. and Phillips, T.D. 2001. Investigation of organophilic montmorillonite clay inclusion in zearalenone-contaminated diets using the mouse uterine weight bioassay. J. Toxicol. Environ. Health A 62: 243–258. https://doi.org/10.1080/009841001459405.

Lerda, D. 2011. Mycotoxins Factsheet, 4th ed. JRC Technical Notes, No. 66956. Available online: ec.europa. eu/jrc/sites/jrcsh/files/Factsheet%20 Mycotoxins_2.pdf.

Li, P., Wei, J., Chiu, Y., Su, H., Peng, F. and Lin, J. 2010. Evaluation on cytotoxicity and genotoxicity of the exfoliated silicate nanoclay. ACS Appl. Mater. Interfaces 2: 1608–1613. https://doi.org/10.1021/am1001162

Li, Y., Tian, G., Dong, G., Bai, S., Han, X., Liang, J., Meng, J. and Zhang, H. 2018. Research progress on the raw and modified montmorillonites as adsorbents for mycotoxins: A review. Appl. Clay Sci. 163: 299–311. https://doi.org/10.1016/j.clay.2018.07.032.

Liew, W.P.P. and Mohd-Redzwan, S. 2018. Mycotoxin: Its impact on gut health and microbiota. Front. Cell. Infect. Microbiol. 8: 1–17. https://doi.org/10.3389/fcimb.2018.00060.

Logrieco, A.F., Miller, J.D., Eskola, M., Krska, R., Ayalew, A., Bandyopadhyay, R., Battilani, P., Bhatnagar, D., Chulze, S., De Saeger, S., Li, P., Perrone, G., Poapolathep, A., Rahayu, E.S., Shephard, G.S., Stepman, F., Zhang, H. and Leslie, J.F. 2018. The mycotox charter: Increasing awareness of, and concerted action for minimizing mycotoxin exposure worldwide. Toxins 10: 149. https://doi.org/10.3390/toxins10040149.

Magnoli, A.P., Cabaglieri, L.R., Magnoli, C.E., Monge, J.C., Miazzo, R.D., Peralta, M.F., Salvano, M.A., Rosa, C.A.R., Dalcero, A.M. and Chiacchiera, S.M. 2008. Bentonite performance on broiler chickens fed with diets containing natural levels of aflatoxin B1. Rev. Bras. Med. Vet. 30(1): 55–60.

Magro, M., Moritz, D.E., Bonaiuto, E., Baratella, D., Terzo, M., Jakubec, P., Malina, O., Cepe, K., Falcao de Aragao, G.M. and Zboril, R.. 2016. Citrinin mycotoxin recognition and removal by naked magnetic nanoparticles. Food Chem. 203: 505–512. https://doi.org/10.1016/j.foodchem.2016.01.147.

Maisanaba, S., Gutiérrez-Praena, D., Puerto, M., Moyano, R., Blanco, A., Jordá, M., Cameán, A., Aucejo, S. and Jos, Á. 2014. Effects of the subchronic exposure to an organomodified clay mineral for food packaging applications on wistar rats. Appl. Clay Sci. 95: 37–40. https://doi.org/10.1016/j.clay.2014.04.006.

Maisanaba, S., Hercog, K., Filipic, M., Jos, Á. and Zegura, B. 2016. Genotoxic potential of montmorillonite clay mineral and alteration in the expression of genes involved in toxicity mechanisms in the human hepatoma cell line HepG2. J. Hazard Mater. 304: 425–433. https://doi.org/10.1016/j.jhazmat.2015.10.018.

Maisanaba, S., Puerto, M., Pichardo, S., Jordá, M., Moreno, F., Aucejo, S. and Jos, Á. 2013. In vitro toxicological assessment of clays for their use in food packaging applications. Food Chem. Toxicol. 57: 266–275. https://doi.org/10.1016/j.fct.2013.03.043.

MambaI, B., NyembeI, B. and Mulaba-Bafubiandi, A. 2010. The effect of conditioning with NaCl, KCl and HCl on the performance of natural clinoptilolite's removal efficiency of Cu^{2+} and Co^{2+} from Co/Cu synthetic solutions. Water SA 36: 437–444. https://doi.org/10.4314/wsa.v36i4.58419.

Marquez, R.N. and Hernandez, I.T.D. 1995. Aflatoxin adsorbent capacity of two Mexican aluminosilicates in experimentally contaminated chick diets. Food Addit. Contam. 431–433. https://doi.org/10.1080/02652039509374326.

Marroquín-Cardona, A., Deng, Y., Taylor, J.F., Hallmark, C.T., Johnson, N.M. and Phillips, T.D. 2009. In vitro and in vivo characterization of mycotoxin-binding

additives used for animal feeds in Mexico. Food Addit. Contam. Part A 26: 733–743. https://doi.org/10.1080/02652030802641872.
Mascolo, N., Summa, V. and Tateo, F. 2004. In vivo experimental data on the mobility of hazardous chemical elements from clays. Appl. Clay Sci. 25: 23–28. https://doi.org/10.1016/j.clay.2003.07.001.
Masimango, N., Remacle, J. and Ramaut, J. 1978. The role of adsorption in the elimination of aflatoxin B1 from contaminated media. Eur. J. Appl. Micorbiol. Biotechnol. 6(1): 101–105. https://doi.org/10.1007/BF00500861.
Masimango, N., Remacle, J. and Ramaut, J. 1979. Elimination of aflatoxin B1 from contaminated media by swollen clays. Ann. Nutr. Aliment. 33(1): 137–147. ISSN : 0003-4037
Mezes, M., Balogh, K. and Toth, K. 2010. Preventive and therapeutic methods against the toxic effects of mycotoxins – A review. Acta Vet. Hung. 58: 1–17. https://doi.org/10.1556/AVet.58.2010.1.1.
Miller, J.D., Schaafsma, A.W., Bhatnagar, D., Bondy, G., Carbone, I., Harris, L.J., Harrison, G., Munkvold, G.P., Oswald, I.P. and Pestka, J.J. 2014. Mycotoxins that affect the North American agri-food sector: State of the art and directions for the future. World Mycotoxin J. 7: 63–82. https://doi.org/10.3920/WMJ2013.1624.
Mitchell, N.J., Xue, K.S., Lin, S., Marroquin-Cardona, A., Brown, K.A., Elmore, S.E., Tang, L., Romoser, A., Gelderblom, W.C., Wang, J.S. and Phillips, T.D. 2014. Calcium montmorillonite clay reduces AFB1 and FB1 biomarkers in rats exposed to single and coexposures of aflatoxin and fumonisin. J. Appl. Toxicol. 34: 795–804. https://doi.org/10.1002/jat.2942.
Mutturee, Y., Tengjaroenkul, B., Pimpukdee, K., Sukon, P. and Tengjaroenkul, U. 2012. Effect of temperature on efficacy of bentonite to adsorb aflatoxin B1. J. Mol. Cat. A Chem. 52: 179–182. http://vmj.kku.ac.th/
Nahm, K.H. 1995. Prevention of aflatoxicosis by addition of antioxidants and hydrated sodium calcium aluminosilicates to the diet of young chicks. Nippon Kakin Gakkaishi 32(2): 117–127.
Nakkeeran, E., Rangabhashiyam, S., Giri Nandagopal, M.S. and Selvaraju, N. 2016. Removal of Cr(VI) from aqueous solution using Strychnos nux-vomica shell as an adsorbent. Desalin. Water Treat. 57(50): 23951–23964. https://doi.org/10.1080/19443994.2015.1137497.
Nakkeeran, E., Saranya, N., Giri Nandagopal, M.S., Santhiagu, A. and Selvaraju, N. 2016. Hexavalent chromium removal from aqueous solutions by a novel powder prepared from Colocasia esculenta leaves. Int. J. Phytoremediation 18(8): 812–821. https://doi.org/10.1080/15226514.2016.1146229.
Nones, J., Nones, J., Riella, H.G., Poli, A., Trentin, A.G. and Kuhnen, N.C. 2015. Thermal treatment of bentonite reduces aflatoxin B1 adsorption and affects stem cell death. Mater. Sci. Eng. C 55: 530–537. https://doi.org/10.1016/j.msec.2015.05.069.
Nones, J., Solhaug, A., Eriksen, G.S., Macuvele, D.L.P., Poli, A., Soares, C., Trentin, A.G., Riella, H.G. and Nones, J. 2017. Bentonite modified with zinc enhances aflatoxin B-1 adsorption and increase survival of fibroblasts (3T3) and epithelial colorectal adenocarcinoma cells (Caco-2). J. Hazard. Mater. 337: 80–89. https://doi.org/10.1016/j.jhazmat.2017.04.068.
Papaioannou, D., Katsoulos, P.D., Panousis, N. and Karatzias, H. 2005. The role of natural and synthetic zeolites as feed additives on the prevention and/or the

treatment of certain farm animal diseases: A review. Micropor. Mesopor. Mat. 84: 161–170. https://doi.org/10.1016/j.micromeso.2005.05.030.

Park, D.L., Lee, L.S., Price, R.S. and Pohland, A.E. 1988. Review of the decontamination of aflatoxins by ammoniation: Current status and regulation. Journal – Association of Official Analytical Chemists 71: 685–703. https://doi.org/10.1093/jaoac/71.4.685.

Park, S., Seo, D. and Lee, J. 2002. Surface modification of montmorillonite on surface acid–base characteristics of clay and thermal stability of epoxy/clay nanocomposites. J. Colloid. Interface Sci. 251: 160–165. https://doi.org/10.1006/jcis.2002.8379

Paterson, R.R.M. and Lima, N. 2010. How will climate change affect mycotoxins in food. Food Research International 42(7): 1902–1914. https://doi.org/10.1016/j.foodres.2009.07.010.

Pfliegler, W.P., Pusztahelyi, T. and Pócsi, I. 2015. Mycotoxins – prevention and decontamination by yeasts. J. Basic Microbiol. 55: 805–818. https://doi.org/10.1002/jobm.201400833.

Phillips, T.D., Afriyie-Gyawu, E., Williams, J., Huebner, H., Ankrah, N.A., Ofori-Adjei, D., Jolly, P., Johnson, N., Taylor, J., Marroquin-Cardona, A., Xu, L., Tang, L. and Wang, J.S. 2008. Reducing human exposure to aflatoxin through the use of clay: A review. Food Addit. Contam. 25: 134–145. https://doi.org/10.1080/02652030701567467.

Phillips, T.D., Kubena, L.F., Harvey, R.B., Taylor, D.R. and Heidelbaugh, N.D. 1988. Hydrated sodium calcium aluminosilicate: A high affinity sorbent for aflatoxin. Poult. Sci. 67: 243–247. https://doi.org/10.3382/ps.0670243

Phillips, T.D., Sarr, A.B. and Grant, P.G. 1995. Selective chemisorption and detoxification of aflatoxins by phyllosilicate clay. Nat. Toxins 3: 204–213. https://doi.org/10.1002/nt.2620030407.

Pinotti, L., Ottoboni, M., Giromini, C., Dell'Orto, V. and Cheli, F. (2016). Mycotoxin contamination in the EU feed supply chain: A focus on cereal byproducts. Toxins 8: 45. https://doi.org/10.3390/toxins8020045.

Pirouz, A.A., Selamat, J., Iqbal, S.Z., Mirhosseini, H., Karjiban, R.A. and Bakar, F.A. 2017. The use of innovative and efficient nanocomposite (magnetic graphene oxide) for the reduction of fusarium mycotoxins in palm kernel cake. Sci. Rep. 7: 12453. https://doi.org/10.1038/s41598-017-12341-3.

Puzyr, A.P., Purtov, K.V., Shenderova, O.A., Luo, M., Brenner, D.W. and Bondar, V.S. 2007. The adsorption of aflatoxin B1 by detonation-synthesis nanodiamonds. Dokl. Biochem. Biophys. 417: 299–301. https://doi.org/10.1134/S1607672907060026.

Ralla, K., Sohling, U., Riechers, D., Kasper, C., Ruf, F. and Scheper, T. 2010. Adsorption and separation of proteins by a smectite clay mineral. Bioprocess. Biosyst. Eng. 33: 847–861. https://doi.org/10.1007/s00449-010-0408-8.

Ramos, A.J. and Hernández, E. 1996. In vitro aflatoxin adsorption by means of a montmorillonite silicate: A study of adsorption isotherms. Anim. Feed Sci. Technol. 62: 263–269. https://doi.org/10.1016/S0377-8401(96)00968-6.

Ramos, A.J., Fink-Gremmels, J. and Hernández, E. 1996a. Prevention of toxic effects of mycotoxins by means of nonnutritive adsorbent compounds. J. Food Prot. 59: 631–641. https://doi.org/10.4315/0362-028X-59.6.631.

Ramos, A.J., Hernández, E., Pla-Delfina, J.M. and Merino, M. 1996b. Intestinal absorption of zearalenone and in vitro study of non-nutritive sorbent materials. Int. J. Pharm. 128: 129–137. https://doi.org/10.1016/0378-5173(95)04239-3.

Rangabhashiyam, S., Anu, N. and Selvaraju, N. 2014. Equilibrium and kinetic modeling of chromium(VI) removal from aqueous solution by a novel biosorbent. Res. J. Chem. Environ. 18: 30–36.

Rangabhashiyam, S., Suganya, E., Varghese Lity, A. and Selvaraju, N. 2016. Equilibrium and kinetics studies of hexavalent chromium biosorption on a novel green macroalgae Enteromorpha sp. Res. Chem. Intermediat. 42(2): 1275–1294. https://doi.org/10.1007/s11164-015-2085-3.

Ringot, D., Lerzy, B., Chaplain, K., Bonhoure, J.P., Auclair, E. and Larondelle, Y. 2007. In vitro biosorption of ochratoxin A on the yeast industry by-products: comparison of isotherm models. Bioresour. Technol. 98: 1812–1821. https://doi.org/10.1016/j.biortech.2006.06.015.

Rosen, M.J. and Kunjappu, J.T. 2004. Surfactants and Interfacial Phenomena, Fourth Edition. John Wiley & Sons, New York. https://doi.org/10.1002/9781118228920.

Ruiz, M.J., Macáková, P., García, A.J. and Font, G. 2011. Cytotoxic effects of mycotoxin combinations in mammalian kidney cells. Food Chem. Toxicol. 49: 2718–2724. https://doi.org/10.1016/j.fct.2011.07.021.

Sabater-Vilar, M., Malekinejad, H., Selman, M.H.J., van der Doelen, M.A.M. and Fink-Gremmels, J. 2007. In vitro assessment of adsorbents aiming to prevent deoxynivalenol and zearalenone mycotoxicoses. Mycopathologia 163: 81–90. https://doi.org/10.1007/s11046-007-0093-6

Santos, R.R., Vermeulen, S., Haritova, A. and Fink-Gremmels, J. 2011. Isotherm modeling of organic activated bentonite and humic acid polymer used as mycotoxin adsorbents. Food Addit. Contam. 28: 1578–1589. https://doi.org/10.1080/19440049.2011.595014.

Saranya, N., Nakeeran, E., Giri Nandagopal, M.S. and Selvaraju, N. 2017. Optimization of adsorption process parameters by response surface methodology for hexavalent chromium removal from aqueous solutions using *Annona reticulata* Linn peel microparticles. Water Sci. Technol. 75(9): 2094–2107. https://doi.org/10.2166/wst.2017.092.

Sassi, A., Flannery, B. and Vardon, P. 2018. Economic impact of mycotoxin contamination in U.S. corn, wheat, and peanuts. Risk Anal., submitted.

Sera, N., Morita, K., Nagasoe, M., Tokieda, H., Kitaura, T. and Tokiwa, H. 2005. Binding effect of polychlorinated compounds and environmental carcinogens on rice bran fiber. J. Nutr. Biochem. 16(1): 50–58. https://doi.org/10.1016/j.jnutbio.2004.09.005.

Sharma, A., Schmidt, B., Frandsen, H., Jacobsen, N., Larsen, E. and Binderup, M. 2010. Genotoxicity of unmodified and organo-modified montmorillonite. Mutat. Res. Genet. Toxicol. Environ. Mutagen. 700: 18–25. https://doi.org/10.1016/j.mrgentox.2010.04.021.

Shetty, P.H. and Jespersen, L. 2006. *Saccharomyces cerevisiae* and lactic acid bacteria as potential mycotoxin decontaminating agents. Trends Food Sci. Technol. 17: 48–55. https://doi.org/10.1016/j.tifs.2005.10.004.

Shetty, P.H., Hald, B. and Jespersen, L. 2007. Surface binding of aflatoxin B1 by *Saccharomyces cerevisiae* strains with potential decontaminating abilities in

indigenous fermented foods. International Journal of Food Microbiology 113: 41–46. https://doi.org/10.1016/j.ijfoodmicro.2006.07.013.
Shi, Y., Xu, Z., Wang, C. and Sun, Y. 2007. Efficacy of two different types of montmorillonite to reduce the toxicity of aflatoxin in pigs. New Zeal. J. Agr. Res. 50: 473–478. https://doi.org/10.1080/00288230709510315.
Smith, A., Panickar, K., Urban, J. and Dawson, H. 2018. Impact of micronutrients on the immune response of animals. Annu. Rev. Anim. Biosci. 6: 227–254. https://doi.org/10.1146/annurev-animal-022516-022914.
Smith, T.K. 1980. Influence of dietary fiber, protein and zeolite on zeralenone toxicosis in rats and swine. J. Anim. Sci. 50(2): 278–285. https://doi.org/10.2527/jas1980.502278x.
Solfrizzo, M., Carratu, M.R., Avantaggiato, G., Galvano, F., Pietri, A. and Visconti, A. 2001a. Ineffectiveness of activated carbon in reducing the alteration of sphingolipid metabolism in rats exposed to fumonisin-contaminated diets. Food and Chemical Toxicology 39: 507–511. https://doi.org/10.1016/S0278-6915(00)00160-5.
Solfrizzo, M., Visconti, A., Avantaggiato, G., Torres, A. and Chulze, S. 2001b. In vitro and in vivo studies to assess the effectiveness of cholestyramine as a binding agent for fumonisins. Mycopathologia 151: 147–153. https://doi.org/10.1023/A:1017999013702.
Stangroom, K.E. and Smith, T.K. 1984. Effect of whole and fractionated dietary alfalfa meal on zearalenone toxicosis and metabolism in rats and swine. Can. J. Physiol. Pharmacol. NRC Research Press 62(9): 1219–1224. https://doi.org/10.1139/y84-203.
Stroka, J. and Maragos, C.M. 2016. Challenges in the analysis of multiple mycotoxins. World Mycotoxin J. 9: 847–861. https://doi.org/10.3920/WMJ2016.2038.
Tranquil, E., Kanarskaya, Z.A., Tikhomirov, D.F. and Kanarsky, A.V. 2013. Compositions and methods for decontamination of animal feed containing mycotoxins typical for both northern and southern climates, US Patent Publication No. 8496984 B2.
Underhill, K.L., Rotter, B.A., Thompson, B.K., Prelusky, D.B. and Trenholm, H.L. 1995. Effectiveness of cholestyramine in the detoxification of zearalenone as determined in mice. Bulletin of Environmental Contamination and Toxicology 54: 128–134. https://doi.org/10.1007/BF00196279.
Van Rensburg, C.J., Van Rensburg, C.E.J., Van Ryssen, J.B.J., Casey, N.H. and Rottinghaus, G.E. 2006. In vitro and in vivo assessment of humic acid as an aflatoxin binder in broiler chickens. Poultry Science 85: 1576–1583. https://doi.org/10.1093/ps/85.9.1576.
Velazquez, A.L.B. and Deng, Y. 2020. Reducing competition of pepsin in aflatoxin adsorption by modifying a smectite with organic nutrients. Toxins 12: 28. doi:10.3390/toxins12010028.
Vila-Donat, P., Marín, S., Sanchis, V. and Ramos, A.J. 2018. A review of the mycotoxin adsorbing agents, with an emphasis on their multi-binding capacity, for animal feed decontamination. Food and Chemical Toxicology 114: 246–259. https://doi.org/10.1016/j.fct.2018.02.044.
Wang, J.S., Luo, H., Billam, M., Wang, Z., Guan, H., Tang, L., Goldston, T., Afriyie-Gyawu, E., Lovett, C., Griswold, J., Brattin, B., Taylor, R.J., Huebner, H.J. and Phillips, T.D. 2005. Short term safety evaluation of processed calcium

montmorillonite clay (NovaSil) in humans. Food Addit. Contam. 22: 270–279. https://doi.org/10.1080/02652030500111129.

Wang, M., Maki, C.R., Deng, Y., Tian, Y. and Phillips, T.D. 2017. Development of high capacity enterosorbents for aflatoxin B1 and other hazardous chemicals. Chem. Res. Toxicol. 30: 1694–1701. https://doi.org/10.1021/acs.chemrestox.7b00154.

Wang, P., Afriyie-Ggawu, E., Tang, Y., Johnson, N.M., Xu, L., Tang, L., Huebner, H.J., Ankrah, N.A., Ofori-Adjei, D., Ellis, W., Jolly, P.E., Williams, J.H., Wang, J.S. and Phillips, T.D. 2008. NovaSil clay intervention in Ghanaians at high risk for aflatoxicosis – II: Reduction in biomarkers of aflatoxin in blood and urine. Food Addit. Contam. Part A 25(5): 622–634. https://doi.org/10.1080/02652030701598694.

WHO. 2005. Bentonite, Kaolin and Selected Clay Minerals. ISBN 9241572310. Available online: https://apps.who.int/iris/handle/10665/43102.

Yao, J., Kang, F. and Gao, Y. 2012. Adsorption of aflatoxin on montmorillonite modified by low-molecular-weight humic acids. Huan Jing Ke Xue 33: 958–964.

Yiannikouris, A., Poughon, L., Cameleyre, X., Dussap, C.-G., François, J., Bertin, G. and Jouany, J.-P. 2003. A novel technique to evaluate interactions between *Saccharomyces cerevisiae* cell wall and mycotoxins: Application to zearalenone. Biotechnology Letters 25: 783–789. https://doi.org/10.1023/A:1023576520932.

Zeng, L., Wang, S., Peng, X., Geng, J., Chen, C. and Li, M. 2013. Al–Fe PILC preparation, characterization and its potential adsorption capacity for aflatoxin B1. Appl. Clay Sci. 83–84: 231–237. https://doi.org/10.1016/j.clay.2013.08.040.

Zhang, Y., Gao, R., Liu, M., Shi, B., Shan, A. and Cheng, B. 2015. Use of modified halloysite nanotubes in the feed reduces the toxic effects of zearalenone on sow reproduction and piglet development. Theriogenology 83: 932–941. https://doi.org/10.1016/j.theriogenology.2014.11.027.

Zhao, Z., Liu, N., Yang, L., Wang, J., Song, S., Nie, D., Yang, X., Hou, J. and Wu, A. 2015. Cross-linked chitosan polymers as generic adsorbents for simultaneous adsorption of multiple mycotoxins. Food Control 57: 362–369. https://doi.org/10.1016/j.foodcont.2015.05.014.

CHAPTER

6

Plants for Plants: Would the Solution Against Mycotoxins be the Use of Plant Extracts?

Asma Chelaghema, Caroline Strub*, Alexandre Colas de la Noue, Sabine Schorr-Galindo and Angélique Fontana

Qualisud, Univ Montpellier, CIRAD, Montpellier Sup Agro, Univ d'Avignon, Univ de La Réunion, Montpellier, France

1. Introduction

Many food commodities are contaminated by filamentous fungi and the toxins produced by some of them, travelling from the field to the plate. These contaminations cause important health risks to humans and animals, and produce losses whether in quality or quantity or both (Hu et al. 2017). The Food and Agriculture Organisation (FAO) estimates that around 1000 million tons of food are damaged each year by mycotoxigenic filamentous fungi (Prakash et al. 2015). For years the only solution used against moulds was chemical treatments, but unfortunately some of them have adverse effects on the health of human and animals (Kiran et al. 2016), and their use could be limited because of their inconveniences such as the resistance of fungi to fungicides, the toxicity of residual fungicides and the pollution of the environment. As a result, alternative strategies can be explored to reduce the occurrence of mycotoxins in food, based on the use of biological methods that do not affect human health or the environment. This strategy consists of the use of physical methods, biocontrol agents and herbal extracts (Mari et al. 2016, Divband et al. 2017). To this end, the use of natural compounds possessing antifungal activities seems interesting, due to their low non-toxic effects for the environment, and their effect on animal and human health (Nicosia et al. 2016). When attacked by pathogens or as part of their metabolism, plants produce a wide variety of compounds with a large diversity in their activities. Among all these molecules, some are particularly efficient at reducing the growth of fungi and their toxins (Divband et al. 2017).

*Corresponding author: caroline.strub@umontpellier.fr

In this context, the aim of this chapter is to describe: (1) the main mycotoxigenic filamentous fungi and their mycotoxins, (2) the plant extracts used for the biocontrol of mycotoxin contamination, (3) the mechanismsof action involved in the inhibition of fungal growth and/or mycotoxin contamination by plant extracts.

2. Main Mycotoxinogenic Species and Their Regulated Mycotoxins

2.1 *Fusarium* and Fusariotoxins

Fusarium species widely contaminates grain products and are one of the most dangerous risks in crop productions (Da Silva Bomfim et al. 2015). The contamination with *Fusarium* species is usually linked to infection and accumulation of harmful mycotoxins, such as trichothecenes (T-2, HT-2, deoxynivalenol and nivalenol), zearalenone and fumonisins. Some species can produce several mycotoxins simultaneously like *Fusarium graminearum*, which is considered to be the most virulent among the *Fusarium* species producing trichothecenes. *F. graminearum* can also produce the zearalenone and some other secondary metabolites such as culmorin, sambucinol, dihydroxy-calconectrin, butenolide and fusarin C which are more or less toxic (Nguyen et al. 2017). Deoxynivalenol is the most reported *Fusarium* toxin in the world (Arraché et al. 2018) and it is the only member of trichothecenes under regulation by the European Commission (European Commission 2006). The main *Fusarium* species involved in zearalenone production is *Fusarium graminearum* but other members such as *Fusarium roseum, Fusarium culmorum, Fusarium oxysporum* and *Fusarium equiseti* are also implied (Yang et al. 2018). Zearalenone is highly prevalent in various crops, particularly maize, but also in barley, wheat, oats, rice and sorghum and is regulated by the European Commission (European Commission 2006). Fumonisins are mainly produced by *Fusarium verticillioides*. Fumonisins have been classified into four groups within which group B is of the most concern. It includes fumonisins B1, B2 and B3. Fumonisin B1represents the greatest health risk due to its high toxicity and is responsible for 70 to 80% of the fumonisin contamination while fumonisin B2 accounts for 15 to 25% and fumonisin B3 is poorly involved with 3 to 8% of contaminated foodstuffs (Da Silva Bomfim et al. 2015). They are classified in group 2B by the International Agency for Research on Cancer because of their potential carcinogenic effect on humans (Ostry et al. 2017). Thus, maximum levels for fumonisins have been established by the European Commission (Shephard et al. 2019).

2.2 *Aspergillus* and Associated Mycotoxins

The genus *Aspergillus* is dominant among food deteriorating moulds. *Aspergillus* species widely occur in foodstuffs, particularly in starchy cereal grains such as wheat, maize, sorghum, rice, barley and millet where they may produce toxic secondary metabolites like aflatoxins and ochratoxin A (Sultana et al. 2015).

Aflatoxins are mainly produced by *Aspergillus flavus* and *Aspergillus parasiticus* which are the main contaminant of peanuts, maize, pistachio and other crops in tropical and subtropical areas, leading to important economic losses. Aflatoxins are divided in four major groups namely B1, B2, G1 and G2 with aflatoxin B1 being the most toxic and classified in Group 1, i.e. carcinogenic to humans, by the International Agency for Research on Cancer (Ibrahim et al. 2017, Ostry et al. 2017). The European Commission set maximum limits for aflatoxinB1 and total aflatoxins (European Commission 2006). Ochratoxin A is usually produced by *Aspergillus* and *Penicillium* species. This toxin can be found in numerous crops, foods and beverages such as cereals, pulses, coffee, beer, wine and grape juice as well as dried vine fruits, nuts, cacao products and spices (El Khoury et al. 2017a). Ochratoxin A has been classified in Group 2B, i.e. possibly carcinogenic to humans, by the International Agency for Research on Cancer (Ostry et al. 2017). The European Union has set regulations for ochratoxin A in many food products (European Commission 2006).

2.3 *Penicillium* and Associated Mycotoxins

The main species of the genus *Penicillium* are *P. chrysogenum, P. expansum, P. verrucosum, P. viridicatum, P. oxalcum, P. brevicompactum, P. cyclopium, P. roqueforti* and *P. velutinum*. They can be present in cereal grains from the field to their storage, as well as in cereal-derived products (Nguyen et al. 2017). Strains of the *Penicillium* genus can produce mycotoxins such as patulin, citrinin and ochratoxin A with potential damage to human health (Cruz et al. 2016).Some of them produce only one mycotoxin while other species can produce several toxins simultaneously. *Penicillium expansum* is known to be the most important producer of patulin (Li et al. 2019). Patulin is found in fruits, fruit products, vegetables, cereals, and cheese. Patulin has become a major concern with regard to public health (Diao et al. 2018). *Penicillium verrucosum* is the main ochratoxin A producer among *Penicillium* sp. (Pittand Hocking 2009). Citrinin is a mycotoxin produced by several species of the genera *Aspergillus* and *Penicillium*, and its presence is generally concomitant with ochratoxin A in food and feed (Degen et al. 2018). Citrinin was firstly isolated from a culture of *Penicillium citrinum* and has been further associated with other species (e.g. *A. ochraceus, P. verrucosum*). Citrinin and ochratoxin A often co-occurs in cereals and plant products. Indeed, some species such as *P. verrucosum* are producing both mycotoxins concomitantly.

3. Plant Extracts Used for the Biocontrol of Mycotoxin Contamination

During the last 25 years, many *in vitro* and *in vivo* investigations have been carried out for the development of new plant-based formulations for their use in agriculture pest management programs. One of the main pros of botanical products resides in their biodegradability which make them good candidates for eco-chemical products and for their use as a bio-rational approach in integrated plant protection programs (Prakash et al. 2015). Into the higher plant reign, a large set of secondary metabolites (e.g. phenolic compounds, tannins, alkaloids etc.) are produced to protect themselves against various kind of aggressions (mechanical, biological or climatic). Many studies have been conducted in the last few years and essential oils, aqueous or alcoholic plant extracts have progressively shown their efficiency against fungal growth and/or mycotoxins production (Ramirez-Mares and Hernandez-Carlos 2015, Moon et al. 2018, Abdullah et al. 2019, Chen et al. 2019).

These researches are summarized in Table 1 and the most recent are detailed below. *Aspergillus* and *Fusarium* are the fungal genera most commonly used to test essential oils and plant extracts, followed by *Penicillium* (Kumar et al. 2007). The production of aflatoxin B1 by *A. flavus* was inhibited using piperine, a main component of black and long peppers (*Piper nigrum* L. and *Piper longum* L.) by 30% at 0.0006 mM of piperine (lyophilized piperine standard, purity ≥97.0%) while no detectable levels of toxin were observed using 0.17 mM of piperine. At the same time, growth inhibition was only of 35% with 0.17 mM of piperine. Using 0.04 mM of piperine allowed an aflatoxin B1 inhibition of 95% with only a growth inhibition of 12% (Caceres et al. 2017). *Cinnamomum zeylanicum* Blume essential oil exhibited high toxicity against all the tested fungi (*Aspergillus flavus, A. niger, A. candidus, A. sydowi, A. fumigatus, Cladosporium cladosporoides, Curvularialunata, Alternaria alternata, Penicillium* sp. and *Mucor* sp.) including an high toxigenic species of *A. flavus* and its minimum inhibitory concentration (MIC) ranged between 0.25 and 6.0 µL/mL culture medium. In addition, *Cinnamomum zeylanicum* Blume essential oil completely inhibited at 0.3 µL/mL the production of aflatoxin B1 (Kiran et al. 2016). Matusinsky et al. (2015) examined essential oils extracted from *Pimpinella anisum, Thymus vulgaris, Pelargonium odoratissimum, Rosmarinus officinalis* and *Foeniculum vulgare* against the fungi *Oculimaculay allundae, Microdochium nivale, Zymoseptoria tritici, Pyrenophora teres* and *Fusarium culmorum*. These essential oils affected the growth of the fungal isolates differently. Indeed, essential oils of *T. vulgaris* and *P. odoratissimum* inhibited the growth of targeted fungal species. *P. odoratissimum* essential oil inhibited the growth of *O. yallundae, P.teres* and *Z. tritici*at 1 µL/mL while *F. culmorum* and *M. nivale* growths were inhibited at 5 µL/mL. Sharma et al. (2017) reported that essential oils obtained from clove (*Syzygium aromaticum*), lemongrass (*Cymbopogon citratus*), mint (*Mentha piperita*) and eucalyptus (*Eucalyptus globulus*) were effective against the wilt causing fungus *Fusarium oxysporum* f.sp. *lycopersici* 1322. The inhibitory effect of oils showed a dose-dependent activity on the tested fungus. Complete inhibition of mycelial growth and spore germination was observed using clove oil at 125 ppm with IC50 value of 18.2 and 0.3 ppm, respectively. Phytochemicals (methyl-trans-p-coumarate, methyl caffeate, syringic acid and ursolic acid) at different concentrations (500, 750 and 1000 ppm) were tested against *Alternaria alternata, Curvularia lunata, Fusarium moniliforme, F. pallidoroseum* and *Helmintho sporium* sp. All tested phytochemicals affected the growth inhibition of all fungi except *A. alternata*, which was found to be resistant to syringic acid (Shaik et al. 2016). Pane et al. (2016) showed an

Table 1. Efficacy of plant extracts against fungal growth and mycotoxin production

Plants	Target	Remarks	References
Myrcia splendens (dichloromethane extract of mature leaves)	*Alternaria alternata*	Inhibitory effect (10.2%) on the mycelial growth at 1 mg/mL.	Pontes et al. (2019)
Micromeria graeca (aqueous extract)	*Aspergillus flavus*	Inhibitory effect on the production of aflatoxin B1 for *Aspergillus flavus* strain E28 (77.7%), strain E73 (70.8%) and strain NRRL 62477 (99.2%) at the dose of 10 mg/mL. At 10 mg/mL, aflatoxin inhibition was accompanied by a mild increase of the colony diameter.	El Khoury et al. (2017)
Curcuma longa L. (turmeric essential oil)	*Aspergillus flavus*	The essential oil decreased the fungal mycelia growth (93.4%) at the concentration of 8 µL/mL The aflatoxinB1 inhibition rate of 8 µL/mL turmeric essential oil was 78.4%.	Hu et al. (2017)
Thymus vulgaris L. (essential oil) *Mentha x piperita* L. (essential oil) *Rosmarinus officinalis* L. (essential oil) *Lavandula angustifolia* Mill. (essential oil)	*Fusarium verticillioides*	100% inhibition of growth at the concentration of 250, 500 and 1000 µL/L after 12 days of incubation. 34.1% inhibition of growth at the concentration of 1000 µL/L after 12 days of incubation. 8.2% inhibition of growth at the concentration of 1000 µL/L after 12 days of incubation. 21.8% inhibition of growth at the concentration of 1000 µL/L after 12 days of incubation.	Vilaplana et al. (2018)
Bay leaves, cumin, fenugreek, melissa, mint and sage (essential oils and phenolic compounds) Anise, chamomile, fennel, rosemary and thyme (phenolic compounds)	*Aspergillus carbonarius*	Inhibitory effect on the production of ochratoxin A between 25% (sage) and 80% (melissa) for all essential oils (5 µL/mL) and between 13% (thyme) and 69% (mint) for phenolic compounds (250 µg GAE/mL). Essential oils had a greater impact than the phenolic extracts on the ochratoxin A production.	El Khoury et al. (2016)

Chamaemelum nobile *Cuminum cyminum* *Zingiber officinale* *Ziziphora clinopodioides*, *Thymus vulgaris* (essential oils)	*A. niger* *A. fumigatus* *A. flavus* *A. ochraceus* *P. citrinum* *P. chrysogenum*	Strong activity (expressed in MIC - minimum inhibitory concentrations - mean values) against tested fungal strains for the oil of *T. vulgaris* (1250 µg/mL), followed by *Cu. Cyminum* (1416 µg /mL), *Zin. officinale* (1833 µg/mL), *Ziz. clinopodioides* (2166 µg/mL) and *Ch. nobile* (3750 µg/mL).	Sharifzadeh et al. (2016)
Rosmarinus officinal (essential oil)	*Aspergillus niger*	From the concentration of 0.50%, a complete absence of growth was observed for the 16 tested isolates.	Baghloul et al. (2016)
Cinnamomum zeylanicum Blume (essential oil)	*A. flavus* *A. niger*	The CZ essential oil caused complete inhibition of all test molds and aflatoxin B1 secretion at 0.25 to 0.6 µg/mL and 0.3 µg/mL respectively.	Kiran et al. (2016)
Parastrephia quadrangularis Extract A: Aerial parts with capitulum in ripe fruits. Extract B: Aerial parts with capitulum in flowering	*Fusarium verticillioides*	*P. quadrangularis* extract A had a significant activity on the growth of *F. verticillioides* M7075 with an MIC of 118.74 µg/mL whereas MIC of extract B was 250 µg/mL.	Abdullah et al. (2017)
Residues from essential oil industry: *Hyssopus officinalis* L., *Thymus mastichina* L., *Lavandula* × *intermedia* var. Super and *Salvia lavandulifolia* Vahl. (ethanolic extracts)	*Penicillium verrucosum*	The most active extracts were those from the solid residues of *Hyssopus officinalis* with 50% inhibition of fungal growth for 92 mg/mL (EC_{50} – median effective concentration)	De Elguea-Culebras et al. (2016)
Piper betle L. var. Tamluk Mitha (essential oil)	*Penicillium expansum*	Inhibition zones of 42.8±1 mm and 53.8±2 mm were observed for 10 and 30 µL of essential oil, respectively, using disc diffusion method and of 28.7±1 mm and 38.5±1 mm for 10 and 30 µL of essential oil, respectively, using disc volatilization method.	Basak and Guha (2015)

(*Contd.*)

Table 1. *(Contd.)*

Plants	*Target*	*Remarks*	*References*
Thymus vulgaris L. (essential oil)	*Aspergillus flavus*	MIC and MFC (minimum fungicidal concentration) were found at 250 µg/mL. Inhibition of the production of both B1 and B2 aflatoxins was observed at 150 µg/mL.	Kohiyama et al. (2015)
Rosmarinus officinalis L. (essential oil)	*Fusarium verticillioides*	The mycelial growth of *Fusarium verticillioides* (Sacc.) Nirenberg was reduced significantly from 150 µg/mL and the production of fumonisins B1 and B2 from the concentration of 300 µg/mL.	Da Silva Bomfim et al. (2015)
Lippia scaberrima, Lippia rehmannii Mentha spicata, Helichrysum splendidum, Cymbopogon citratus, Thymus vulgaris, Syzygium aromaticum, Cinnamon zeylanicum and Cinnamomum camphora (essential oils) R-(−)-carvone, S-(+)-carvone, eugenol, limonene, neral, citral and thymol) (pure compounds)	*Fusarium oxysporum*	Inhibition of growth of *F. oxysporum* S-1187 was obtained from 500 µL/L with *Thymus vulgaris, Syzygiumaro maticum* essential oil, as well as pure citral, eugenol and thymol.	Manganyi et al. (2015)
Punicagranatum L. (peel methanol extract)	*F. sambucinum*	MIC and MFC were found to be 20 and 120 mg/mL, respectively. 75.5% inhibition on mycelial growth and complete inhibition on spore germination was observed at the concentration of 20 mg/mL.	Elsherbiny et al. (2016)
Punica granatum L. (peel concentrated aqueous extract)	*Penicillium digitatum* *Penicillium expansum* *Botrytis cinerea*	Inhibition of conidia germination was significantly inhibited by 100, 91.0 and 82.7% for *B. cinerea, P. digitatum* and *P. expansum*, respectively, using 12 g/L extract. Viability of conidia was 0%, 20% and 42.3% for *B. cinerea, P. digitatum* and *P. expansum*, respectively, using 12 g/L extract.	Nicosia et al. (2016)

Parastrephia lepidophylla (ethanolic extract)	*Penicillium digitatum*	MIC and MFC were found to be 150 mg of GAE/mL and 350 mg GAE/mL. Conidia germination was inhibited from 200 mg GAE/mL.	Ruiz et al. (2016)
Basil, cinnamon, eucalyptus, mandarin, oregano, peppermint, tea tree and thyme (essential oils)	*Aspergillus niger* *Aspergillus flavus* *Aspergillus parasiticus* *Penicillium chrysogenum.*	Oregano, thyme and cinnamon were the best effective essential oil considering MIC. Combination of oregano and thyme showed in a synergistic effect against *A. flavus*, *A. parasiticus* and *P. chrysogenum* while association of peppermint and tea tree resulted in a synergistic effect against *A. niger*	Hossain et al. (2016)
Curcuma longa (ethanol extract) and its major compounds (curdione, isocurcumenol, curcumenol, curzerene, β-elemene, curcumin, germacrone and curcumol)	*Fusarium graminearum*	*Curcuma longa* had a strong inhibitory effect on *Fusarium graminarium.* Eight chemical constituents of *Curcuma longa* had inhibitory effects on the mycelia growth of *F. graminearum*. Among the eight chemicals, curdione had the best inhibitory effect on the growth of *F. graminearum* with an inhibitory rate of 52.9%.	Chen et al. (2018)
Borreria verticillata, *Lactuca capensis*, *Colocasiae sculenta*, *Jatropha gossy* and *Ageratum conyzoide* extracts (ethanolic, hot and cold-water extracts)	*Aspergillus flavus*, *A. parasiticus* and *A. ochraceus*	The extracts of *Borreria verticillata* inhibited the growth of the tested fungal strains. *Borreria verticillata* appeared to be more effective as an antifungal agent than the other plant extracts. The ethanolic extracts were more effective than the cold and hot water extracts.	Jeff-Agboola and Onifade (2016)

dose-dependent inhibition of the mycelial growth and conidia germination of *Alternaria alternate* using crude foliar extracts of a wild *Capsicum annuum* accession. Furthermore, a reduced soft rot disease severity was observed on artificially infected ripe cherry tomatoes treated with these extracts.The methanol extract of pomegranate peels at the concentration of 20 mg/mL was tested against *Fusarium sambucinum*. The results showed that mycelial growth of *F. sambucinum* was inhibited at 75.5%, while spore germination was completely inhibited. The extract exhibited a minimum inhibitory concentration (MIC) of 20 mg/mL and a minimum fungicidal concentration (MFC) of 120 mg/mL (Elsherbiny et al. 2016). Chen et al. (2018) observed that *Curcuma longa* alcohol extract had a strong inhibitory effect on various pathogenic fungi including *Fusarium graminearum, Fusarium chlamydosporum, Alternaria alternata, Fusarium tricinctum, Sclerotinia sclerotiorum, Botrytis cinerea, Fusarium culmorum, Rhizopusoryzae, Cladosporium cladosporioides, Fusarium oxysporum* and *Colletotrichum higginsianum*. Another study by Jeff-Agboola and Onifade (2016) showed that the extracts of *Borreria verticillata, Lactuca capenisi, Colocasiaes culenta* and *Jatropha gossy* exhibited antifungal activity against *A. flavus* and *A. parasiticus*, while only the extracts of *Ageratum conyzoide* had inhibitory effect on *A. ochraceus*. An aqueous extract of the medicinal plant *Micromeria graeca* was shown to completely inhibit the aflatoxin production by *Aspergillus flavus* without reducing fungal development. The essential oils of bay leaves, cumin, fenugreek, melissa, mint, and sage showed a great impact on the ochratoxin A production of *Aspergillus carbonarius* on synthetic grape medium at 28°C for four days. Reduction levels ranged between 25% (sage) and 80% (melissa) at 5 µL/mL. Phenolic compounds extracted from bay leaves, cumin, fenugreek, melissa, mint, sage, anise, chamomile, fennel, rosemary and thyme reduced ochratoxin A production and reduction levels ranged between 13% (thyme) and 69% (mint). Although they did not affect the growth of *A. carbonarius* (El Khoury et al. 2017a).

4. Mechanisms of Action Involved in the Inhibition of Fungal Growth and/or Mycotoxin Contamination by Plant Extracts

The action of plant extracts against mycotoxin contamination of food products involves three main mechanisms (anti-fungal and/or anti-mycotoxin synthesis and/or detoxifying activities) that can happen alone or in combination and involve a reduction of mycotoxin levels (Fig. 1).

4.1 Antifungal Activity

The mechanism of action of the plant-based extracts is described less. Their composition rich in bioactive molecules allow for the plant extracts to act simultaneously on several cellular targets. Some of them have been identified: cell membrane, cell wall, mitochondria and efflux pump (Nazzaro et al. 2017).

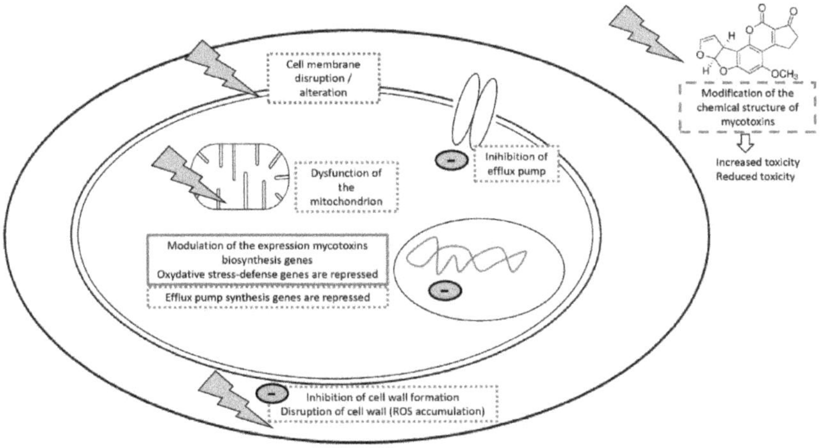

Figure 1. Schematic representation of mechanisms involved in the reduction of mycotoxin contamination (dotted line: fungicide/fungistatic actions, solid line: anti-mycotoxin synthesis activity, dash line: detoxifying activity)

4.1.1 Cell Membrane Disruption, Alteration, and Inhibition of Cell Wall Formation

Plants extracts like essential oils are mixtures of molecules characterized by their poor solubility in water and high hydrophobicity (Jing et al. 2014). This hydrophobicity facilitates their incorporation into plasma membranes and those of intracellular organelles (especially mitochondria). These compounds can then act by altering the lipid membrane composition, like decreasing the ergosterol levels (Kedia et al. 2015). Ergosterol is the main constituent of the fungal cell membrane which is specific and important for fungi and its presence is very important for maintaining the cell function and integrity (Hu et al. 2017). Reduction or absence of ergosterol in fungal membranes leads to a modification in the cell permeability and disruption in cell organelles (Kiran et al. 2016). *Cinnamomum zeylanicum* Blume essential oil in the concentration of 0.5 µg/mL was capable to reduce with a percentage of 100% the ergosterol in the plasma membrane of *Aspergillus flavus*. Such a reduction resulted in a significant leakage of vital cellular ions (Ca^{+2}, K^{+} and Mg^{+2}) in the *Cinnamomum zeylanicum* Blume essential oil fumigated hyphae compared to control. A significant deformation in conidiophores and conspicuous depressions in conidia of *Cinnamomum zeylanicum* Blume essential oil fumigated hyphae compared to control were also observed (Kiran et al. 2016). Essential oil derived from turmeric (*Curcuma longa* L.) was shown to significantly reduce the level of ergosterol of *Aspergillus flavus* (Hu et al. 2017). Inhibition of ergosterol biosynthesis was also detected at a concentration of 100 µg/mL of essential oil of *Thymus vulgaris* L. in *Aspergillus flavus* by Kohiyama et al. (2015). In a research work performed by Da Silva Bomfim et al. (2015), essential oil of *Rosmarinus officinalis* L. at the concentration of 600 µg/mL was capable to inhibit ergosterol production in *Fusarium verticillioides*. Chen et al. (2018) demonstrated that the utilisation of the extract of *Curcuma longa*, curdione and curcumenol, decreased the ergosterol content in *Fusarium graminearum,* resulting in the inhibition of the growth of the fungal strain. This result is further proof that ergosterol is one of the most important compounds to ensure essential cellular functions, e.g. the integrity of the membrane structure, membrane binding enzyme activity, cell viability and cellular transport. The absence of ergosterol or its decrease cause unnatural function of the fungal cell membrane or yet cell break. Moreover, treatments of hyphae with extracts of *Parastrephia quadrangularis* induced distortion in permeability of plasma membrane in *Fusarium verticillioides* (Abdullah et al. 2017). The fungal cell wall of fungi is mainly composed of different polysaccharides according to taxonomic groups (chitin, glucans, chitosan, chitin, cellulose), with smaller quantities of proteins and glycoproteins (Lagrouh et al. 2017). Chitin and glucans have been used as potential targets to search for new antifungals and many natural products have been identified as inhibitors through stimulating plant defence mechanisms against fungal pathogens which cause destruction of the fungal cell wall (Vicente et al. 2003). Thus, thyme essential oils were capable to induce the activities of chitinase, β-1,3-glucanase in avocado fruit which lead to the degradation of the cell wall of the fungal pathogens (Sivakumar and Bautista-Baños 2014). Anethole, a major component of anise oil, causes a change on the hyphal morphology such as swollen hyphae at the tips and inhibits chitin synthase activity in permeabilized hyphae of *Mucormucedo* IFO 7684 (Yutani et al. 2011). Methanol extract of pomegranate peels inhibited the growth of *Fusarium sambucinum in vitro*. This inhibition was correlated with some morphological modifications in *F. sambucinum* hyphae including curling, twisting and collapse. Alterations of the fungal ultrastructure were also observed including cell empty cavity and the disintegration of cytoplasmic organelles (Elsherbiny et al. 2016).

4.1.2 Dysfunction of the Fungal Mitochondria

Mitochondria are organelles, which are a central part needed to maintain the viability of eukaryotic cells due to their enzyme activities. Mitochondrial enzymes (lactate dehydrogenase, malate dehydrogenase, succinate dehydrogenase, citrate synthetase, isocitrate dehydrogenase, α-ketoglutarate dehydrogenase) are key enzymes in respiratory chains of the cells which produce ATP (Siahmoshteh et al. 2018). Some plant extracts can inhibit the action of these enzymes. Hu et al. (2017) have shown that turmeric essential oils could inhibit the activities of mitochondrial ATPase and dehydrogenases. The ATPase inhibitory rate at the concentration of 8 μL/mL of turmeric essential oil was 74.6% in *Aspergillus flavus.* The turmeric essential oil also exhibited inhibition on both succinate dehydrogenase (SDH) and malate dehydrogenase (MDH) activities, which are key enzymes in tricarboxylic acid (TCA) cycle, which is greatly important for energy metabolism of *Aspergillu sflavus.* The citral is a naturally occurring isoprenoid with two isomers (geranial and neral). This compound was found to cause the loss of matrix and increase of irregular mitochondria resulted in the change of the mitochondrial morphology and function of *Penicillium digitatum* (Zheng et al. 2015). The deformation of the mitochondriawas correlated with a decrease in intracellular ATP content and an increase in extracellular ATP content in *Penicillium digitatum* cells. This result confirmed that citral had a detrimental effect on mitochondrial membrane permeability. Citral was also found to decrease the activities of citrate synthetase, isocitrate dehydrogenase, α-ketoglutarate dehydrogenase, and succinate dehydrogenase and increase citric acid content (Zheng et al. 2015). Moreover, inhibitory effect in the tricarboxylic acid cycle (TCA) pathway of *Penicillium digitatum* cells was observed. These results showed that in addition to damaging the permeability of the mitochondrial membrane, citral could disrupt the TCA cycle of *Penicillium digitatum.*

4.1.3 Inhibition of Efflux Pumps

The fungal plasma membrane (PM) H^+-ATPase enzyme plays an important role in eukaryotic cell physiology in maintaining the transmembrane electrochemical proton gradient necessary for nutrient uptake (Kongstad et al. 2015). The fungal PM H^+-ATPase enzyme also allows the export of the toxic substances out of the cell (Lagrouh et al. 2017) and its inhibition could cause the fungal cell death. *Lippiarugosa* essential oil at the concentration of 2000 mg/L completely inhibited the activity of the PM H^+-ATPase in *Aspergillus flavus.* This inhibition was accompanied by an acidification of the external medium (Tatsadjieu et al. 2009).

4.1.4 Reactive Oxygen Species (ROS) Production

Reactive oxygen species (ROS) are natural by-products of cellular metabolism originating from NADPH oxidase located in the plasma membrane. They are also one of the major factors inducing the lipid peroxidation of eukaryotic cell membrane under various environmental stimuli (Gao et al. 2016). Accumulation of intracellular ROS results in oxidative damages affecting important biomolecules such as proteins, DNA and lipids, and can be associated with cell death (Almshawit and Macreadie 2017). Among ROS, NO (nitrous oxide) is a highly active molecule recognized as an intra and inter cellular signalling molecule and its role in fungi remain unclear in literature. The efficiency of dill oil, curcuma longa oil and thymol have been related to an accumulation of ROS in the cytoplasm of fungal cells (OuYang et al. 2018). Shen et al. (2016) showed that thymol provoke the lysis of *Aspergillus flavus* spore via the induction of nitric oxide. OuYang et al. (2018) also demonstrated that citral increased the accumulation of ROS in *Penicillium digitatum.* As well, *Rosmarinus officinalis* L. essential oil inhibits fumonisin production by *Fusarium verticillioides* as a consequence of cell wall perturbation by ROS leading to the loss of cellular components (Da Silva Bomfim et al. 2015).

4.2 Mycotoxin Synthesis Inhibition Using Plants Extracts

The frequency of contamination of global crops by mycotoxins shows that the strategies currently used are insufficient to ensure the safety of food and that it is necessary to develop others, in addition to, or replacing existing ones. In this context, strategies based on the use of natural compounds, generally recognized as not harmful to the environment and to health, seem interesting. Caceres et al. (2017) demonstrated that piperine inhibited aflatoxin production correlated with an enhancement of antioxidant status in *A. flavus*. This inhibition was relying on the dose of piperine. They showed that aflatoxin reduction was correlated to the down regulation of genes belonging to the aflatoxin's biosynthetic pathway and of gene encoding veA, a global regulator. Piperine also causes an over-expression of genes coding for several basic leucine zipper (bZIP) transcription factors such as *atfA, atfB* and *ap-1* and genes encoding enzymes belonging to superoxide dismutase and catalase's families. Hu et al. (2017) suggested that the down-regulation of the transcription level of aflatoxin biosynthetic genes, i.e. *aflM, aflO, aflP,* and *aflQ* may be responsible of the reduction in aflatoxin B1 production by *Aspergillus flavus* exposed to turmeric essential oil. El Khoury et al. (2016) showed that essential oils (fennel, cardamom, anise, chamomile, celery, cinnamon, thyme, taramira, oregano and rosemary) used principally to inhibit the production of ochratoxin A have an effect on the expression levels of the genes responsible for the ochratoxin biosynthesis in *Aspergillus carbonarius* (*acOTApks* and *acOTAnrps* along with the *acpks* gene) and the two regulatory genes *laeA* and *veA*. According to the results obtained by El Khoury et al. (2017b), the use of aqueous extract of *Micromeria graeca* (Hyssop) inhibited aflatoxin B1 synthesis. This inhibition was synchronized with 3-fold down-regulation of *aflR* and *aflS* genes, the two internal cluster co-activators, causing a radical repression of all aflatoxin biosynthesis genes. In addition, fifteen regulating genes including *veA* and *mtfA* and two major global-regulating transcription factors was impacted by the extract. The effect of Hyssop was also related to an alteration of the expression of a number of oxidative stress-defense genes such as *msnA, srrA, catA, cat2, sod1, mnsod* and *stuA*.

4.3 Mycotoxin Detoxifying Using Plants Extracts

Mycotoxins are resistant to food processing and the preservation processes because they are chemically stable and heat resistant. Various physical, chemical and biological methods have been described for mycotoxin detoxification in food (Vijayanandraj et al. 2014). In any case, each treatment suffers from some limitations. Physical methods like pressure-cooking cause the destruction of certain nutrients. Chemical methods can be a risk for the food safety due to chemical residues. The use of microorganisms, such as a biological detoxification method, is frequently used, but microorganisms use the food nutrients for their own growth and thus produce several undesirable compounds (Iram et al. 2015). Natural plants extracts can be an alternative to synthetic chemicals for detoxification of mycotoxins. Velazhahan et al. (2010) demonstrated that the use of the seed extract of Ajowan (*Trachyspermum ammi* (L.) Sprague ex Turrill) provoked the degradation of aflatoxin G1. The dialyzed *T. ammi* extract was found more effective than the crude extract, it was capable of degrading >90% of the aflatoxin G1. Another study carried out by Vijayanandraj et al. (2014) proved the effectiveness of the leaf extract of Vasaka (*Adhatodavasica* Nees) to degrade the aflatoxin B1 (≥98%) after incubation for 24 h at 37 °C.

Corymbia citriodora aqueous extracts were also tested for their detoxification potential against aflatoxin B1 and B2 (aflatoxin B1 100 µg/L and aflatoxin B2 50 µg/L). Results indicated that *C. citriodora* leaf extract was more effective to degrade aflatoxin B1 and aflatoxin B2 (95.21% and 92.95%, respectively) than *C. citriodora* branch extract, under optimized conditions (Iram et al. 2015). Aqueous extract of seeds and leaves of *Trachyspermum ammi* were also evaluated for their ability to detoxify aflatoxin B1 and B2 (aflatoxin B1: 100 µg/L and aflatoxin B2: 50 µg/ L) by *in vitro* and *in vivo* assays. Results showed that *T. ammi* seeds extract was found to be significantly effective in degrading aflatoxin B1 and aflatoxin B2 (92.8 and 91.9% respectively) (Iram et al. 2016). The treatment with *Ocimumtenui florum* extract at 85°C for 4 h led to 74.7% degradation of aflatoxin B1 (Panda and Mehta 2013). Perczak et al. (2016) studied the ability of essential oils to reduce zearalenone toxin level. They showed that the type of essential oil influenced the effectiveness of zearalenon toxin level reduction. Indeed, lemon peel *(Citrus limonum,* Italy*)*, grapefruit peel (*Citrus paradisi,* Argentina*)*, eucalyptus leaf oil *(Eucalyptus radiata,* China), and palmarosa leaf oil *(Cymbopogon martinii)* were the most effective, while lavender, thyme flower and leaf (*Thymus vulgaris,* Spain), and rosemary flower and leaf *(Rosmarinus officinalis,* Spain*)* did not degrade zearalenone.

5. Essential Oils Safety

Essential oils are complex mixtures of many chemical compounds among which terpenes and their oxygenated compounds are the principal components. They are usually used as aroma additives in food, pharmaceuticals and cosmetics (Von Fraunhofer and Joshi 2019). Since they are molecules of natural origin, biodegradable, they are therefore considered as a possible alternative to fungicide (El Ouadi et al. 2017). In addition to these characteristics, essential oils have proven to be effective in the treatment of many incurable diseases in humans like cancer (Hanif et al. 2019). Most essential oils and their main compounds are considered non-mutagenic or genotoxic. The negative effects generally occur at high doses and most essential oils have low acute toxicity, more than 2 g/kg for both oral and dermal application (Pavela and Benelli 2016). However, that does not mean that the use of essential oils is completely safe. Indeed, numerous studies have shown that the use of essential oils could have harmful effects on human health, such as asthma exacerbation and the decrease of pulmonary function (Levy et al. 2019). Moreover, some volatile compounds of essential oils can be classified as potentially hazardous (Nematollahi et al. 2018). That does not mean that essential oils have to be avoided but they must be used with caution and according to regulations. Essential oil safety and its regulation is of world concern (Es et al. 2017). World Health Organisation (WHO), Food and Agriculture Organisation of the United Nations (FAO), International Organisation for Standardisation (ISO), Food and Drug Administration (FDA) as well as EU Commission or American Essential Oil Trade Association (AEOTA) contribute to ensure the control of safe and legal use of essential oils. Their use as food additives or flavourings has to be authorized by food safety authorities. For this purpose, many studies are carried out on essential oils in order to set their safety limits through evaluation of DL50 on mice or rat (Prakash et al. 2015). European Food Safety Authority (EFSA) assesses risks before allowing their use in food (Barbieri and Borsotto 2018). EU legislation regulates essential oils through the general regulations REACH (Registration, Evaluation, Authorization and Restriction of Chemicals) and CLP (Classification, Labelling and Packaging of substances and mixtures). In USA, essential oils have to be recognized as GRAS (Generally Recognized As Safe) and approved by Food and Drug Administration (FDA) to be used as food additives. Although they showed low toxicity on non-target vertebrates, their use in crops as biopesticides remains low and includes regulatory requirements which can vary considerably from one country to another (Prakash et al., 2015, Isman 2016, Pavela and Benelli 2016, Dosoky and Setzer 2018, Isman 2020).

6. Conclusion

The use of plant extracts as antifungal and anti-mycotoxin agents seems to be promising and safe for human and environmental health compared to chemicals, as long as the regulatory doses established by the health authorities are respected. Many studies have shown that plant extracts inhibited mycotoxinogenesis at lower concentrations than those reducing fungal growth. Thus, their use as food bio-preservation agents for mycotoxin control in the agri-food chain is attractive because at these low concentrations, they have less impact on the organoleptic quality of foodstuffs. However, studies need to be conducted to assess consumer acceptability. Plant extracts are complex and contain many different compounds. So, data on their composition and their active molecules have still to be collected. Their absence of toxicity must be checked as well as the possible synergistic effects between the plant extracts and/or their constituents. Concerning the degradation of mycotoxins by plant extracts, chemical reactions are poorly known and studies to decipher them must continue as well as the *in vivo* evaluation of toxicity of the molecules resulting from mycotoxin degradation. The assessment of the impact of extracts on the organoleptic, technological and nutritional quality of foodstuffs must not be omitted.

References

Abdullah, E., Idris A. and Saparon, A. 2017. Antifungal activity of *Parastrephia quadrangularis* (Meyen) Cabrera extracts against *Fusarium verticillioides*. ARPN Journal of Engineering and Applied Sciences 12: 3218–3221. doi.org/10.1111/lam.12844

Abdullah, B.A., Salih, D.T. and Saadullah, A.A. 2019. Antifungal activity of some medicinal plant extracts against some fungal isolates. Revista Innovaciencia 7: 1–7. https://revistas.udes.edu.co/innovaciencia/article/view/671

Almshawit, H. and Macreadie, I. 2017. Fungicidal effect of thymoquinone involves generation of oxidative stress in *Candida glabrata*. Microbiological Research 195: 81–88. doi.org/10.1016/j.micres.2016.11.008

Arraché, E.R.S., Fontes, M.R.V., Buffon J.G. and Badiale-Furlong, E. 2018. Trichothecenes in wheat: Methodology, occurrence and human exposure risk. Journal of Cereal Science 82: 129–137. doi.org/10.1016/j.jcs.2018.05.015

Baghloul, F., Mansori, R. and Djahoudi, A. 2016. In vitro antifungal effect of *Rosmarinus officinalis* essential oil on *Aspergillus niger*. National Journal of Physiology, Pharmacy and Pharmacology 7: 285–289. doi.org/10.5455/njppp.2017.7.7021513102016

Barbieri, C. and Borsotto, P. 2018. Essential oils: Market and legislation. pp. 107–127. *In*: H. El-Shemy (ed.). Potential of Essential Oils. IntechOpen. www.intechopen.com/books/potential-of-essential-oils/essential-oils-market-and-legislation

Basak, S. and Guha, P. 2015. Modelling the effect of essential oil of betel leaf (*Piper betle* L.) on germination, growth, and apparent lag time of *Penicillium expansum* on semi-synthetic media. International Journal of Food Microbiology 215: 171–178. doi: 10.1016/j.ijfoodmicro.2015.09.019

Caceres, I., El Khoury, R., Bailly, S., Oswald, I.P., Puel, O. and Bailly, J.D. 2017. Piperine inhibits aflatoxin B1 production in *Aspergillus flavus* by modulating fungal oxidative stress response. Fungal Genetics and Biology 107: 77–85. doi: 10.1016/j.fgb.2017.08.005

Chen, C., Long, L., Zhang, F., Chen, Q., Chen, C., Yu, X., Liu, Q., Bao, J. and Long, Z. 2018. Antifungal activity, main active components and mechanism of *Curcuma longa* extract against *Fusarium graminearum*. PLoS One 13: 1–19. doi: 10.1371/journal.pone.0194284

Chen, J., Shen, Y., Chen, C. and Wan, C. 2019. Inhibition of key citrus postharvest fungal strains by plant extracts in vitro and in vivo: A review. Plants 8: 26–45. doi.org/10.3390/plants8020026

Cruz, J.S., Costa, G.L.D. and Villar, J.D.F. 2016. Biological activities of crude extracts from *Penicillium waksmanii* isolated from mosquitoes vectors of tropical diseases. Journal of Pharmaceutical Science 5: 90–102. www.arca.fiocruz.br/handle/icict/14418

Da Silva Bomfim, N., Nakassugi, L.P., Oliveira, J.F.P., Kohiyama, C.Y., Mossini, S.A.G., Grespan, R., Nerilo, S.B., Mallmann, C.A., Filho, B.A. and Machinski Jr, M. 2015. Antifungal activity and inhibition of fumonisin production by *Rosmarinus officinalis* L. essential oil in *Fusarium verticillioides* (Sacc.) Nirenberg. Food Chemistry 166: 330–336. doi: 10.1016/j.foodchem.2014.06.019

De Elguea-Culebras, G.O., Sánchez-Vioque, R., Santana-Méridas, O., Herraiz-Peñalver, D., Carmona, M. and Berruga, M.I. 2016. In vitro antifungal activity of residues from essential oil industry against *Penicillium verrucosum*, a common contaminant of ripening cheeses. LWT-Food Science and Technology 73: 226–232. doi: 10.1016/j.lwt.2016.06.008

Degen, G.H., Ali, N. and Gundert-Remy, U. 2018. Preliminary data on citrinin kinetics in humans and their use to estimate citrinin exposure based on biomarkers. Toxicology Letters 282: 43–48. doi: 10.1016/j.toxlet.2017.10.006

Diao, E., Hou, H., Hu, W., Dong, H. and Li, X. 2018. Removing and detoxifying methods of patulin: A review. Trends in Food Science and Technology 81: 139–145. doi: 10.1016/j.tifs.2018.09.016

Divband, K., Shokri, H. and Khosravi, A.R. 2017. Down-regulatory effect of *Thymus vulgaris* L. on growth and Tri4 gene expression in *Fusarium oxysporum* strains. Microbial Pathogenesis 104: 1-5. doi: 10.1016/j.micpath.2017.01.011

Dosoky, N.S. and Setzer, W.N. 2018. Biological activities and safety of *Citrus* spp. essential oils. International Journal of Molecular Sciences 19: 1966–1991. doi:10.3390/ijms19071966

El Khoury, R., Atoui, A., Verheecke, C., Maroun, R., El Khoury, A. and Mathieu, F. 2016. Essential oils modulate gene expression and ochratoxin A production in *Aspergillus carbonarius*. Toxins (Basel) 8: 242–256. doi: 10.3390/toxins8080242

El Khoury, R., Atoui, A., Mathieu, F., Kawtharani, H., El Khoury, A., Maroun, R.G. and El Khoury, A. 2017a. Antifungal and antiochratoxigenic activities of essential oils and total phenolic extracts: A comparative study. Antioxidants 6: 44. doi: 10.3390/antiox6030044

El Khoury, R., Caceres, I., Puel, O., Bailly, S., Atoui, A., Oswald, I.P., El Khoury, A. and Bailly, J.D. 2017b. Identification of the anti-aflatoxinogenic activity of *Micromeria graeca* and elucidation of its molecular mechanism in *Aspergillus flavus*. Toxins (Basel) 9: 87–102. doi: 10.3390/toxins9030087

El Ouadi, Y., Manssouri, M., Bouyanzer, A., Majidi, L., Bendaif, H., Elmsellem, H., Shariati, M.A., Melhaoui, A. and Hammouti, B. 2017. Essential oil composition and antifungal activity of *Melissa officinalis* originating from north-east Morocco, against postharvest phytopathogenic fungi in apples. Microbial Pathogenesis 107: 321–326. doi.org/10.1016/j.micpath.2017.04.004

Elsherbiny, E.A., Amin, B.H. and Baka, Z.A. 2016. Efficiency of pomegranate (*Punica granatum* L.) peels extract as a high potential natural tool towards *Fusarium* dry rot on potato tubers. Postharvest Biology and Technology 111: 256–263. doi: 10.1016/j.postharvbio.2015.09.019

Es, I., Khaneghah, A.M. and Akbariirad, H. 2017. Global regulation of essential oils. pp. 327–338. *In*: S.M.B. Hashemi, A.M. Khaneghah and A.S. Sant'Ana (eds.). Essential Oils in Food Processing: Chemistry, Safety and Applications. Wiley & Sons. doi: 10.1002/9781119149392.ch11

European Commission. 2006. Commission Regulation (EC) No 1881/2006 of 19 December 2006 setting maximum levels for certain contaminants in foodstuffs. Official Journal of the European Union L364:5–24.data.europa.eu/eli/reg/2006/1881/oj

Gao, T., Zhou, H., Zhou, W., Hu, L., Chen, J. and Shi, Z. 2016. The fungicidal activity of thymol against *Fusarium graminearum* via inducing lipid peroxidation and disrupting ergosterol biosynthesis. Molecules 21: 1–13. doi: 10.3390/molecules21060770

Hanif, M.A., Nisar, S., Khan, G.S., Mushtaq, Z. and Zubair, M. 2019. Essential oils. pp. 3–17. *In*: S. Malik (ed.). Essential Oil Research. Springer, Cham. https://doi.org/10.1007/978-3-030-16546-8_1

Hossain, F., Follett, P., Vu, K.D., Harich, M., Salmieri, S. and Lacroix, M. 2016. Evidence for synergistic activity of plant-derived essential oils against fungal pathogens of food. Food Microbiology 53: 24–30. doi: 10.1016/j.fm.2015.08.006

Hu, Y., Zhang, J., Kong, W., Zhao, G. and Yang, M. 2017. Mechanisms of antifungal and anti-aflatoxigenic properties of essential oil derived from turmeric (*Curcuma longa* L.) on *Aspergillus flavus*. Food Chemistry 220: 1–8. doi: 10.1016/j.foodchem.2016.09.179

Ibrahim, F., Asghar, M.A., Iqbal, J., Ahmed, A. and Khan, A.B. 2017. Inhibitory effects of natural spices extracts on *Aspergillus* growth and aflatoxin synthesis. Australian Journal of Crop Science 11: 1553–1558. doi: 10.21475/ajcs.17.11.12.pne709

Iram, W., Anjum, T., Iqbal, M., Ghaffar, A. and Abbas, M. 2015. Mass spectrometric identification and toxicity assessment of degraded products of aflatoxin B1 and B2 by *Corymbia citriodora* aqueous extracts. Scientific Reports 5: 14672–14698. doi: 10.1038/srep14672

Iram, W., Anjum, T., Iqbal, M., Ghaffar, A. and Abbas, M. 2016. Structural elucidation and toxicity assessment of degraded products of aflatoxin B1 and B2 by aqueous extracts of *Trachyspermum ammi*. Frontiers in Microbiology 7: 346. doi: 10.3389/fmicb.2016.00346

Isman, M.B. 2016. Pesticides based on plant essential oils: Phytochemical and practical considerations. pp. 13–26. *In*: V.D. Jeliazkov and C.L. Cantrell (eds.). Medicinal and Aromatic Crops: Production, Phytochemistry, and Utilization. American Chemical Society. doi: 10.1021/bk-2016-1218.ch002

Isman, M.B. 2020. Botanical insecticides in the twenty-first century – fulfilling their promise? Annual Review of Entomology 65: 233–249. doi.org/10.1146/annurev-ento-011019-025010

Jeff-Agboola, Y. and Onifade, A. 2016. In vitro efficacies of some Nigerian medicinal plant extracts against toxigenic *Aspergillus flavus*, A. parasiticus and A. ochraceus. British Journal Pharmaceutical Research 9: 169. doi: 10.9734/BJPR/2016/20390

Jing, L., Lei, Z., Li, L., Xie, R., Xi, W., Guan, Y., Sumnar, L.Z. and Zhou, Z. 2014. Antifungal activity of citrus essential oils. Journal of Agricultural and Food Chemistry 62: 301163033. doi: 10.1021/jf5006148

Kedia, A., Jha, D.K. and Dubey, N.K. 2015. Plant essential oils as natural fungicides against stored product fungi. pp. 208–214. *In*: A. Mendes-Vilas (ed.). Battle Against Microbial Pathogens; Basic Science Technological Advances and Eductional Programs. Badajoz: Formatex Research Center. api.semanticscholar.org/CorpusID:19960256

Kiran, S., Kujur, A. and Prakash, B. 2016. Assessment of preservative potential of *Cinnamomum zeylanicum* Blume essential oil against food-borne molds, aflatoxin B1 synthesis, its functional properties and mode of action. Innovative Food Science and Emerging Technologies 37: 184–191. doi: 10.1016/j.ifset.2016.08.018

Kohiyama, C.Y., Ribeiro, M.M.Y., Mossini, S.A.G., Bando, E., da Silva Bomfim, N., Nerilo, S.B., Rocha, G.H.O., Renta, G., Mikcha, J.M.G. and Machinski Jr, M. 2015. Antifungal properties and inhibitory effects upon aflatoxin production of *Thymus vulgaris* L. by *Aspergillus flavus* Link. Food Chemistry 173: 1006–1010. doi: 10.1016/j.foodchem.2014.10.135

Kongstad, K.T., Wubshet, S.G., Kjellerup, L., Winther, A.M.L. and Staerk, D. 2015. Fungal plasma membrane H^+-ATPase inhibitory activity of o-hydroxybenzylated flavanones and chalcones from *Uvaria chamae* P. Beauv. Fitoterapia 105: 102–106. doi: 10.1016/j.fitote.2015.06.013

Kumar, R., Kumar, A., Dubey, N.K. and Tripathi, Y.B. 2007. Evaluation of *Chenopodium ambrosioides* oil as a potential source of antifungal, antiaflatoxigenic and antioxidant activity. International Journal of Food Microbiology 115: 159–164. doi: 10.1016/j.ijfoodmicro.2006.10.017

Lagrouh, F., Dakka, N. and Bakri, Y. 2017. The antifungal activity of Moroccan plants and the mechanism of action of secondary metabolites from plants. Journalde Mycologie Médicale 27: 303–311. doi: 10.1016/j.mycmed.2017.04.008

Levy, J., Neukirch, C., Larfi, I., Demoly, P. and Thabut, G. 2019. Tolerance to exposure to essential oils exposure in patients with allergic asthma. Journal of Asthma 56: 853–860. doi.org/10.1080/02770903.2018.1493601

Li, B., Chen, Y., Zong, Y., Shang, Y., Zhang, Z., Xu, X., Wang, X., Long, M. and Tian, S. 2019. Dissection of patulin biosynthesis, spatial control and regulation mechanism in *Penicillium expansum*. Environmental Microbiology 21: 1124–1139. doi: 10.1111/1462-2920.14542

Manganyi, M.C., Regnier, T. and Olivier, E.I. 2015. Antimicrobial activities of selected essential oils against *Fusarium oxysporum* isolates and their biofilms. South African Journal of Botany 99: 115–121. doi: 10.1016/j.sajb.2015.03.192

Mari, M., Bautista-Baños, S. and Sivakumar, D. 2016. Decay control in the postharvest system: Role of microbial and plant volatile organic compounds. Postharvest Biology and Technology 122: 70–81. doi: 10.1016/j.postharvbio.2016.04.014

Matusinsky, P., Zouhar, M., Pavela, R. and Novy, P. 2015. Antifungal effect of five essential oils against important pathogenic fungi of cereals. Industrial Crops and Products 67: 208–215. doi: 10.1016/j.indcrop.2015.01.022

Moon, Y., Lee, H. and Lee, S. 2018. Inhibitory effects of three monoterpenes from ginger essential oil on growth and aflatoxin production of *Aspergillus flavus* and their gene regulation in aflatoxin biosynthesis. Applied Biological Chemistry 61: 243–250. doi.org/10.1007/s13765-018-0352-x

Nazzaro, F., Fratianni, F., Coppola, R. and Feo, V.D. 2017. Essential oils and antifungal activity. Pharmaceuticals 10: 1–20. doi: 10.3390/ph10040086

Nematollahi, N., Kolev, S.D. and Steinemann, A. 2018. Volatile chemical emissions from essential oils. Air Quality, Atmosphere & Health 11: 949–954. doi.org/10.1007/s11869-018-0606-0

Nguyen, P.A., Strub, C., Fontana, A. and Schorr-Galindo, S. 2017. Crop molds and mycotoxins: Alternative management using biocontrol. Biological Control 104: 10–27. doi: 10.1016/j.biocontrol.2016.10.004

Nicosia, M.G.L.D., Pangallo, S., Raphael, G., Romeo, F.V., Strano, M.C., Rapisarda, P., Drobay, S. and Schena, L. 2016. Control of postharvest fungal rots on citrus fruit and sweet cherries using a pomegranate peel extract. Postharvest Biology and Technology 114: 54–61. doi: 10.1016/j.postharvbio.2015.11.012

Ostry, V., Malir, F., Toman, J. and Grosse, Y. 2017. Mycotoxins as human carcinogens – The International Agency for Research on Cancer Monographs classification. Mycotoxin Research 33: 65–73. doi.org/10.1007/s12550-016-0265-7

OuYang, Q., Tao, N. and Zhang, M. 2018. A damaged oxidative phosphorylation mechanism is involved in the antifungal activity of citral against *Penicillium digitatum*. Frontiers in Microbiology 9: 1–13. doi: 10.3389/fmicb.2018.00239

Panda, P. and Mehta, A. 2013. Aflatoxin detoxification potential of *Ocimum tenuiflorum*. Journal of Food Safety 33: 265–272. doi: 10.1111/jfs.12048

Pane, C., Fratianni, F., Parisi, M., Nazzaro, F. and Zaccardelli, M. 2016. Control of *Alternaria* post-harvest infections on cherry tomato fruits by wild pepper phenolic-rich extracts. Crop Protection 84: 81–87. doi: 10.1016/j.cropro.2016.02.015

Pavela, R. and Benelli, G. 2016. Essential oils as ecofriendly biopesticides? Challenges and constraints. Trends in Plant Science 21: 1000–1007. https://doi.org/10.1016/j.tplants.2016.10.005

Perczak, A., Juś, K., Marchwińska, K., Gwiazdowska, D., Waśkiewicz, A. and Goliński, P. 2016. Degradation of zearalenone by essential oils under in vitro conditions. Frontiers in Microbiology 7: 1–11. doi: 10.3389/fmicb.2016.01224

Pitt, J.I. and Hocking, A.D. 2009. Fungi and Food Spoilage (Vol. 519). Springer, New York. doi: 10.1007/978-0-387-92207-2

Pontes, F.C., Abdalla, V.C.P., Imatomi, M., Fuentes, L.F.G. and Gualtieri, S.C.J. 2019. Antifungal and antioxidant activities of mature leaves of *Myrcia splendens* (Sw.) DC. Brazilian Journal of Biology 79: 127–132. doi: 10.1590/1519-6984.179829

Prakash, B., Kedia, A., Mishra, P.K. and Dubey, N.K. 2015. Plant essential oils as food preservatives to control moulds, mycotoxin contamination and oxidative deterioration of agri-food commodities – Potentials and challenges. Food Control 47: 381–391. doi: 10.1016/j.foodcont.2014.07.023

Ramirez-Mares, M.V. and Hernandez-Carlos, B. 2015. Plant-derived natural products from the American continent for the control of phytopathogenic fungi: A review. Journal of Global Innovations in Agricultural and Social Science 3: 96–118. doi: 10.17957/JGIASS/3.4.721

Ruiz, M.D.P., Ordóñez, R.M., Isla, M.I. and Sayago, J.E. 2016. Activity and mode of action of *Parastrephia lepidophylla* ethanolic extracts on phytopathogenic fungus strains of lemon fruit from Argentine Northwest. Postharvest Biology and Technology 114: 62–68. doi: 10.1016/j.postharvbio.2015.12.003

Shaik, A.B., Ahil, S.B., Govardhanam, R., Senthi, M., Khan, R., Sojitra, R., Kumar, S. and Srinivas, A. 2016. Antifungal effect and protective role of ursolic acid and three phenolic derivatives in the management of sorghum grain mold under field conditions. Chemistry & Biodiversity 13: 1158–1164. doi: 10.1002/cbdv.201500515

Sharifzadeh, A., Javan, A.J., Shokri, H., Abbaszadeh, S. and Keykhosravy, K. 2016. Evaluation of antioxidant and antifungal properties of the traditional plants against foodborne fungal pathogens. Journal de Mycologie Médicale 26: e11–e17. doi: 10.1016/j.mycmed.2015.11.002

Sharma, A., Rajendran, S., Srivastava, A., Sharma, S. and Kundu, B. 2017. Antifungal activities of selected essential oils against *Fusarium oxysporum* f. sp. lycopersici 1322, with emphasis on *Syzygium aromaticum* essential oil. Journal of Bioscience and Bioengineering 123: 308–313. doi: 10.1016/j.jbiosc.2016.09.011

Shen, Q., Zhou, W., Li, H., Hu, L. and Mo, H. 2016. ROS involves the fungicidal actions of thymol against spores of *Aspergillus flavus* via the induction of nitric oxide. PLoS One 11: 1–14. doi.org/10.1371/journal.pone.0155647

Shephard, G.S., Burger, H.M., Rheeder, J.P., Alberts, J.F. and Gelderblom, W.C. 2019. The effectiveness of regulatory maximum levels for fumonisin mycotoxins in commercial and subsistence maize crops in South Africa. Food Control 97: 77–80. doi.org/10.1016/j.foodcont.2018.10.004

Siahmoshteh, F., Hamidi-Esfahani, Z., Spadaro, D., Shams-Ghahfarokhi, M. and Razzaghi-Abyaneh, M. 2018. Unraveling the mode of antifungal action of *Bacillus subtilis* and *Bacillus amyloliquefaciens* as potential biocontrol agents against aflatoxigenic *Aspergillus parasiticus*. Food Control 89: 300–307. doi: 10.1016/j.foodcont.2017.11.010

Sivakumar, D. and Bautista-Baños, S. 2014. A review on the use of essential oils for postharvest decay control and maintenance of fruit quality during storage. Crop Protection 64: 27–37. doi: 10.1016/j.cropro.2014.05.012

Sultana, B., Naseer, R. and Nigam, P. 2015. Utilization of agro-wastes to inhibit aflatoxins synthesis by *Aspergillus parasiticus*: A biotreatment of three cereals for safe long-term storage. Bioresource Technology 197: 443–450. doi: 10.1016/j.biortech.2015.08.113

Tatsadjieu, N.L., Dongmo, P.J., Ngassoum, M.B., Etoa, F.X. and Mbofung, C.M.F. 2009. Investigations on the essential oil of *Lippia rugosa* from Cameroon for its potential use as antifungal agent against *Aspergillus flavus* Link ex. Fries. Food Control 20: 161–166. doi: 10.1016/j.foodcont.2008.03.008

Velazhahan, R., Vijayanandraj, S., Vijayasamundeeswari, A., Paranidharan, V., Samiyappan, R., Iwamoto, T., Friebe, B. and Muthukrishnan, S. 2010. Detoxification of aflatoxins by seed extracts of the medicinal plant, *Trachyspermum ammi* (L.) Sprague ex Turrill – Structural analysis and biological toxicity of degradation product of aflatoxin G1. Food Control 21: 719–725. doi: 10.1016/j.foodcont.2009.10.014

Vicente, M.F., Basilio, A., Cabello, A. and Peláez, F. 2003. Microbial natural products as a source of antifungals. Clinical Microbiology and Infection 9: 15–32. doi: 10.1046/j.1469-0691.2003.00489.x

Vijayanandraj, S., Brinda, R., Kannan, K., Adhithya, R., Vinothini, S., Senthil, K., Chinta, R.R., Paranidharan, V. and Velazhahan, R. 2014. Detoxification of aflatoxin B1 by an aqueous extract from leaves of *Adhatoda vasica* Nees. Microbiological Research 169: 294–300. doi: 10.1016/j.micres.2013.07.008

Vilaplana, R., Pérez-Revelo, K. and Valencia-Chamorro, S. 2018. Essential oils as an alternative postharvest treatment to control fusariosis, caused by *Fusarium verticillioides*, in fresh pineapples (*Ananas comosus*). Scientia Horticulturae 238: 255–263. doi: 10.1016/j.scienta.2018.04.052

Von Fraunhofer, J.A. and Joshi, R.K. 2019. Essential oils and the legislative landscape. American Journal of Essential Oils and Natural Products 7: 01–06. www.essencejournal.com/pdf/2019/vol7issue1/PartA/5-5-27-895.pdf

Yang, D., Jiang, X., Sun, J., Li, Xia, Li, Xusheng, Jiao, R., Peng, Z., Li, Y. and Bai, W. 2018. Toxic effects of zearalenone on gametogenesis and embryonic development: A molecular point of review. Food and Chemical Toxicology 119: 24–30. doi.org/10.1016/j.fct.2018.06.003

Yutani, M., Hashimoto, Y., Ogita, A., Kubo, I., Tanaka, T. and Fujita, K.I. 2011. Morphological changes of the filamentous fungus *Mucor mucedo* and inhibition of chitin synthase activity induced by anethole. Phytotherapy Research 25: 1707–1713. doi: 10.1002/ptr.3579

Zheng, S., Jing, G., Wang, X., Ouyang, Q., Jia, L. and Tao, N. 2015. Citral exerts its antifungal activity against *Penicillium digitatum* by affecting the mitochondrial morphology and function. Food Chemistry 178: 76–81. doi: 10.1016/j.foodchem.2015.01.077

CHAPTER

7

Binders Used in Feed for Their Protection against Mycotoxins

Abderahim Ahmadou[1,2,4*], Nicolas Brun[5], Alfredo Napoli[3], Noel Durand[1,2] and Didier Montet[1,2]

[1] Centre de Coopération Internationale en Recherche Agronomique pour le Développement (CIRAD), UMR Qualisud, F-34398 Montpellier, France

[2] Qualisud, Univ Montpellier, CIRAD, Montpellier SupAgro, Univ d'Avignon, Univ de La Réunion, Montpellier, France

[3] CIRAD, UR BioWooEB, 73 rue Jean-François Breton, 34398 Montpellier Cedex 5, France

[4] Institut Polytechnique Rural de Formation et Recherche Appliquée (IPR/IFRA) de Katibougou, Unité de Technologie Alimentaire, 45 Route de Koulouba, Annexe de Dar-Salam Bamako, Mali

[5] ICGM, Univ Montpellier, CNRS, ENSCM, Montpellier, France

1. Introduction

The immobilization of a xenobiotic by non-covalent bonding to adsorbents constitutes an interesting method of "decontamination" increasingly used when mycotoxins are present in feed. This technique can be very effective for certain toxins, subject to the appropriate choice of adsorbent. It can also have very little effectiveness in terms of "adsorption of toxins", but still be used in animal feed for other reasons: for example the fluidifying properties of clays and the beneficial effects of certain adsorbents for commercial reasons (Guerre 2016).

Adsorption technique could decrease mycotoxins levels in foods by trapping mycotoxins to decrease their bioavailability during digestion in the intestinal tract. Some mycotoxins have the ability to adsorb on materials. The physical structure of the adsorbent, i.e. the total load and its distribution, the pores size and surface accessibility, are the most important parameters for the efficiency of adsorption. The characteristics of mycotoxins such as their polarity, their solubility, their size, their shape and, in the case of ionized compounds, their distribution charge and their dissociation constants are also very important.

*Corresponding author: abderahim.ahmadou@mesrs.ml

Adding adsorbents to animal feed is the most common method of countering the harmful effects of mycotoxins. These adsorbents are only effective if the complex formed is stable in the digestive system. Usually, the complex mycotoxins-absorbent will be found in the urine and feces (Jard 2011).

The object of this chapter was to clarify the interest effects that adsorbents can have when they are used with raw materials contaminated by mycotoxins.

2. Mycotoxins Removal Methods

Monitoring preventive practices is not enough to completely eliminate the risk of contamination. It is therefore also necessary to develop strategies to decontaminate raw products. Reducing mycotoxin levels can be done during the food manufacturing process (Bullerman and Bianchini 2007) or by adding additives to food that remove or deactivate mycotoxins in the body. In all cases, the decontamination process must destroy or inactivate the toxin, must not generate toxic residues, must maintain food nutritional quality and its digestibility and must not modify the technological properties of the product.

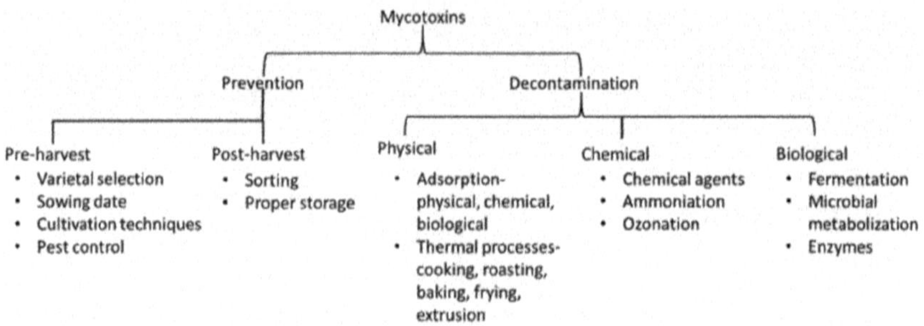

Figure 1. Conventional prevention and decontamination strategies for mycotoxins (Pankaj et al. 2018)

Different physical methods can be used; we have classified them according to whether they lead to the elimination or denaturation of toxins (Guerre 2016).

3. Mycotoxin Binders

Mycotoxin binders are considered a physical technique used principally for feed decontamination but can also be used in human intervention. The most important feature of the adsorption, a solid compound, is the physical structure of the adsorbent, i.e. the total charge and charge distribution, the size of the pores and the accessible surface area. On the other hand, the properties of the absorbed molecules, the mycotoxins, like polarity, solubility, size, shape and in the case of ionized compounds, charge distribution and dissociation constants also play a significant role.

Therefore, the efficacy of every adsorption process has to be investigated in with regard to the particular properties of the adsorbate (Huwig et al. 2001).

The efficiency of binders in mitigating adverse effects of mycotoxins in food was demonstrated in a randomized and double-blinded clinical trial (Wang et al. 2005, Afriyie-Gyawu et al. 2008, Wang et al. 2008).

Several solid compounds have been reported in literature and used as binders in food for the adsorption of mycotoxins, including silicate minerals (zeolites, clays and other phyllosilicates) and carbonaceous materials (activated carbons and more recently biochars).

3.1 Mycotoxins Decontamination with Clays

The term clay corresponds to compounds consisting of more or less hydrated lamellar silicates originating from the alteration of three-dimensional framework silicates such as feldspar (Ward 1991). When aluminum is present, these compounds are called aluminosilicates.

The very strong adsorption of aflatoxins by these compounds, associated with a marked decrease in the toxicity of contaminated food, is probably at the origin of the current craze for adsorbents. The beneficial effects of clays in animal feed, in the absence of contamination by mycotoxins, have been widely studied. Since these compounds interfere with the absorption of trace elements, harmful effects have also been reported (Moshtaghian et al. 1991).

3.2 Hydrated Sodium Calcium Aluminosilicate

Better known as HSCAS (hydrated sodium calcium aluminosilicate), sodium and calcium aluminosilicates are the most widely studied class of aluminosilicate for its adsorbent properties. These compounds belong to the family of zeolites; having a deficit in positive charge they constitute excellent adsorbents of cations.

The HSCAS reveals remarkable adsorbent properties with respect to aflatoxins, the complexes formed being stable for a pH range from 2.5 to 10 (Phillips et al. 1988). The aflatoxins thus adsorbed are "immobilized", less than 10% being extractable by organic solvents. This adsorption, the mechanism of which remains to be discussed, is rapid and intense. A plateau has been noted 30 min after the toxin has been dissolved, more than 200 nmoles of AFB1 can be fixed per mg of HSCAS (Phillips et al. 1990). This efficiency through its strong fixation and surface area explains why only small amounts of HSCAS are necessary in food to reduce the deleterious effects of aflatoxins (0.1 to 0.5%). This protective effect has been observed in chicken, turkey and pigs. It is accompanied by a decrease in AFM1 levels in the milk of ruminants (Smith 1980).

The adsorbent properties of HSCAS have also been explored with other toxins. The results obtained with respect to zearalenone or ochratoxin A are partial, those obtained for trichothecenes are disappointing (Galvano et al. 1998, Kubena et al. 1998). It may be explained by the ochratoxin size difference and its polarity.

Phillips et al. (1990) interpreted the binding mechanism as the formation of a complex by the β-carbonyl function of the aflatoxin withaluminium ions. Thus, HSCAS can be used as an 'inorganic sponge' sequestering aflatoxins in the gastro-intestinal tract of farm animals. Ramos et al. (1996) investigated the adsorption of aflatoxins to montmorillonite according to Freundlich and Langmuir isotherm calculations. They obtained a better fit of their adsorption data employing the Freundlich isotherm and suggested, therefore, the presence of a heterogeneous surface with different adsorption centers having different affinities for the adsorbate or the co-existence of different adsorption mechanisms or both.

3.3 Zeolites

Zeolites are crystallized substances having a structure made up of interconnected tetrahedra of SiO_4 and AlO_4. To be part of the zeolite family, the aluminosilicate must conform to a ratio (Si + Al) / O equal to 0.5. The aluminosilicate tetrahedron is negatively charged, leaving large spaces between molecules, thus trapping cations (usually Na^+ and / or Ca^{2+}).

In the zeolites used as adsorbents, the free spaces are interconnected, constituting vast spaces capable of adsorbing compounds of size much greater than sodium or calcium. In addition, these compounds hydrate and dehydrate very easily, without any modification of their structure. Regarding the adsorption of mycotoxins, the effects of zeolites have been mainly explored in the presence of aflatoxins. The *in vitro* adsorption of AFB1 in solution in different media would be close to 60% (Dvorak 1989). This adsorption, although lower in the presence of nitrogen compounds, would decrease the toxicity of feed containing 2.5 ppm of aflatoxins and 5% of zeolite in chicken (Scheideler 1993).

The origin of zeolite is crucial; a comparative study of 5 forms revealed significant differences in protection against the toxicity of aflatoxin-contaminated food in chicken. Synthetic zeolites, especially sodium zeolites, would be the most active (Miazzo et al. 2000). Finally, adding 5% of synthetic anionic zeolite to a food containing 250 ppm of zearalenone would prevent the effects of the toxin in rats (weight gain and quantity of food consumed). No protection would be observed with synthetic cationic zeolite.

3.4 Bentonite

Bentonite is a clay of the "montmorillonite" type, formed by the aging of volcanic ash. The term bentonite therefore groups together different products of the same origin but of different compositions. Composed of aluminum and silica, they also belong to the large category of aluminosilicates. Some are rich in sodium, others in calcium, potassium or magnesium; all contain trace elements and traces of toxic metals. The origin of bentonite and its physical and chemical properties will be very important in its ability to adsorb toxins.

Sodium bentonite is the most common and mostly used in animal feed. Due to its large internal surface, this compound adsorbs approximately 6 to 7 times its weight in water. Its cation exchange capacity is close to 80-85 meq/100 g (Ramos et al. 1996).

The uses of bentonite are multiple. In animal feed it is mainly used for its properties as:

- binding agent in food (pellets), its incorporation rate varies from 1.5 to 3% of the ration,
- anti-caking agent in feeds to avoid the formation of lumps,
- water absorbent to reduce liquid losses from silages with low dry matter content, and
- source of trace elements, mainly selenium and magnesium.

In therapy, bentonite has been used as an adsorbent in the treatment of paraquat poisoning in cats and rats (Meredith and Vale 1987). Its adsorbent properties vis-à-vis aflatoxins have been explored *in vitro* and *in vivo*:

- *In vitro*, 2% of bentonite adsorb 94 to 100% of AFB1 (400 pg) in solution in phosphate buffer at pH 6.5. The adsorbent capacities vary according to the nature of the bentonite used. Chloroform is a perfect extractor for the complexes formed and provides a recovery percentage varying from 5 to 25%. In milk, 2% of bentonite adsorbs 89% of the AFM1 dose (3 to 6 ppb). Similar adsorption percentages are observed at pH 2, and on complex biological mixtures representative of intestinal fluids. Studies on moldy foods show that 10% bentonite adsorbs 70% of AFB1 (44.6 ppb) present (Ramos et al. 1996).
- *In vivo*, adding sodium bentonite to aflatoxin-contaminated foods decreases their toxicity in pigs and poultry. An incorporation rate of 0.5% (bentonite) seems to give optimal effect.

Protection against aflatoxins teratogenic effects is also observed in rats. It should be emphasized that, for bentonite as for other clays, not all the deleterious effects of aflatoxins are systematically inhibited (Smith and Carson 1984).

Montmorillonite itself allows, in a 2% solution, an adsorption of the order of 95% of the AFB1 contents present in phosphate buffer at pH 6.5. A chloroformic extraction of the complexes formed provides a recovery percentage varying from 10 to 57%, depending on the montmorillonite tested. Similar results are obtained on liquid media whose composition is close to intestinal fluids (Ramos et al. 1996). The adsorption seems saturable around 99% and the further addition of montmorillonite does not increase the amount adsorbed. The equilibrium is reached in one hour and the complex formed is stable for pH between 2.5 and 7. One gram of montmorillonite would allow the adsorption of more than one mg of aflatoxins at pH 6 which is similar to the intestinal acidity found in the body.

3.5 Uses of Other Phyllosilicates for Mycotoxins Decontamination

Kaolin ($Al_2Si_2O_5(OH)_4$) or aluminum silicate, is a phyllosilicate used in the treatment of gastric ulcers; it is also endowed with adsorbent properties. Kaolin allows the *in vitro* absorption of 87% of the AFB1 content (ppm) present in phosphate buffer (pH 6.5). However, adsorption is easily reversible, with almost 77% of the toxin remaining extractable by chloroform. It nevertheless allows the decontamination of peanuts and is sufficient to prevent the toxin effects in rats and ducks during contamination of food with 5 ppm of AFB1 (addition of 0.2 to 1% of kaolin). This effect turns out to be insufficient if the level of contamination is higher (20 ppm) (Masimango et al. 1979).

Sepiolite ($(MgO)_2$ $(SiO_2)_3$, $2H_2O$) is a magnesium silicate and also a phyllosilicate. Used at 2%, it is capable of absorbing almost 87% of the AFB1 content (8 ppm) present in phosphate buffer (pH 6.5) (Masimango et al. 1979).

This adsorption is reversible and almost 77% of the toxin remain extractable by chloroform. However, it explains the protective effects of 0.5% sepiolite in food against the toxicity of AFB1 (800 ppm) in pigs (Schell et al. 1993).

3.6 Mycotoxins Decontamination with Carbonaceous Binders

Activated carbon is a porous black water-insoluble powder obtained by pyrolysis of different types of biomass in which activation process has been done to increase its specific surface area. Activation process can be done after the pyrolysis process by physical way (i.e. thermal treatment of charcoal under CO_2 and/or steam atmosphere) or before the pyrolysis process by the chemical way (impregnation of biomass with a chemical, acid, base or salt).

"Officinal" activated carbon is used *in vivo* for the adsorption of toxins and toxics in humans and animals.

Its adsorbent capacities vary according to its porosity (contact surface) and the medium in which it is found (aqueous / non-aqueous, concentration of organic matter and salts, pH ...).

The adsorbent capacities of activated carbon vis-à-vis aflatoxins were demonstrated very early (Decker and Corby 1980). 100 mg of activated carbon are capable of absorbing 1 mg of toxin in the presence of 2% bovine serum albumin and 0.5% corn oil at pH 7 (Decker and Corby 1980). The complex formed appears stable, and protects against the toxicity of AFB1: reduction in the signs of hepatotoxicity (plasma markers and liver damage), increase in the percentage of survival, and increase in the fecal excretion of mycotoxin and its metabolites. These effects have been observed in both acute and chronic poisoning, in non-ruminant mammals, ruminants, and poultry, when administered AFB1 or AFM1.

Activated carbon was used to remove patulin from naturally contaminated cider and bentonite removed AFM1 from naturally contaminated milk (Jans et al. 2014). De Nijs et al. (2012) discussed the efficiency of mycotoxin mitigation and food safety aspects of such techniques.

Regarding ochratoxin A, although activated carbon is an effective adsorbent *in vitro*, it is not able to decrease the toxicity of mycotoxin *in vivo* in chicken. This effect could be due to a decrease in the adsorption capacities of carbon in food, due to their high content of organic matter. A clear protective effect would be observed when administering large doses of carbon in the feed (5-10%) of piglets contaminated with ochratoxin A (1 ppm); it is however accompanied by a decrease in plasma concentrations of vitamin E (Plank et al. 1990).

As is activated carbon, biochar is a pyrogenic black carbon produced by pyrolysis conversion, in an inert atmosphere, of biomass feedstock, including agricultural and forest residuals. Biochar may be activated or not depending on the final use. Biochar differs from activated carbon by its end use in soils. As such, it must respond to various constraints related to the physical, chemical and biological balances to be achieved with soils and plants, in addition to interactions with mycotoxins. Thus, biochar has attracted great attention because of its potential to help mitigate climate change and improve soil fertility (Lehmann et al. 2007). In addition, many researchers have found that biochar can be used as an alternative adsorbent to remove different kinds of contaminants, including heavy metals, nutrients, and pharmaceuticals, from aqueous solutions (Zhou et al. 2013).

When biomass is pyrolyzed in the aim of producing biochars, slow pyrolysis with moderate temperature (350–800°C) is normally adopted (Qiet al. 2017). Within this temperature range, most of biochars are carbonized from biomass feedstocks.

For instance, biochars were produced from cashew nut shell at 400, 600 and 800 °C. The influence of pyrolysis temperatures on the adsorption characteristics and mechanisms of ochratoxin A on cashew nut shell biochars were investigated by Ahmadou et al. (2019b). Biochars produced at higher temperatures have higher specific surface areas, resulting in higher OTA adsorption capacities and faster adsorption kinetics.

With 1000 mg of each biochar in 5 mL of water-methanol solution (w/w) containing OTA at 38 ng/mL; results showed a difference in OTA adsorption rates which was mainly related to the increase in the pyrolysis temperature. The OTA adsorption rates for the biochars produced at 400, 600 and 800°C were respectively as follows: 66.21%, 78.98% and 92.35%.

The increase of biochar production temperature allowed an increase in the OTA adsorption rate.

Most of the OTA was adsorbed by the different biochars in the first 20 min and equilibrium was reached after 30 min of contact (Ahmadou et al. 2019b).

Aflatoxins adsorption rate was 100% at 180 ng/mL with 25 mg of all biochars (400, 600 and 800°C). The pH and temperature of pyrolysis had not effects on aflatoxins adsorption by the different biochars. The adsorption of the totality of aflatoxins can be explained by a high availability of specific surface area and stirring promotes contact between biochar adsorption sites and the aflatoxins molecules. Phillips et al. (1995) suggest that one adsorbed AFB1 molecule occupied a surface area of about 1.38 nm^2; the cashew nut shell biochars specific surface areas are between 151 and 306 m^2/g, which is dramatically larger than the required specific surface area for aflatoxins sorption. That could explain why the pyrolysis temperature has few effects on biochars adsorption capacity and biochars could adsorb more aflatoxins because the adsorption limits are not reached. Biochar adsorbs more aflatoxins compared to OTA. It means that the molecules of aflatoxins had a better affinity for all biochars than OTA (Ahmadou et al. 2019a).

3.7 Uses of Resins for Mycotoxins Decontamination

Resins are synthetic polymers made up of a three-dimensional macromolecular framework with active groups capable of fixing organic or mineral compounds by low energy bonds. The most common resins are cation exchange resins, of the $R\text{-}SO_3^-$, $R\text{-}PO_3^-$, R-COO type and anion exchange resins, of the [R-N (CH_3) 3] exponent type. These compounds are often used in the separation of mycotoxins with a view to their subsequent determination (Shetty and Bhat 1999).

The existence of aluminosilicates makes the use of cation exchange resins unattractive; on the other hand, anion exchange resins are used for their adsorbent properties in vivo. Colestyramine is marketed in human medicine (Questran ND) for the treatment of high cholesterol and pruritus accompanying incomplete hepatic cholestasis.

Its adsorbent properties have also been explored with respect to zearalenone *in vitro* (concentration of 1%). 1 g of colestyramine is capable of absorbing nearly 2 ng of toxin in media with a composition close to gastric or intestinal content (Ramos et al. 1996). In rats, used at a rate of 5% in food, it decreases the urinary excretion of the toxin and its metabolites, as well as the level of residues present in the liver and the kidneys (Smith 1982). In mice, the resin (2.5% in food) also prevented increases in body weight and uterine weight following the ingestion of a food contaminated with 6 ppm of zearalenone (Underhill et al. 1995).

Adsorbent properties of colestyramine have also been demonstrated during administration of ochratoxin A (OTA). In rats, the resin (2% of the food) is likely to decrease plasma OTA concentrations (incorporated in the food at a concentration of 1 ppm) while increasing its fecal excretion and decreasing its urinary excretion. These effects, less marked for a food rich in saturated lipids (Madhyastha et al. 1992), could correspond to a decrease in the enterohepatic cycle of the toxin (Roth et al. 1988). The fixation of OTA on the resin would compete with that of bile salts, but with greater affinity. This fixation is accompanied by a decrease in the nephrotoxicity (enzymuria) of ochratoxin A in rats (Creppy et al. 1995, Kerkadi et al. 1998).

Thus, colestyramine seems to have interesting adsorbent properties vis-à-vis anionic mycotoxins, its high cost however makes it difficult to imagine its systematic use in animal feed.

The adsorbent properties of polymers of divinyl benzene-styrene, another resin capable of fixing anions, have also been explored. The *in vivo* addition of this adsorbent in the rat food (5%) modified the toxicokinetics of zearalenone administered orally at a dose of 100 mg/kg (decrease in urinary excretion and residues in the liver) (Smith 1982). It reduces also the deleterious effects of a food contaminated with 3 ppm of T2 toxin administered for two weeks (Carson and Smith 1983). The results obtained in this last experiment are similar to those obtained with bentonite, but superior to those obtained with a cation exchange resin (also administered at 5% in the food).

Finally, polyvinylpyrrolidone, used at a rate of 0.2% in food, was unable to protect from the toxic effects of deoxynivalenol in pigs (5 to 14 ppm of toxin for five weeks) (Friend et al. 1984). Used at a concentration of 0.3% in food, it decreased the toxicity of aflatoxins in chicken (2.5 ppm of a mixture containing more than 80% of AFB1 given for three weeks) (Kececi et al. 1998).

4. Conclusion

The limitations of other mycotoxins elimination methods in food or feed have prompted scientists to turn to the use of binders. Many studies have been carried out in recent years on adsorbents as binders, seeking to obtain good efficiency, and better adsorption specificity while reducing the impact on food nutritional quality compared to other mycotoxins elimination methods.

However, the adsorption efficiency of the various adsorbents is strongly linked to the adsorbent itself, to the mycotoxin tested and to the conditions of use.

Most of the adsorbents described in this chapter have shown good results for mycotoxins elimination with certain limits, and this pushes scientists to deepen research around the use of adsorbents. This research is mainly focused on the mechanisms explaining the adsorption phenomenon, determining the limits of adsorbents and optimizing their adsorption capacities. The preferred area for using these binders for mycotoxin removal appears to be animal feed.

References

Afriyie-Gyawu, E., Ankrah, N.A., Huebner, H.J., Ofosuhene, M., Kumi, J., Johnson, N.M., Tang, L., Xu, L., Jolly, P.E., Ellis, W., Ofori-Adjei, D., Williams, J.H., Wang, J.S. and Phillips, T.D. 2008. NovaSil clay intervention in Ghanaians at high risk for aflatoxicosis. I. Study design and clinical outcomes. Food Addit Contam. 25: 76–87.

Ahmadou, A., Napoli, A., Durand, N. and Montet, D. 2019a. High physical properties of cashew nut shell biochars in the adsorbtion of mycotoxins. Inter. Journal of Food Research 6: 18–28.

Ahmadou, A., Brun N., Napoli, A., Durand, N. and Montet, D. 2019b. Effect of pyrolysis temperature on ochratoxin A adsorption mechanisms and kinetics by cashew nut shell biochars. J. of Food Sci. and Technology 7: 877–888.

Bullerman, L.B. and Bianchini, A. 2007. Stability of mycotoxins during food processing. Int. J. Food Microbiol. 119: 140–146.

Carson, M.S. and Smith, T.K. 1983. Role of bentonite in prevention of T-2 toxicosis in rats. J. Anim. Sci. 57: 1498–1506.

Creppy, E.E., Baudrimont, I. and Betbeder, A.M. 1995. Prevention of nephrotoxicity of ochratoxin A: A food contaminant. Toxicol. Lett. 82–83: 869–877.

De Nijs, M., van Egmond, H.P., Nauta, M., Rombouts, F.M. and Notermans, S. 2012. Mycotoxin binders in feed: Food safety aspects. In report 2011.019 of RIKILT, Institute of Food Safety. Wageningen University (University& Research Centre), Wageningen.

Decker, W.J. and Corby, D.G. 1980. Activated charcoal adsorbs aflatoxin B1. Vet. Hum. Toxicol. 22: 388–389.

Dvorak, M. 1989. Ability of bentonite and natural zeolite to adsorb aflatoxin from liquid media. Vet. Med. (Praha) 34: 307–316.

Friend, D.W., Trenholm, H.L., Young, J.C., Thompson, B.K. and Hartin, K.E. 1984. Effect of adding potential vomitoxin (deoxinivalenol) detoxicants for a *F. graminearum* inoculated corn supplement to wheat diets fed to pigs. Can. J. Anim. Sci. 64: 733–741.

Galvano, F., Pietri, A., Bertuzzi, T., Piva, A., Chies, L. and Galvano, M. 1998. Activated carbons: In vitro affinity for ochratoxin A and deoxynivalenol and relation of adsorption ability to physicochemical parameters. J. Food Prot. 61: 469–475.

Guerre, P. 2016. Worldwide mycotoxins exposure in pig and poultry feed formulations. Toxins 8, 12: 350. DOI: https://doi.org/10.3390/toxins8120350"10.3390/toxins8120350

Hutwig, A., Freimund, S., Kappeli, O. and Dutler, H. 2001. Mycotoxin detoxication of animal feed by different adsorbents. Toxicology Letters 122: 179–188.

Jans, D., Pedrosa, K., Schatzmayr, D., Bertin, G. and Grenier, B. 2014. Mycotoxin reduction in animal diets. pp. 101–110. *In*: Leslie, J.F. and Logrieco, A.F. (eds.). Mycotoxin Reduction in Grain Chains. Wiley, Oxford.

Jard, G., Liboz, T., Mathieu, F., Guyonvarc'h, A. and Lebrihi, A. 2011. Review of mycotoxin reduction in food and feed: From prevention in the field to detoxification. Food Additives and Contaminants 20: 11–28.

Kececi, T., Oguz, H., Kurtoglu, V. and Demet, O. 1998. Effects of polyvinylpolypyrrolidone, synthetic zeolite and bentonite on serum biochemical and haematological characters of broiler chickens during aflatoxicosis. Br. Poult. Sci. 39: 452–458.

Kerkadi, A., Barriault, C., Tuchweber, B., Frohlich, A.A., Marquardt, R.R., Bouchard, G. and Youssef, I.M. 1998. Dietary cholestyramine reduces ochratoxin A-induced nephrotoxicity in the rat by decreasing plasma levels and enhancing fecal excretion of the toxin. J. Toxicol. Environ. Health 53: 231–250.

Kubena, L.F., Harvey, R.B., Bailey, R.H., Buckley, S.A. and Rottinghaus, G.E. 1998. Effects of a hydrated sodium calcium aluminosilicate (T-Bind) on mycotoxicosis in young broiler chickens. Poult. Sci. 77: 1502–1509.

Lehmann, J., Gaunt J. and Rondom M. 2007. Biochar sequestration in terrestrial. Mitigation and Adaptation Strategies for Global Change 20: 7–11.

Madhyastha, M.S., Frohlica, A.A. and Marquardt, R.R. 1992. Effect of dietary cholestyramine on the elimination pattern of ochratoxin A in rats. Food Chem. Toxicol. 30: 709–714.

Masimango, N., Remacle, J. and Ramaut, J. 1979. Elimination of aflatoxin B1 by clays from contaminated substrates. Ann. Nutr. Aliment. 33: 137–147.

Meredith, T.J. and Vale, J.A. 1987. Treatment of paraquat poisoning in man: Methods to prevent absorption. Hum. Toxicol. 6: 49–55.
Miazzo, R., Rosa, C.A., De Queiroz Carvalho, E.C., Magnoli, C., Chiacchiera, S.M., Palacio, G., Saenz, M., Kikot, A., Basaldella, E. and Dalcero, A. 2000. Efficacy of synthetic zeolite to reduce the toxicity of aflatoxin in broiler chicks. Poult. Sci. 79: 1–6.
Moshtaghian, J., Parsons, C.M., Leeper, R.W., Harrison, P.C. and Koelkebeck, K.W. 1991. Effect of sodium aluminosilicate on phosphorus utilization by chicks and laying hens. Poult. Sci. 70: 955–962.
Pankaj, S.K., Shi, H. and Keener, K.M. 2018. A review of novel physical and chemical decontamination technologies for aflatoxin in food. Trends in Food Science & Technology 73: 1–11. Elsevier.
Phillips, T.D., Kubena, L.F., Harvey, R.B., Taylor, D.R. and Hedelbaugh, N.D. 1988. Hydrated sodium calcium aluminosilicate: A high affinity sorbent for aflatoxin. Poult. Sci. 67: 243–247.
Phillips, T.D., Clement, B.A., Kubena, L.F. and Harvey, R.B. 1990. Detection and detoxification of aflatoxins: Prevention of aflatoxicosis and aflatoxin residues with hydrated sodium calcium aluminosilicate. Vet. Hum. Toxicol. 32: 15–19.
Phillips, T.D., Sarr, A.B. and Grant, P.G. 1995. Selective chemisorption and detoxification of aflatoxins by phyllosilicate clay. Nat. Toxins 3: 204–213.
Plank, G., Bauer, J., Grunkemeier, A., Fishcer, S., Gedek, B. and Berner, H. 1990. The protective effect of adsorbents against ochratoxin A in swine. Tierarztl. Prax. 18: 483–489.
Qi, F., Dong, Z., Lamb, D., Naidu, R., Bolan, N.S., Ok, Y.S., Liu, C., Khan, N. and Johir, M.A.H. 2017. Effects of acidic and neutral biochars on properties and cadmium retention of soils. Chemosphere 180: 564–573.
Ramos, A.-J., FinkK-Gremmels, J. and Hernandez, E. 1996. Prevention of toxic effects of mycotoxins by means of nonnutritive adsorbent compounds. J. Food Protect. 59: 631–641.
Roth, A., Chakor, K., Creppy, E.E., Kane, A., Roschenthaler, R. and Dirheimer, G. 1988. Evidence for an enterohepatic circulation of ochratoxin A in mice. Toxicology 48: 293–308.
Scheideler, S.E. 1993. Effects of various types of aluminosilicates and aflatoxin B1 on aflatoxin toxicity, chick performance, and mineral status. Poult. Sci. 72: 282–288.
Schell, T.C., Lindemann, M.D., Kornegay, E.T., Blodgett, D.J. and Doerr, J.A. 1993. Effectiveness of different types of clay for reducing the detrimental effects of aflatoxin-contaminated diets on performance and serum profiles of weanling pigs. J. Anim. Sci. 71: 1226–1231.
Shetty, P.H and Bhat, R.V. 1999. A physical method for segregation of fumonisin-contaminated maize. Food Chemistry 66: 371–374.
Smith, T.K. 1980. Influence of dietary fiber, protein and zeolite on zeralenone toxicosis in rats and swine. J. Anim. Sci. 50: 278–285.
Smith, T.K. and Carson, M.S. 1984. Effect of diet on T-2 toxicosis. Adv. Exp. Med. Biol. 177: 153–167.
Smith, T.K. 1982. Dietary influences on excretory pathways and tissue residues of zearalenone and zearalenols in the rat. Can. J. Physiol. Pharmacol. 60: 1444–1449.

Underhill, K.L., Rotter, B.A., Thmpson, B.K., Prelusky, D.B. and Trenholm, H.L. 1995. Effectiveness of cholestyramine in the detoxification of zearalenone as determined in mice. Bull. Environ. Contam. Toxicol. 54: 128–134.

Wang, P., Afriyie-Gyawu, E., Tang, Y., Johnson, N.M., Xu, L., Tang, L., Huebner, H.J., Ankrah, N.A., Ofori-Adjei, D., Ellis, W., Jolly, P.E., Williams, J.H., Wang, J.S. and Phillips, T.D. 2008. NovaSil clay intervention in Ghanaians at high risk for aflatoxicosis: II. Reduction in biomarkers of aflatoxin exposure in blood and urine. Food Addit Contam. 25: 622–634.

Ward, T.L., Watkins, K.L., Southern, L.L., Hoyt, P.G. and French, D.D. 1991. Interactive effects of sodium zeolite-A and copper in growing swine: Growth, and bone and tissue mineral concentrations. J. Anim. Sci. 69: 726–733.

Zhou, Y., Gao, B., Zimmerman, A., Fang, J., Sun, Y. and Cao, X. 2013. Sorption of heavy metals on chitosan-modified biochars and its biological effects. Chem. Eng. 231: 512–518.

CHAPTER

8

Toxicology of Mycotoxins: Overview and Challenges

Nolwenn Hymery[1]* and Isabelle P. Oswald[2]

[1] Laboratoire de Biodiversité et d'Écologie Microbienne, Brest University Parvis Blaise Pascal 29280 Technopole Plouzané, France

[2] Toxalim (Research Center in Food Toxicology), Université de Toulouse, INRAE, ENVT, INP-Purpan, UPS, 180 chemin de Tournefeuille, BP9317 331027 Toulouse Cedex 03, France

1. Mycotoxins in Food and Beverages

Aflatoxins (AFs) are produced by *Aspergillus* species; Ochratoxin A(OTA) are produced by both *Aspergillus* and *Penicillium*; trichothecenes (TCTs) [HT-2, T2 toxin and deoxynivalenol (DON)], nivalenol (NIV), zearalenone (ZEN), fumonisins type B (FBs), emerging mycotoxins such as fusaproliferin (FUS), moniliformin (MON), beauvericin (BEA), and enniatins (ENNs) [enniatin A (ENNA), and enniatin B (ENNB)], altenuene (ALT), alternariol (AOH), alternariol methyl ether (AME), altertoxin (ATX), and tenuazonic acid (TeA) are produced mainly by *Fusarium* and *Aternaria* toxins (Cigi□ and Prosen 2009, Yang et al. 2014, Marin et al. 2013). The presence of mycotoxins has been thoroughly investigated in different foodstuffs such as cereal products (Juan et al. 2017, Saladino et al. 2017), vegetables (Dong et al. 2019, Rodríguez-Carrasco et al. 2016) and some studies have also been carried out in cooked food (Sakuma et al. 2013, Carballo et al. 2018). Concerning beer, coffee and tea beverages some literature does exist (Campone et al. 2020, Pallarés et al. 2017, García-Moraleja et al. 2015). Negative results regarding the consumption of herbal tea beverages and the presence of AFs, 3-acetyl DON, 15-acetyl DON, NIV, HT-2, T-2, OTA, ZEN, ENNA, ENNA1, ENNB, ENNB1, and BEA have been described (Pallarès et al. 2019).

There is less data concerned with the mycotoxin contamination of dietary supplements which represents a possible risk for human health, especially in the case of products intended for people suffering from certain health problems. Veprikova et al. (2015) described an analytical method including 57 mycotoxins based on a QuEChERS-like (quick, easy, cheap, effective, rugged, safe) approach and ultrahigh performance liquid chromatography

*Corresponding author: nolwenn.hymery@univ-brest.fr

Table 1. Total diet studies revised for mycotoxin analysis (Carballo et al. 2019)

Country	*Analyzed food*	*Composite (n)*	*Individual food (n)*	*Analyzed mycotoxins*	*Reference*
Canada	Cereal and cereal products; alcoholic beverages; coffee; tea; beans; fruits; sugars; chocolate; cheese; milk; eggs; dessert; meat; herb and spices; dried fruits; soya products; mixed dishes	140	–	OTA	Tam et al. (2011)
China	Cereal and their products; vegetables; legumes, nuts, seeds; fruits; meat and poultry; fats, oils; alcoholic and non-alcoholic beverages; mixed dishes; snacks; sugars; condiments, sauces	600	1800	AFB_1, AFB_2, AFG_1, AFG_2, PAT, ZEN, DON, FB_1, FB_2, FB_3, α-ZOL, βZOL, DON, 3AcDON, 15AcDO	Yau et al. (2016)
China	Cereal products; beans; potatoes; meat; eggs; aquatic products; milk; vegetables; fruits; saccharides; beverages; condiments	240	–	STC, CIT, CPA, MON, GLIO, MPA, Verru	Qiu et al. (2017)
France	Vegetarian food; biscuits; breakfast cereals; breads; pasta; rice; cakes; chocolates; desserts; nuts and oilseeds; vegetables; pulses; eggs; sugars; breads, buns; butter; dairy products; coffee; meat; offal; fruits; soft drinks; alcoholic beverages; pizzas, salt cakes, quiches; sandwiches; soup; prepared dishes; salads; compotes	456	2280	AFB_1, AFB_2, AFG_1, AFG_2, AFM_1, OTA, PAT, ZEN, FB_1, FB_2, DON, NIV, 3AcDON, 15AcDON, T-2 and HT-2, NEO, FUS-X, DAS, MAS	Leblanc et al. (2005)
France	Breads; breakfast cereals; pasta; rice; croissants; pastries; biscuits; cakes; milk; dairy products; eggs; butter; offal; delicatessen meat; vegetables; fruits; dried	577	1319	AFB_1, AFB_2, AFG_1, AFG_2, AFM_1, FB_1, FB_2, OTA, PAT, ZEN, DON, NIV, 3AcDON, 15AcDON, T-2, HT-2	Sirot et al. (2013)

	fruits; nuts and seeds; chocolate; alcoholic and non-alcoholic beverages; coffee; pizzas; sandwiches; snacks; mixed dishes; desserts; compotes				
Ireland	Cereals; milk; dairy products; eggs; meat; fish; potatoes; vegetables; fruits; dried fruits; nuts; seeds; herbs spices; soups; sauces; sugars; beverages; fats; oils; snacks; pizza	141	1043	AFB_1, AFB_2, AFG_1, AFG_2, AFM_1, OTA, PAT, ZEN, FB_1, FB_2, DON, DAS, NIV, 3AcDON, 15AcDON, T-2, HT-2	FSAI (2016)
Lebanon	Breads and toasts; biscuits; croissants; cakes; pastries; pasta; pizza; pies; rice; pulses; olive oils, sesame oils; nuts, seeds, olives and dried dates; cheese; milk and milk-based beverages; milk-based ice cream and pudding; yogurt and yogurt-based products; caffeinated beverages; alcoholic beverages	47	705	AFB_1, AFM_1, OTA, DON	Raad et al. (2014)
New Zealand	Alcoholic and non-alcoholic beverages; cereal and cereal products; condiments; dairy products; eggs; fats and oils; fish; fruits; meat; nuts; seeds; snacks; sugars; vegetables; infant food	48	–	AFB_1, AFB_2, AFG_1, AFG_2, AFM_1, AFM_2	FSANZ (2001)
New Zealand	Alcoholic and non-alcoholic beverages; cereal and cereal products; condiments; dairy products; eggs; fats; oils; fish; seafood; fish products; fruits; meat products; nuts and seeds; snacks; sugars; vegetables; infant food	65	–	AFB_1, AFB_2, AFG_1, AFG_2, OTA	FSANZ (2003)
New Zealand	Alcoholic and non-alcoholic beverages; cereal products; condiments; dairy products; eggs; fats; oils; fish; fruits; meat; nuts; seeds; snacks; sugars; vegetables; infant food, fast food	570	–	AFB_1, AFB_2, AFG_1, AFG_2, AFM_1, OTA, PAT, ZEN, FB_1, FB_2, DON	FSANZ (2011)

(Contd.)

Table 1. *(Contd.)*

Country	*Analyzed food*	*Composite (n)*	*Individual food (n)*	*Analyzed mycotoxins*	*Reference*
Spain	Cereal and cereal products; milk; cheeses; dried fruits; sweet corns; breakfast cereals; corn snacks; alcoholic beverages; coffee; vegetables; baby food; apple juice; jams and apple sauce; ethnic food; gluten-free food	1690	3447	AFB_1, AFB_2, AFG_1, AFG_2, AFM_1, OTA, PAT, ZEN, FBs, DON, T-2, HT-2	Cano-Sancho et al. (2012)
Spain	Milk; dairy products	60		AFM_1	Urieta et al. (1991); Urieta et al. (1996)
Spain	Cereal and cereal products; olives; pickles; apple; pear; eggs; milk; milk shakes; custards; soya products; cheeses; grapes; alcoholic beverages; juices; oils	240	–	AFB_1, AFB_2, AFG_1, AFG_2, OTA, ZEN, FB_1, FB_2, DON, NIV, 3AcDON, 15AcDON, T-2, HT-2, T-2 triol, NEO, FUS-X, DAS	Beltrán et al. (2013)
The Netherlands	Alcoholic and non-alcoholic beverages; sugars; dairy products; eggs; fish; fruits; cereal products; legumes; meat; offal; nuts; seeds; oils; fats; soy products; tuber; vegetables	88	–	AFB_1, AFB_2, AFG_1, AFG_2, AFM_1, AOH, AME, BEA, ENN A, ENN A_1, ENN B, ENN B1, OTA, PAT, ZEN, α-ZOL, β-ZOL, STE, FB_1, FB_2, FB_3, DON, DON 3G, FUS-X, NEO, DAS, NIV, 3AcDON, 15AcDON, T-2, HT-2, MON, MPA, NPA, PeA, ROC,	López et al. (2016); Sprong et al. (2016a) Sprong et al. (2016b)
Viet Nam	Rice; wheat and products; tubes root and products; beans products; tofu; oily seeds; vegetables; sugars; seasoning; fats; oils; meat products; egg and milk; fish; other aquatic products	42	1134	AFB_1, FBs, OTA	Huong et al. (2016)

coupled with tandem mass spectrometry. The main mycotoxins determined were Fusarium trichothecenes, ZEN and ENNs, and Alternaria mycotoxins. Co-occurrence of enniatins, HT-2/T-2 toxins, and Alternaria toxins were observed in many cases. The highest mycotoxin concentrations were found in milk thistle-based supplements (up to 37 mg/kg in the sum).

2. Regulated Mycotoxins

2.1 Toxicology of Regulated Mycotoxin: *In Vitro* Data

Out of the several hundred mycotoxins identified, only a small proportion has been explored in chemical structures and toxicity in animals and cell lines. Mycotoxins cause adverse effects in host cells, such as oxidative stress, apoptosis, cell death, DNA damage, cell cycle arrest, and so on, which varies with respect to different mycotoxins (Table 2).

Table 2. Cytotoxicity of regulated mycotoxins (Wen et al. 2016)

Adverse effects in cells	*Major mycotoxins*
Oxidative stress	AFs, OTA, FB_1, ZEN, T-2, DON and PAT
Immune suppress	AFs, OTA and DON
Apoptosis	AFB_1, AFG_1, OTA, FB_1, ZEN, T-2, DON and PAT
Cell death/necrosis	AFB_1, FB_1, T-2, OTA, DON, ZEN and PAT
Lipid peroxidation	AFB_1, FB_1, OTA, DON and ZEN
DNA damage	AFB_1, FB_1, OTA, T-2, DON and PAT
Cell cycle arrest	AFB_1, FB_1, DON, OTA, and PAT

Concerning the individual toxicology of major mycotoxins, much data exists. But, one mycotoxin can be produced by several fungi, and a fungus can produce several mycotoxins. Single mycotoxin contamination is not the norm but rather the exception. Nevertheless, the current regulations were established on toxicological data from studies taking into account only one mycotoxin exposure at a time, and do not consider the combined effects of mycotoxins. However, the natural co-occurrence of mycotoxins in food and beverages is well-known but not enough toxicological data exists about their combined effect. It is therefore of the utmost importance to evaluate the toxicological impact of mycotoxin combinations to better reflect feed and food contamination and their associated animal and human health risks.

A deeper view of the different interactions between mycotoxins can be found in the review by Grenier and Oswald (2011). When focused on the *in vitro* effects of fusariotoxin mixtures on cell viability using mammalian cell models, we can conclude that mycotoxin toxicological combined effects are unpredictable based on their individual effects, despite an increasing number of co-exposure studies (Table 3). For example, concerning ZEN+FB1 mixture,

Table 3. *In vitro* interactions between fusariotoxins on cell viability (Smith et al. 2016)

Mycotoxin couples/cells	*Doses (μM)*	*Exposure*	*Toxicological effect*	
		Interaction between TCT		
DON+15-ADON	DON: 0.25–4	48 h	Synergistic	At low inhibitory concentration levels ($IC_{10, 20, 30}$)
Human epithelial colorectal adenocarcinoma cells: **Caco-2**	15-ADON: 0.25–4		Additive	At medium inhibit concentration levels ($IC_{40, 50}$)
DON+15-ADON	DON: 0.2–15	24 h	Synergistic	From IC_{10} to IC_{80}
Intestinal porcine epithelial cells (ileum + jejunum): **IPEC-1**	15-ADON: 0.2–15			
DON+3-ADON	DON: 0.25–4	48 h	Synergistic	At low and medium inhibitory concentration levels ($IC_{10, 20, 30, 40}$)
Human epithelial colorectal adenocarcinoma cells: **Caco-2**	3-ADON: 0.42–6.67		Additive	At the 50% growth inhibition level (IC_{50})
DON+3-ADON Intestinal porcine epithelial cells (ileum + jejunum): **IPEC-1**	DON: 0.2–15 3-ADON: 2–150	24 h	Antagonistic	At low inhibitory concentration levels (IC_{10}–IC_{30})
			Additive	At medium inhibitory concentration levels (IC_{30}–IC_{60})
			Synergistic	At high inhibitory concentration levels (IC_{60}–IC_{80})
15-ADON+3-ADON	15-ADON: 0.25–4	48 h	Synergistic	At low cytotoxicity levels ($IC_{10, 20, 30}$)
Human epithelial colorectal adenocarcinoma cells: **Caco-2**	3-ADON: 0.42–6.67		Additive	At medium inhibitory concentration levels ($IC_{40, 50}$)
15-ADON+3-ADON	15-ADON: 0.2–15	24 h	Synergistic	At all cytotoxicity levels (IC_{10}–IC_{80})
Intestinal porcine epithelial cells (ileum + jejunum): **IPEC-1**	3-ADON: 2–150			

DON+15-ADON+3-ADON Human epithelial colorectal adenocarcinoma cells: **Caco-2**	DON: 0.25–4	48 h	Synergistic	At low cytotoxicity levels ($IC_{10, 20, 30}$)
	15-ADON: 0.25–4		Additive	At the 40% growth inhibition level (IC_{40})
	3-ADON: 0.42–6.67		Antagonistic	From the 50% growth inhibition level (IC_{50})
DON+NIV	DON: 10–100	24 h, 48 h and 72 h	Additive	At 50% growth inhibition level (IC_{50})
Murine monocyte macrophage cells: **J774A.1**	NIV: 10–100			
DON+NIV	DON: 0.5–2	48 h	Antagonistic	At the lowest dose
Intestinal porcine epithelial cells (jejunum): **IPEC-J2**	NIV: 0.5–2		Synergistic	At the highest dose
DON+NIV	DON: 0.25–4	48 h	Synergistic	At all cytotoxicity levels (from IC_{10} to IC_{50})
Human epithelial colorectal adenocarcinoma cells: **Caco-2**	NIV: 0.2–3.2			
DON+NIV	DON: 0.2–15	24 h	Synergistic	At all cytotoxicity levels (from IC_{10} to IC_{80})
Intestinal porcine epithelial cells (ileum + jejunum): **IPEC-1**	NIV: 0.2–15			
DON+FX	DON: 0.25–4	48 h	Synergistic	At all cytotoxicity levels (from IC_{10} to IC_{50})
Human epithelial colorectal adenocarcinoma cells: **Caco-2**	FX: 7.5–120			
DON+FX	DON: 0.2–15	24 h	Antagonistic	At all inhibitory concentration levels (IC_{10}-IC_{80})
Intestinal porcine epithelial cells (ileum + jejunum): **IPEC-1**	FX: 0.12–9			

(*Contd.*)

Table 3. (*Contd.*)

Mycotoxin couples/cells	*Doses (μM)*	*Exposure*	*Toxicological Effect*	
NIV+FX	NIV: 0.2–3.2	48 h	Synergistic	At low cytotoxicity levels ($IC_{10, 20}$)
Human epithelial colorectal adenocarcinoma cells: **Caco-2**	FX: 7.5–120		Additive	At medium cytotoxicity levels ($IC_{30, 40, 50}$)
NIV+FX	NIV: 0.2–15	24 h	Additive	At all cytotoxicity levels (IC_{10}-IC_{80})
Intestinal porcine epithelial cells (ileum + jejunum): **IPEC-1**	FX: 0.16–12			
DON+NIV+FX Human epithelial colorectal adenocarcinoma cells: **Caco-2**	DON: 0.25–4	48 h	Antagonistic	At low cytotoxicity levels ($IC_{10, 20}$)
	NIV: 0.2–3.2		Additive	At medium cytotoxicity levels ($IC_{30, 40, 50}$)
	FX: 7.5–120			
DON+T2	DON: 0.25–4	24 h, 48 h and 72 h	Antagonistic	
Chinese hamster ovary cells: **CHO-K1**	T2: 0.006–0.1			
DON+T2	DON: 0.25–8	24 h, 48 h and 72 h	Antagonistic	
Monkey kidney epithelial cells: **Vero**	T2: 0.001–0.05			
DON+T2	DON: 0.04–0.1	14 days	Additive	
Hematopoietic progenitors granulo monocyte: **CFU-GM**	T2: 0.0005–0.0016			

Kouadio et al. (2007) and Wan et al. (2013) observed antagonistic effects on Caco-2 and IPEC-J2. The Caco-2 cell line, first isolated in the 70s from a human colon adenocarcinoma, forms a monolayer of highly polarized cells, joined by functional tight junctions, with well-developed and organized microvilli on the apical (AP) membrane. This cell line has proved to be a good choice for studies of intestinal absorption and toxicity of xenobiotics. The second cell model, IPEC-J2 cells are intestinal porcine enterocytes isolated from the jejunum of a neonatal un-suckled piglet. This cell line is unique as it is derived from the small intestine and is neither transformed nor tumorigenic in nature. IPEC-J2 cells mimic the human physiology more closely than any other cell line of non-human origin. Therefore, it is an ideal tool to study epithelial transport, interactions with enteric bacteria, and the effects of probiotic microorganisms and the effect of nutrients and other feedstuffs on a variety of widely used parameters reflecting epithelial functionality.

2.2 Toxicology of Regulatory Mycotoxin: *In Vivo* Data

The *in vivo* effects of the regulated mycotoxins have been investigated in rodents as well as in several farm animal species. Human intoxication has also been described and health based guidance values have been established. These effects are well characterized and they range from acute mortality to slowed growth and reproductive deficiency. Consumption of a low amount of toxins can also lead to impaired immunity (Bennet and Klich 2003). The main adverse effect and health based guidance values of the regulated mycotoxins are summarized in Table 4.

3. Emerging Mycotoxins

Moreover, several fungal species can produce other mycotoxins with toxicological properties such as beauvericin (BEA), enniatins (ENNs), and monoliformin (MON), a group of lesser-studied toxins called emerging mycotoxins (Jestoi 2008) (a non-exhaustive list of mycotoxin producing *Aspergillus*, *Penicillium* and *Fusarium* species, split into eight groups, is provided in Table 2). Their frequent detection is partly a result of the continuous development of sensitive analytical methods with low detection limits (LODs) (Fredlund et al. 2013, Lindblad et al. 2013, Stanciu et al. 2017). Even if these mycotoxin-producing fungi differ according to ecological conditions, it is important to emphasize that mycotoxins are found all over the world in foodstuffs and feedstuffs due to trade in these commodities that contributes to their worldwide dispersal. Moreover, Table 4 shows that one mycotoxin can be produced by several fungi, and that a fungus can produce several mycotoxins.

3.1 Toxicology of Emerging Mycotoxin: *In Vitro* Data

To follow *in vitro* toxicity mechanisms, common cells parameters can be monitored. First, cell viability and IC50 (inhibiting concentration 50%)

Table 4. Main adverse effect and health based guidance values of the regulated mycotoxins

Mycotoxin	*Main adverse effects*	*Health-based guidance value*	*Reference (EFSA opinion)*
Deoxynivalenol (DON)	Feed refusal, emesis Reduction of growth Immunotoxicity Not classifiable as to carcinogenic to humans	ARfD[a] 8 μg/kg bw[C]/day Group-TDI[b] (for DON and its acetylated and modified forms) 1 μg/kg bw[C]/day	Knutsen et al. 2017
Zearalenone (ZEN)	Endocrine disruptor (interaction with estrogen-receptors)	TDI[c] 0. μg/kg bw[C]/day Group TDI[b](for ZEN and modified form) 0.25 μg/kg bw[C]/day	EFSA 2016
Ochratoxin A (OTA)	Nephrotoxic (renal tumors) Carcinogenic to animals and possibly to humans	Because of carcinogenicity, exposure should be kept as low as reasonably achievable. No health-based guidance value are set up by EFSA	Schrenk et al. 2020b
Fumonisin B (FB)	Induction of apoptosis in liver Tumorigenic in rodents Possibly carcinogenic to humans	Group TDI[b](for FB and modified form) 1 μg/kg bw[C]/day	Kutsen et al. 2018
Aflatoxin B1 (AFB1)	Genotoxic carcinogen Carcinogenic to humans	Because of carcinogenicity, exposure should be kept as low as reasonably achievable. No health-based guidance value are set up by EFSA	Schrenk et al. 2020a

[a]: ARfD: Acute Reference Dose; [b]: TDI: Tolerable Daily Intake; [c]: bw: Body weight

Table 5. Some emerging mycotoxins of interest and their fungal sources, with primary food and feed hosts and endemic regions in all cereals and cereals-based products in temperate regions (Europe) (Smith et al. 2016)

BEA*	*Fusarium (acuminatum, armeniacum, anthophilum, avenaceum, beomiforme, dlamini, equiseti, fujikuroi, globosum, langsethiae, longipes, nygamai, oxysporum, poae, proliferatum, pseudoanthophilum, sambucinum, semitectum, sporotrichioides, subglutinans)*
ENNs* (A, A1, B, B1)	*Fusarium (acuminatum, avenaceum, langsethiae, lateritium, poae, proliferatum, sambucinum, sporotrichioides, tricinctum)*
MON*	*Fusarium (acuminatum, avenaceum, culmorum, equiseti, fujikuroi, napiforme, nygamai, oxysporum, proliferatum, pseudonygamai, sporotrichioides, subglutinans, thapsinum, tricinctum, verticillioides)*

*Beauvericin (BEA); enniatins (ENNs); moniliformin (MON).

values can be determined with MTT or neutral red assays. Measurement of the trans-epithelial electrical resistance (TEER) was performed to evaluate monolayer integrity and possible damage of the cellular monolayer during the experiments. The production of reactive oxygen species (ROS) – including superoxide radicals ($O_2{\bullet}^-$), hydrogen peroxide (H_2O_2), and hydroxyl radicals (•OH) – injures the cell by (among other mechanisms) damaging membrane integrity and cellular metabolism. One indicator of oxidative stress is malondialdehyde, a lipid peroxidation product (LPO) responsible for altering the permeability of the cell membrane. Furthermore, enzymes such as superoxide dismutase (SOD), catalase (CAT), glutathione peroxidase (GPx), and glutathione reductase (GR), as well as non-enzymatic components such as glutathione (GSH), are activated following the generation of ROS. This is related to the activation of the antioxidant protective capacity of cells.

Concerning emerging mycotoxins (Table 5), some data is recently available. Moreover, these limited toxicity studies are mainly *in vitro* studies and suggest an effect on the gastro-intestinal tract, immunity and steroidogenesis (Fraeyman et al. 2017, Prosperini et al. 2017 , Mallebrera et al. 2018). Fraeyman et al. recently reviewed *in vitro* studies concerning *in vitro* toxicity of emerging mycotoxins. Results depended on cell models and concentration. For example, in the same order of concentration (1.5-5 µM) incubation in the presence of BEA or ENNs on human colon adenocarcinoma (Caco-2) cells for 24-72 h resulted in generation of reactive oxygen species (ROS), lipid peroxidation (LPO), loss of mitochondrial membrane potential, cell cycle arrest, apoptosis and necrosis. Likewise, incubation of human cells with BEA and ENN B caused apoptosis and necrosis. The cytotoxic effect was more pronounced for BEA. Moreover, BEA but not ENN B was recently described to be genotoxic for Jurkat T-cells (a T cell acute leukemia human cell line) after incubation with 3–5 µM for 24 h. Both BEA and ENN B exert *in vitro* immunomodulatory effects, such as an increased IL-10 secretion by human mature dendritic cells, at concentrations as low as 1.6 µM BEA and 2 µM ENN B and after exposure for 48 h. Major studies were cited in Table 6.

Table 6. *In vitro* toxicity of emerging mycotoxins from *Fusarium* and *Alternaria* species (Fraeyman et al. 2017)

Cell line	*Mycotoxin*	*Exposure time*	*Exposure dose (μM)*	*Effect*
Caco-2 [a]	BEA	0 min	1.5	ROS[b] generation
		24–72 h		IC_{50}: 20.6–3.2 μM (MTT[c]); IC_{50}: 8.8–1.9 μM (NR[d])
		24–72 h	1.5–3.0	LPO[e], ↓ GSH, ↑ GSSG, loss of mitochondrial membrane potential, cell cycle arrest in S and G2/M, apoptosis and necrosis
		24 h	12	DNA damage
	ENN A	<1 h	1.5–3.0	ROS generation
		24–72 h		IC_{50}: 9.3–0.46 μM
		24–72 h	1.5–3.0	LPO, loss of mitochondrial membrane potential, cell cycle arrest in SubG0/G1 and (Sub)G2/M, DNA damage, apoptosis and necrosis
	ENN A1	10 min	1.5	ROS generation
		24–72 h		IC_{50}: 12.3–0.46 μM
		24–72 h	1.5–3.0	LPO, loss of mitochondrial membrane potential, DNA damage, cell cycle arrest in (Sub)G0/G1 and G2/M, apoptosis, necrosis
	ENN B	10 min	3.0	ROS generation
		48–72 h		IC_{50}: 10.7–1.4 μM
		24–72 h	1.5–3.0	LPO, loss of mitochondrial membrane potential, cell cycle arrest in (Sub)G0/G1, and G2/M, apoptosis, necrosis
	ENN B1	5–10 min	1.5–3.0	ROS generation
		48–72 h		IC_{50}: 10.8–0.8 μM

		24–74 h	1.5–3.0	LPO, loss of mitochondrial membrane potential, DNA damage, cell cycle arrest in (Sub)G0/G1, G2/M and S, apoptosis, necrosis
	MON	72 h		IC_{50}: 30.9 µg/mL
	AOH	24 h	15–30	Changes in MMP [f], ↓ G1 phase, ↑ S and G2/M phase, apoptosis, necrosis
HT-29 [g]	ENN A	24–48 h		IC_{50}: 9.3–8.2 µM
	ENN A1	24–48 h		IC_{50}: 9.1–1.4 µM
	ENN B	24–48 h		IC_{50}: ≥2.8 µM
	ENN B1	24–48 h		IC_{50}: 16.8–3.7 µM
HCT116 [g]	AOH			$IC_{50, 24h}$: 65 µM ↓ Early apoptotic and late apoptotic/necrotic cells, ROS generation PTP[h]-dependent MMP Caspase-cascade activation, activation of p53 protein expression
	AME			$IC_{50, 24h}$: 120 µM Apoptotic cell death, PTP-opening, induction of MMP, cytochrome c release Caspase-cascade activation, ↑ p53 protein, ROS generation
IPEC-J2 [i]	BEA	24–72 h	5–10	TEER[j] reduction (between –59% and –80%), no reduction of cell viability
	ENN A	72 h	5	TEER reduction (–70%), no reduction of Cell viability
	ENN A1	24–72 h	10	TEER reduction (between –29% and –74%), no reduction of cell viability
	ENN B	48–72 h	2.5	TEER reduction (between –55% and –68%), no reduction of cell viability

(Contd.)

Table 6. *(Contd.)*

Cell line	*Mycotoxin*	*Exposure time*	*Exposure dose (µM)*	*Effect*
	ENN B1	48–72 h	5	TEER reduction (between –44% and –58%), no reduction of cell viability
	ENN combinations		1.5	Additive effect on TEER reduction
	MON	72 h	5–10	No effect on TEER or viability
Hep-G2 [k]	ENN A	24–48 h		IC_{50}: 26.2–11.4 µM
	ENN A1	24–48 h		IC_{50}: 11.6–2.6 µM
	ENN B	24–48 h		IC_{50}: >30 µM
	ENN B1	24–48 h		IC_{50}: 24.3–8.5 µM
	MON	48–72 h		IC_{50}: 39.5–24.1 µg/mL
H295R [l]	ENN B	72 h	10–100	↓ Viability by 37%, ↑ S-phase, ↓ G0/G1phase, ↑ apoptosis ↓ HMGR, STAR, CYP11A, HSD3B2, CYP17A1 ↑ CYP1A1, MC2R, NR0B1, CYP21A2, CYP11B1, CYP19 ↓ Progesterone, testosterone and cortisol; estradiol unaffected
	AOH		3.87	No influence on viability ↑ 7 proteins (FDX1, HSD3B, CYP21A2, SCAMP3, SOAT1, ARF6, RRP15) ↓ 15 proteins (ACTBL2, NUCKS1, EIF2B5, COX2, CRMP1, ABHD14A-ACY1, ATP5J, ACSF2, HN1, ETHE1, HIST1H1E, ACBD5, NPC1, NR5A1, TOMM7) Upregulation mRNA for CYP21A2 and HSD3B ↑ G0/G1 and ↑ G2/M phase

H29R[1]	AOH		No effect on testosterone and cortisol levels ↑ Progesterone and estradiol levels ↓ NR0B1 gene ↑ CYP1A1, MC2R, HSD3B2, CYP17, CYP21, CYP11B2, CYP19
Neonatal Leydig cells	ENN B	10–100	↓ Viability by 20%, ↓ estradiol in unstimulated cells ↓ Estradiol and testosterone in LH stimulated cells, probably due to cytotoxicity
Human breast adenocarcinoma RGA cell line	AOH		Agonistic estrogen response, relative estrogenic potential: 0.0004% and equivalent estrogenic quantity of 17β-estradiol: 2.9 fg/mL
Cell free buffer	AOH		Binding affinity to ERα: 10,000 × lower compared to 17β-estradiol Binding affinity to ERβ: 2500 × lower compared to 17β-estradiol Similar EC_{50}
Ishikawa human endometrial adenocarcinoma cell line	AOH	2.5–10	↑ Alkaline phosphatase mRNA and activity ↓ G1 phase and ↑ S and G2/M phase ↓ Cell number due to inhibition of proliferation
Porcine oocytes and embryos	BEA	>0.5	↓ Rate of development of maturing oocyte and 2–4 cell stage embryo, activated oocytes and 2–4 cell stage embryos more sensitive than maturing oocytes, compromised cytoplasmic maturation and abnormal meiosis in oocytes, ↓ Cumulus viability and progesterone synthesis, cumulus cells control intracellular BEA through MDR1 activity, in oocytes mitochondrial function was altered, altered gene expression in cumulus cells and oocytes, altered MDR1 activity in activated oocytes, ↓ viability embryo

(Contd.)

Table 6. (*Contd.*)

Cell line	*Mycotoxin*	*Exposure time*	*Exposure dose (μM)*	*Effect*
Pig granulosa cells	AOH		0.8–1.6	↓ Cell viability, ↓ progesterone levels, ↓ P450scc ↓ α-tubulin, actin and EIF4a
	AME		0.8–1.6	↓ Cell viability, ↓ progesterone levels, ↓ P450scc
	TeA		6.4–100	No influence on viability No influence on progesterone concentrations
Bovine granulosa cells	BEA		3	↓ estradiol and progesterone production ↓ CYP11A1 and CYP19A1 mRNA
			6–10	↓ (fetal calf serum-induced) proliferation
CHO-K1[m]	BEA	24–72 h		IC_{50}: 10.7–2.2 μM Combination of BEA + PAT[n], BEA + STG[o], BEA + PAT + STG: synergistic effect at low (IC < 1), additive effect at higher (IC 0.6–5.9) doses
	ENN A	24–72 h		>7.5–2.83 μM
	ENN A1	24–72 h		8.8–1.65 μM
	ENN B	24–72 h		11.0–2.44 μM
	ENN B1	24–72 h		4.53–2.47 μM
	ENN combinations	24 h		Additive effects: A + B1, A1 + B, B + B1 Synergistic effects: A + A1, A + B, A1 + B1, A1 + B1, A + A1 + B, A + A1 + B1, A1 + B + B1 (higher concentrations) Antagonistic effects: A + A1 + B1, A1 + B +B1 (lower concentrations)
	MON			IC_{50}: >100 μg/mL
THP-1 [p] monocyte	AOH	24–48 h	7.5–15	Cell cycle arrest in S- and G2/M-phase

				↓ CD14 and CD11b upregulation during macrophage differentiation ↓ Downregulation of CD71 during macrophage differentiation, ↓ TNF-α secretion due to ↓ gene expression +DON: additive effect +ZEN: synergistic effect on macrophage differentiation
CCRF-CEM [q]	BEA	24 h	1	Cytotoxicity, apoptosis
Human lymphocytes	MON	48 h	10–25 15–25	Chromosome breaks, chromatid breaks and exchanges, polyploidy, increase in sister chromatid exchanges and micronuclei frequency All effects were dose-dependent
Human immature dendritic cells	BEA			IC_{50}: 1.0 μM
	ENN B			IC_{50}: 1.6 μM
	MON		80	20% mortality, ↓ endocytosis, ↓ CD1a expression
Human mature dendritic cells	BEA			IC_{50}: 2.9 μM, ↓ CCR7 expression, ↑ IL-10 concentration
	ENN B			IC_{50}: 2.6 μM, ↓ CD80, CD86 and CCR7 expression, ↑ IL-10
	MON		80	20% mortality
Human macrophages	BEA		≥0.5	IC_{50}: 2.5 μM, ↓ endocytosis
	ENN B			IC_{50}: 2.5 μM, ↓ endocytosis, ↑ CD71
	MON			↓ Endocytosis, ↓ CD71, ↓ HLA-DR
	AOH	24 h	30	Changed morphology: from round to elongated with dendrite-like protrusions ↑ CD83 and CD86 ↓ HLA-DR and CD68 ↑ Secretion of TNFα and IL-6 ↓ Endocytosis and ↓ autophagy Double DNA strand breaks

(*Contd.*)

Table 6. (*Contd.*)

Cell line	*Mycotoxin*	*Exposure time*	*Exposure dose (μM)*	*Effect*
RAW 2654.7 mouse macrophage	AOH	24–48 h	30	Changed morphology: from round to flattened, star-shaped or elongated spindle-shaped cells Micronuclei, polyploidy, ↑ CD86, CD80, MHCII (T cell activation), ↑ CD11b ↑ mRNA of TNFα and IL-6, but only ↑ TNFα secretion, ↑ endocytosis
Mouse hemidiaphragm preparation	BEA		5	Inhibition (in) directly elicited tetanic muscle contraction; inhibition nerve-evoked and directly elicited muscle twitches, reduction amplitude and frequency of miniature endplate potentials
		1 h	7.5	Inhibition directly elicited twitches, induction contracture, decrease resting membrane potential
		1 h	10	Complete block of (in) directly elicited isometric muscle contraction, amplitude reduction of directly elicited muscle twitch, decrease resting membrane potential
C5-O [r]	MON	72 h		IC_{50}: 34.2 μg/mL
V79 [s]	MON	72 h		IC_{50}: >100 μg/mL
	AOH		5–50	Induction of micronuclei cell cycle arrest in G2 and S phase

↓ decrease; ↑ increase; [a]human adenocarcinoma colon cells; [b]reactive oxygen species; [c]tetrazolium salt reduction assay; [d]Neutral Red assay; [e]lipid peroxidation; [f]mitochondrial membrane permeabilization; [g]human colon carcinoma cells; [h]permeability transition pore; [i]intestinal porcine epithelial cells from the jejunum; [j]transepithelial electrical resistance; [k]human hepatocellular carcinoma cells; [l]human adrenocortical carcinoma cells; [m]Chinese hamster ovary cells; [n]patulin; [o]sterigmatocystin; [p]human acute monocyte leukemia cell line; [q]human leukemia cells; [r]Balb/c mice keratinocyte cells; [s]Chinese hamster lung fibroblast.

The extrapolation of *in vitro* data to the *in vivo* situation is not straightforward and should hence should be done with caution especially since different animal species are considered. However, the low cytotoxicity of ENN B towards IPEC-J2 in Fraeyman et al. (2018) *in vitro* study corroborates the absence of major *in vivo* effects on the intestinal health of broiler chickens.

The additive, synergic or antagonist toxic effects of different mycotoxins should attract more attentions, which is also a new tendency in mycotoxins toxicity research. Several authors were interested in mixtures of TCT and other fusariotoxins, such as FB1, ZEN, and the emerging mycotoxin BEA (Smith et al. 2016). Conclusions of the different authors and studies were species- and organ-dependent. Opposite observations highlight the complexity of the mycotoxin interactions, with the influence of the used cell models. For example, Ruiz et al. (2011a, b) observed antagonistic effects on hamster CHO-K1 and monkey Vero cells with DON+BEA co-exposure (Ruiz et al. 2011a, b), whereas T2+BEA showed opposite cytotoxic effect on CHO-K1 and Vero cells (synergism and antagonism respectively) despite the similar mycotoxin doses, the same time of exposure (24 to 72 h) and the same used assessment to measure cell viability (neutral red assay). Concerning the ternary mixture DON+T2+BEA studied the effects were the same as those observed for T2+BEA on CHO-K1 and Vero cells.

Concerning the mixtures involving ZEN, FB1 and emerging mycotoxins, major results presented antagonistic or additive cytotoxic effects. In particular, ZEN and its derivatives α- and β-zearalenol (α-ZOL and β-ZOL) in binary and ternary mixtures were studied by Wang et al. (2014) on HepG2 and Tatay et al. (2014) on CHO-K1. Wang et al. (2014) showed mainly an antagonistic effect of ZEN+α-ZOL, whereas Tatay et al. (2014) mostly observed additivity between ZEN and its derivatives on CHO-K1. Additivity effects were observed for FB1+BEA at the lowest concentration (about 0.06 μM BEA and FB1) and synergism at the highest dose (about 6 μM BEA and FB1) on PK15 cells (Klarić et al. 2006). Several authors studied binary, ternary, and quaternary EN mixtures (ENA, ENA1, ENB, and ENB1) and in similar concentrations, with the same cell viability assessment and time of exposure (MTT assay, during 24 h) (Lu et al. 2013, Prosperini et al. 2014). Observed effects are not necessarily dose- and time-dependent. Prosperini et al. (2014) indicated antagonism at low cytotoxicity levels (IC_5–IC_{25}) and additivity at medium and high inhibitory concentration levels (IC_{50}–IC_{90}) on Caco-2 cells. Lu et al. (2013) observed synergistic effects at low cytotoxicity levels (IC_{25}) and additivity at medium and high inhibitory concentration levels (IC_{50}–IC_{90}) on CHO-K1.

Additionally to cocktail effect, the concerns regarding the contamination of foods by masked mycotoxins have been raised due to their re-convertion to its native form in the digestive tract (Freire and Sant'Ana 2018). At the same time, the masked mycotoxins were easily eluded by the conventional screening methods due to biotransformation, leading to underreporting. The lack of conclusive toxicokinetic and toxicodynamic studies, in addition to the limited understanding of the transformation mechanisms of mycotoxins also hinders the determination of the maximum tolerable level of these metabolites in food, being a major future challenge.

3.2 Toxicology of Emerging Mycotoxins: *In Vivo* Data

In vivo data on the toxicity of emerging mycotoxins are very scarce; they have been obtained on rodents and on farm animals, especially poultry but never on humans.

Enniatin B (ENN B)

Few toxicological studies of ENN B have been performed *in vivo*. A toxicokinetic trial using pigs demonstrated a high bioavailability for ENN B. Although ENN B bioaccumulation in the lipophilic tissues was observed no acute toxicity was observed in mice after intraperitoneal administration (Prosperini et al. 2017).

Beauvericin (BEA)

In acute *in vivo* genotoxicity studies performed in mice (oral exposure for three days, at doses ranging from 50 to 200 mg/kg bw/day), BEA showed no effects in terms of mortality, clinical signs and body weight decrease. Comet assay did not show significant difference with the control in the various tissues analyzed (liver, duodenum, kidney, colon, spleen and bone marrow) (Caloni et al. 2020).

Several trials were performed in birds; however due to the use of naturally contaminated grains, other mycotoxins such as DON and/or MON were present in combination with BEA. These experiments showed that the contaminated feed had no influence on growth performances, carcass characteristics and blood parameters of the animals, even with high levels of mycotoxin contamination (Caloni et al. 2020).

Moniliformin (MON)

The toxicokinetic behavior of MON is largely unknown. After single oral gavage, 42% of the administered MON was excreted in the urine of rats within 24 h post administration, and less than 1% was excreted in the feces (Fraeyman et al. 2017). MON is strikingly toxic *in vivo*. It targets the heart, causing acute heart failure, but the mycotoxins can also cause muscle weakness, respiratory distress and negatively affect immunity and animal performance. Accordingly, the massive increase of the number of mitochondria associated with the disruption of heart muscle fibres observed in Japanese quail and in broilers appears to be a compensatory event for the MON-induced decrease of cellular energy with 27 to 154 mg MON/kg feed (Fremy et al. 2019).

Toxins from Alternaria

In vivo studies on the effects of Alternaria mycotoxins on reproductive and developmental health are limited. In hamsters, AME (200 mg/kg bw, intraperitoneal administration) was toxic to fetus but did not cause teratogenic malformations. In a chicken embryo assay, AOH, AME and ALT did not cause mortality, difference in weight of the hatched chicks or teratogenic effects at doses up to 1000, 500 and 1000 μg/egg, respectively. TeA caused emesis, salivation, tachycardia, hemorrhages and hemorrhagic gastro-enteropathy in rats, mice, chickens, dogs and monkeys (Fraeyman et al. 2017).

4. Conclusion

Foods that were commonly analyzed were cereals, meat, vegetables, fruits, nuts and beverages. The findings in food were below the current European legislation, except for some sporadic samples of wine and milk meaning less than 1% of total analyzed samples. Dietary exposure was evaluated, through the estimated daily intake mycotoxin evaluation and risk assessment concluded that relatively scarce toxicological concern was associated with mycotoxins exposure. However, a special attention should be paid to meat and cereal products high percentile consumers (Carballo et al. 2019). This study characterizes the health for OTA, DON, NIV, T-2, HT-2, AFs, ZEN FBs and PAT by comparing the estimate dietary intake (EDI) previously calculated with the tolerable daily intake (TDI) established (ng kg^{-1} bw day^{-1}). Despite the relatively high frequency of mycotoxins detected in the different food of habitual consumption, calculated EDIs were generally below the established TDIs, except for DON and OTA who amounted in some TDSs, appreciable ratio of their respective TDI through cereal and meat products.

References

Beltrán, E., Ibáñez, M., Portolés, T., Ripollés, C., Sancho, J.V., Yusà V., Marín, S. and Hernández, F. 2013. Development of sensitive and rapid analytical methodology for food analysis of 18 mycotoxins included in a total diet study. Analytica Chimica Acta 783: 39–48.

Bennett, J.W. and Klich, M. 2003. Mycotoxins. Clinical Microbiology Review 16: 497–516. doi.org/10.1128/cmr.16.3.497-516.2003

Caloni, F., Fossati, P., Anadón, A. and Bertero, A. 2020. Beauvericin: The beauty and the beast. Environmental Toxicology and Pharmacology 75: 103349. doi: 10.1016/j.etap.2020.103349.

Campone, L., Rizzo, S., Piccinelli, A.L., Celano, R., Pagano, I., Russo, M., Labra, M. and Rastrelli, L. 2020. Determination of mycotoxins in beer by multi heart-cutting two-dimensional liquid chromatography tandem mass spectrometry method. Food Chemical 318: 126496.

Cano-Sancho, G., Ramos, V., Marín, S. and Sanchis. V. 2012. Estudio de dieta total en Cataluña. 2008-2009 Generalitat de Cataluña, Barcelona, España. Micotoxinas 135–140.

Carballo, D., Moltó, J.C., Berrada, H. and Ferrer. E. 2018. Presence of mycotoxins in ready-to-eat food and subsequent risk assessment. Food and Chemical Toxicology 121: 558–565. doi.org/10.1016/j.fct.2018.09.054

Carballo, D., Tolosa, J., Ferrer, E. and Berrada, H. 2019. Dietary exposure assessment to mycotoxins through total diet studies: A review. Food and Chemical Toxicology 128: 8–20. doi.org/10.1016/j.fct.2019.03.033

Cigić, I.K. and Prosen, H. 2009. An overview of conventional and emerging analytical methods for the determination of mycotoxins. Int. J. Mol. Sci. Jan. 10(1): 62–115. doi: 10.3390/ijms10010062

Dong, H., Xian, Y., Xiao, K., Wu, Y., Zhu, L. and He, J. 2019. Development and comparison of single-step solid phase extraction and QuEChERS clean-up for the analysis of 7 mycotoxins in fruits and vegetables during storage by UHPLC-MS/MS. Food Chemistry 274: 471–479.

EFSA Panel on Contaminants in the Food Chain (CONTAM). 2016. Scientific opinion on the appropriateness to set a group health-based guidance value for zearalenone and its modified forms. EFSA Journal 14: 4425.

Fraeyman, S., Croubels, S., Devreese, M. and Antonissen, G. 2017. Emerging Fusarium and Alternaria mycotoxins: Occurrence, toxicity and toxicokinetics. Toxins 18: 9. doi.org/10.3390/toxins9070228

Fraeyman, S., Meyer, E., Devreese, M., Antonissen, G., Demeyere, K., Haesebrouck, F. and Croubels, S. 2018. Comparative in vitro cytotoxicity of the emerging Fusarium mycotoxins beauvericin and enniatins to porcine intestinal epithelial cells. Food and Chemical Toxicology 121: 566–572. doi.org/10.1016/j.fct.2018.09.053

Fredlund, E., Gidlund, A., Sulyok, M., Börjesson, T., Krska, R., Olsen, M. and Lindblad, M. 2013. Deoxynivalenol and other selected Fusarium toxins in Swedish oats – Occurrence and correlation to specific Fusarium species. Int. J. Food Microbiol. 167: 276–283.

Freire, L. and Sant'Ana, A.S. 2018. Modified mycotoxins: An updated review on their formation, detection, occurrence, and toxic effects. Food and Chemical Toxicology 111: 189–205. doi.org/10.1016/j.fct.2017.11.021

Fremy, J.M., Alassane–Kpembi, I., Oswald, I.P, Cottril, B. and van Egmond, H.P. 2019. A review on combined effects of moniliformin and co-occurring Fusarium toxins in farm animals. World Mycotoxin Journal 12: 281–291. doi.org/10.3920/WMJ2018.2405

FSANZ – Food Standards Australia New Zealand. 2001. The 19th Australian total diet study. Canberra: Food standards Australia New Zealand Available from: http://www.foodstandards.gov.au/publications/Pages/19thaustraliantotal dietsurveyapril2001/ 19thaustraliantotaldietsurvey/Default.aspx, 2001.

FSANZ – Food Standards Australia New Zealand. 2003. The 20th Australian total diet study. Canberra: Food standards Australia New Zealand Available from: http://www.foodstandards.gov.au/publications/Pages/20thaustraliantot aldietsurveyjanuary2003/ 20thaustraliantotaldietsurveyfullreport/Default.aspx, 2003.

FSAI-Food Safety Authority of Ireland. Report on a Total Diet Study Carried Out by the Food Safety Authority of Ireland in the Period 2012–2014. 2016, https://www.fsai.ie/publications_TDS_2012-2014/9.7.18

García-Moraleja, A., Font, G., Mañes, J. and Ferrer, E. 2015. Analysis of mycotoxins in coffee and risk assessment in Spanish adolescents and adults. Food and Chemical Toxicology 86: 225–233. doi.org/10.1016/j.fct.2015.10.014

Grenier, B. and Oswald, I. 2011. Mycotoxin co-contamination of food and feed: Meta-analysis of publications describing toxicological interactions. World Mycotoxin Journal 4: 285–313. doi.org/10.3920/WMJ2011.1281

Huong, B.T.M., Brimer, L. and Dalsgaard, A. 2016. Dietary exposure to aflatoxin B 1, ochratoxin A and fuminisins of adults in Lao Cai province, Viet Nam: A total dietary study approach. Food and Chemical Toxicology 98: 127–133. doi.org/10.1016/j.fct.2016.10.012

Jestoi, M. 2008. Emerging Fusarium—Mycotoxins fusaproliferin, beauvericin, enniatins, and moniliformin—A review. Critical Reviews of Food Science and Nutrition 48: 21–49. doi.org/10.1080/10408390601062021

Juan, C., Berrada, H., Mañes, J. and Oueslati. S. 2017. Multi-mycotoxin determination in barley and derived products from Tunisia and estimation of their dietary intake. Food and Chemical Toxicology 103: 148–156. doi.org/10.1016/j.fct.2017.02.037

Klarić, M.Š., Pepeljnjak, S., Domijan, A.M. and Petrik, J. 2006. Lipid peroxidation and glutathione levels in porcine kidney PK15 cells after individual and combined treatment with fumonisin B1, beauvericin and ochratoxin A. Basic Clinical Pharmacology and Toxicology 100: 157–164.

Knutsen, H.K., Alexander, J., Barregard, L., Bignami, M., Bruschweiler, B., Ceccatelli, S., Cottrill, B., Dinovi, M., Grasl-Kraupp, B., Hogstrand, C., Hoogenboom, L., Nebbia, C.S., Oswald, IP., Petersen, A., Rose, M., Roudot, A.C., Schwerdtle, T., Vleminckx, C., Vollmer, G., Wallace, H., De Saeger, S., Sundstøl Eriksen, G., Farmer, P., Fremy, J.M., Gong, Y.Y., Meyer, K., Naegeli, H., Parent-Massin, D., Rietjens, I., van Egmond, H., Altieri, A., Eskola, M., Gergelova, P., Ramos Bordajandi, L., Benkova, B., Dörr, B., Gkrillas, A.,Gustavsson, N., van Manen, M. and Edler, L. 2017. Risks to human and animal health related to the presence of deoxynivalenol and its acetylated and modified forms in food and feed. EFSA Journal 15: 4718.

Knutsen, H.K., Barregard, L., Bignami, M., Bruschweiler, B., Ceccatelli, S., Cottrill, B., Dinovi, M., Edler, L., Grasl-Kraupp, B., Hogstrand, C., Hoogenboom, L., Nebbia, C.S., Petersen, A., Rose, M., Roudot, A.C., Schwerdtle, T., Vleminckx, C., Vollmer, G., Wallace, H., Dall'Asta, C., Humpf, H.U., Galli, C, Gutleb, A., Metzler, M., Oswald, I.P., Parent-Massin, D., Binaglia, M., Steinkellner, H. and Alexander, J. 2018. Appropriateness to set a group health-based guidance value for fumonisins and their modified forms. EFSA Journal 16: 5172.

Kouadio, J.H., Dano, S.D., Moukha, S., Mobio, T.A. and Creppy, E.E. 2007. Effects of combinations of Fusarium mycotoxins on the inhibition of macromolecular synthesis, malondialdehyde levels, DNA methylation and fragmentation, and viability in Caco-2 cells. Toxicon 49: 306–317. doi.org/10.1016/j.toxicon.2006.09.029

Leblanc, J.C., Tard, A., Volatier, J.L. and Verger. P. 2005. Estimated dietary exposure to principal food mycotoxins from the first French Total Diet Study. Food and Additive Contaminants 22: 652–672. doi.org/10.1080/02652030500159938

Lindblad, M., Gidlund, A., Sulyok, M., Börjesson, T., Krska, R., Olsen, M. and Fredlund, E. 2013. Deoxynivalenol and other selected Fusarium toxins in Swedish wheat – Occurrence and correlation to specific Fusarium species. Int. J. Food Microbiol. 167: 284–291.

López, P., De Rijk, T., Sprong, R.C., Mengelers, M.J.B., Castenmiller, J.J.M. and Alewijn. M.A. 2016. Mycotoxin-dedicated total diet study in The Netherlands in 2013: Part II – Occurrence. World Mycotoxin Journal 9: 89–108. doi.org/10.3920/WMJ2015.1906

Lu, H., Fernández-Franzón, M., Font, G. and Ruiz, M.J. 2013. Toxicity evaluation of individual and mixed enniatins using an in vitro method with CHO-K1 cells. Toxicology in Vitro 27: 672–680. doi.org/10.1016/j.tiv.2012.11.009

Mallebrera, B., Prosperini, A., Font, G. and Ruiz, M.J. 2018. In vitro mechanisms of Beauvericin toxicity: A review. Food and Chemical Toxicology 111: 537–545. doi.org/10.1016/j.fct.2017.11.019

Marin, S., Ramos, A.J., Cano-Sancho, G. and Sanchis, V. 2013. Mycotoxins: Occurrence, toxicology, and exposure assessment. Food Chem. Toxicol. 60: 218–237. doi: 10.1016/j.fct.2013.07.047

Pallarés, N., Font, G., Mañes, J. and Ferrer, E. 2017. Multimycotoxin LC–MS/MS analysis in tea beverages after dispersive liquid–liquid microextraction (DLLME). Journal of Agricultural and Food Chemistry 65: 10282–10289. doi.org/10.1021/acs.jafc.7b03507

Pallarés, N., Carballo, D., Ferrer, E., Fernández-Franzón, M. and Berrada, H. 2019. Mycotoxin dietary exposure assessment through fruit juices consumption in children and adult population. Toxins 22: 684–696. doi.org/10.3390/toxins11120684

Prosperini, A., Font, G. and Ruiz, M.J. 2014. Interaction effects of Fusarium enniatins (A, A1, B and B1) combinations on in vitro cytotoxicity of Caco-2 cells. Toxicology in Vitro 28: 88–94. doi.org/10.1016/j.tiv.2013.06.021

Prosperini, A., Berrada, H., Ruiz, M.J., Caloni, F., Coccini, T., Spicer, L.J. and Perego, M.C. 2017. Lafranconi, A: A Review of the mycotoxin enniatin B. Front Public Health. 16: 5–304. doi.org/10.3389/fpubh.2017.00304

Qiu, N.N., Lyu, B., Zhou, S., Zhao, Y.F. and Wu, Y.N. 2017. The contamination and dietary exposure analysis for seven mycotoxins in the Fifth Chinese Total Diet Study. Chinese Journal of Preventive Medicine 51: 943–948. doi.org/10.3760/cma.j.issn.0253-9624.2017.10.014

Raad, F., Nasreddine, L., Hilan, C., Bartosik, M. and Parent-Massin, D. 2014. Dietary exposure to aflatoxins, ochratoxin A and deoxynivalenol from a total diet study in an adult urban Lebanese population. Food and Chemical Toxicology 73: 35–43. doi.org/10.1016/j.fct.2014.07.034

Rodríguez-Carrasco, Y., Mañes, J., Berrada, H. and Juan, C. 2016. Development and validation of a LC-ESI-MS/MS method for the determination of Alternaria toxins alternariol, alternariol methyl-ether and tentoxin in tomato and tomato-based products. Toxins 8: 328. doi.org/10.3390/toxins8110328

Ruiz, M.J., Franzova, P., Juan-García, A. and Font, G. 2011a. Toxicological interactions between the mycotoxins beauvericin, deoxynivalenol and T-2 toxin in CHO-K1 cells in vitro. Toxicon 58: 315–326. doi.org/10.1016/j.toxicon.2011.07.015

Ruiz, M.J. Macáková, P. Juan-García, A. and Font, G. 2011b. Cytotoxic effects of mycotoxin combinations in mammalian kidney cells. Food and Chemical Toxicology 49: 2718–2724. doi.org/10.1016/j.fct.2011.07.021

Sakuma, H., Watanabe, Y., Furusawa, H., Yoshinari, T., Akashi, H., Kawakami, H., Saito, S. and Sugita-Konishi, Y. 2013. Estimated dietary exposure to mycotoxins after taking into account the cooking of staple foods in Japan. Toxins 5: 1032–1042. doi.org/10.3390/toxins5051032

Saladino, F., Quiles, J.M., Mañes, J., Fernández-Franzón, M., Luciano, F.B. and Meca, G. 2017. Dietary exposure to mycotoxins through the consumption of commercial bread loaf in Valencia, Spain. LWT-Food Science and Technology 75: 697–701. doi.org/10.1016%2Fj.lwt.2016.10.029

Schrenk, D., Bignami, M., Bodin, L., Chipman, J.K., del Mazo, J., Grasl-Kraupp, B., Hogstrand, C., Hoogenboom, L., Leblanc, J.C., Nebbia, C.S., Nielsen, E., Ntzani, E., Petersen, A., Sand, S., Schwerdtle, T., Vleminckx, C., Marko, D., Oswald, I.P., Piersma, A., Routledge, M., Schlatter, J., Baert, K., Gergelova, P. and Wallace, H. 2020a. Risk assessment of aflatoxins in food. EFSA Journal 18: 6040

Schrenk, D., Bodin, L., Chipman, J.K., del Mazo, J., Grasl-Kraupp, B., Hogstrand, C., Hoogenboom, L., Leblanc, J.C., Nebbia, C.S., Nielsen, E., Ntzani, E., Petersen, A., Sand, S., Schwerdtle, T., Vleminckx, C., Wallace, H., Alexander, J., Dall'Asta, C., Mally, A., Metzler, M., Binaglia, M., Horvath, Z., Steinkellner, H. and Bignami, M. 2020b. Scientific opinion on the risks to public health related to the presence of ochratoxin A in food. EFSA Journal 18: 6113.

Sirot, V., Fremy, J. and Leblanc. J. 2013. Dietary exposure to mycotoxins and health risk assessment in the second French total diet study. Food and Chemical Toxicology 252: 1–11. doi.org/10.1016/j.fct.2012.10.036

Smith, M.C., Madec, S., Coton, E. and Hymery, N. 2016. Natural Co-occurrence of mycotoxins in foods and feeds and their in vitro combined toxicological effects. Toxins 8: 94. doi.org/10.3390/toxins8040094

Sprong, R.C., De Wit-Bos, L., Zeilmaker, M.J., Alewijn, M., Castenmiller, J.J.M. and Mengelers, M.J.B. 2016a. A mycotoxin-dedicated total diet study in The Netherlands in 2013: Part I – Design. World Mycotoxin Journal 9: 73–88. doi.org/10.3920/WMJ2015.1904

Sprong, R.C., De Wit-Bos, L., Te Biesebeek, J.D., Alewijn, M., Lopez, P. and Mengelers. M.J.B. 2016b. A mycotoxin-dedicated total diet study in The Netherlands in 2013: Part III – Exposure and risk assessment. World Mycotoxin Journal 9: 109–128. doi.org/10.3920/WMJ2015.1905

Stanciu, O., Juan, C., Miere, D., Loghin, F. and Mañes, J. 2017. Presence of Enniatins and Beauvericin in Romanian wheat samples: From raw material to products for direct human consumption. Toxins (Basel). 12; 9(6): 189. doi: 10.3390/toxins9060189

Tam, J., Pantazopoulos, P., Scott, P.M., Moisey, J., Dabeka, R.W. and Richard. I.D.K. 2011. Application of isotope dilution mass spectrometry: Determination of ochratoxin A in the Canadian Total Diet Study. Food and Additive Contaminants 28: 754–761. doi.org/10.1080%2F19440049.2010.504750

Tatay, E., Meca, G., Font, G. and Ruiz, M.J. 2014. Interactive effects of zearalenone and its metabolites on cytotoxicity and metabolization in ovarian CHO-K1 cells. Toxicology in Vitro 28: 95–103. doi.org/10.1016/j.tiv.2013.06.025

Urieta, I., Jalon, M., Garcia, J.L. and De Galdeano, G. 1991. Food surveillance in the Basque country (Spain) I: The design of a total diet study. Food and Additive Contaminants 8: 371–380.

Urieta, I., Jalon, M. and Eguileor. I. 1996. Food surveillance in the Basque country (Spain). II: Estimation of the dietary intake of organochlorine pesticides, heavy metals, arsenic, aflatoxin M1, iron and zinc through the Total Diet Study, 1990/91. Food and Additive Contaminants 13: 29–52.

Veprikova, Z., Zachariasova, M., Dzuman, Z., Zachariasova, A., Fenclova, M., Slavikova, P., Vaclavikova, M., Mastovska, K., Hengst, D. and Hajslova, J. 2015. Mycotoxins in plant-based dietary supplements: Hidden health risk for consumers. Journal of Agricultural of Food Chemistry 63: 6633–6643. doi.org/10.1021/acs.jafc.5b02105

Wan, L.Y.M., Turner, P.C. and El-Nezami, H. 2013. Individual and combined cytotoxic effects of Fusarium toxins (deoxynivalenol, nivalenol, zearalenone and fumonisins B1) on swine jejunal epithelial cells. Food and Chemical Toxicology 57: 276–283. doi.org/10.1016/j.fct.2013.03.034

Wang, H.W., Wang, J.Q., Zheng, B.Q. Li, S.L., Zhang, Y.D., Li, F.D. and Zheng, N. 2014. Cytotoxicity induced by ochratoxin A, zearalenone, and α-zearalenol: Effects of individual and combined treatment. Food and Chemical Toxicology 71: 217–224. doi.org/10.1016/j.fct.2014.05.032

Wen, J., Mu, P. and Deng, Y. 2016. Mycotoxins: Cytotoxicity and biotransformation in animal cells. Toxicological Research 5: 377–387. doi.org/10.1039%2Fc5tx00293a

Yang, J., Li, J., Jiang, Y., Duan, X., Qu, H., Yang, B., Chen, F. and Sivakumar, D. 2014. Natural occurrence, analysis, and prevention of mycotoxins in fruits and their processed products. Crit. Rev. Food Sci. Nutr. 54(1): 64–83. doi: 10.1080/10408398.2011.569860

Yau, A.T.C., Chen, M.Y.Y., Lam, C.H., Ho, Y.Y., Xiao, Y. and Chung, S.W.C. 2016. Dietary exposure to mycotoxins of the Hong Kong adult population from a Total Diet Study. Food and Additive Contaminants 33: 1026–1035. doi.org/10.1080/19440049.2016.1184995

CHAPTER

9

Gut Microbiome and Their Possible Roles in Combating Mycotoxins

Anil Kumar Anal*, Sushil Koirala and Smriti Shrestha
Department of Food Agriculture and Bioresources, Asian Institute of Technology, Pathum Thani 12120, Thailand

1. Introduction

Over the past decade, immense interest in food safety has been a rising concern for both human, animal and plant health as well as the key towards access to the international food markets. Awareness in the global health burden of food-borne diseases has led to the interest in food safety to all including producers, processors, suppliers, retailers and consumers globally (Hoffmann et al. 2019). Foods contaminated with any type of hazards (physical, chemical and biological) cause food-borne illness to the consumers. Mycotoxins which occur mostly as a natural contaminant in food and feed, are responsible for autoimmune diseases, allergies, mutations and cancers (García-Cela et al. 2012). Moreover, mycotoxin contamination of food and feed is responsible for a huge loss of human life, livestock, forage crops and feeds, and the overall economy of the country every year (Ji et al. 2016).

Mycotoxins, the secondary metabolites of fungal origins are very small molecules (MW < 700 Da) and highly toxic in nature, causing severe health effects. Till date, more than 400 mycotoxins have been identified, among which aflatoxins (AFs), ochratoxin A (OTA), deoxynivalenol (DON) and zearalenone (ZEN) are strictly regulated for agro-products safety (Li et al. 2014). The highly toxic and thermally stable mycotoxins can enter the human food supply chain *via* bioaccumulation in food products that harbour fungal growth. Humans are at the risk of exposure to mycotoxins either through the environment (food, air) or biological such as residues in food materials, metabolites and adducts in tissues, fluids, and excreta. Among the food sources, toxin contaminated plant-derived foods and animal products with residual mycotoxins are the significant sources of mycotoxin exposure to humans (Ladeira 2016).

Several studies have revealed that along with mammalian metabolism, human microbiome in the gut plays an essential role in the metabolic fate

*Corresponding author: anilkumar@ait.ac.th; anil.anal@gmail.com

of mycotoxins. The metabolic transformation of mycotoxin can substantially affect its toxicity level depending on the microbiota composition in the gastrointestinal (GI) tract. Moreover, bilateral interaction takes place between both gut microbiome and mycotoxins in the GI tract with a high potential to induce the compositional changes in gut microbiota (Marko 2017). The GI tract microbial community is commonly referred to as gut microbiota, has coevolved with the host and exhibits a homeostatic community structure amongst each other and with the host. The gut microbiota are responsible for numerous health impacts including metabolism, immunity, physiology and nutrient absorption (Konstantinov 2017).

2. Human Gut Microbiome

The human microbiome is a collective term for all genomes of microbial community, also known as microbiota inhabiting the whole body. These include from protozoa to archaea, eukaryotes, and overwhelming microscopic organisms that live symbiotically inside different parts of the human body including oral cavity, genital organs, respiratory tract, skin and gastrointestinal tract (Lloyd-Price et al. 2016). The microbial population of human microbiota is approximately ~10^{13}-10^{14} microbial cells with microbial cell to human cell ratio being 1:1. The estimation of these numbers is derived from the total microbial cells in the colon (3.8×10^{13}), the organ which pools the densest and maximum number of microbes (Sender et al. 2016).

The microbes present in the human gastrointestinal tract are known as gut flora or gut microbiota. Gut is the one niche that the human microbiota inhabits. They consist of both commensal (mutual) and pathogenic microorganisms in the gastrointestinal tract (GIT), including but not limited to bacteria, fungi, viruses, protozoa and their total genetic components (Cresci and Bawden 2015). With over 1000 bacterial species and three million genes, GIT is a complex microbial ecosystem governing human health. Gut microbiota, which is often termed as a human bioreactor, is an intricate component of the GIT with an estimated total microbial population of around 100 trillion microorganisms that when weighed would account around a 2 kg mass in the gut alone. The human gut microbiota is classified into four main or phyla, predominated in order of *Firmicutes* (~64%), *Bacteroidetes* (~23%), *Actinobacteria* (~3%) and *Proteobacteria* (~2%) (Sánchez-Tapia et al. 2019). Gut microbiota is referred to as an invisible organ due to its metabolic activity being 100 times higher than that of the liver and also as the symbiotic bacteria due to its symbiotic relationship with the host (An et al. 2019). This microbial community in the gut is active, not passive, having communication with the host via gut-brain signalling and mediating a range of other health benefits (Mohajeri et al. 2018).

2.1 Gastrointestinal (GI) Tract as Human Bioreactor

Gut microbiota essentially provides favourable conditions (pH, temperature, nutrients) for the breakdown of dietary fibres and other non-digestible portions of food until the GI tract. These dietary fibres are usually non-digestible portions of food. However, gut flora converts them *via* fermentation to produce short-chain fatty acids (SCFAs) and gases which are useful for gut health. Acetate, Butyrate, Propionate and Formate are major SCFAs produced as a result of fermentation or breakdown of these non-digestible portions of food (Wong et al. 2006).

For epithelial cells of the colon, **butyrate** is the primary energy source that triggers apoptosis of colon cancer cells and intestinal gluconeogenesis which has a positive effect on glucose and overall energy homeostasis. It promotes growth of colonocytes through β-oxidation by consuming oxygen, which then induces hypoxia maintaining the balance of oxygen levels in the gut. It substantially prevents gut microbiota dysbiosis (Byndloss et al. 2017). Another product of fermentation in the human gut is **propionate**. Propionate carries significance by channelling gluconeogenesis as it is transferred to the liver from the gut. It also mediates satiety signalling via communication with the receptors from gut fatty acid (De Vadder et al. 2014). Among all SCFAs, **acetate** is the most abundant. It is considered useful for promoting the growth of commensal bacteria in the gut. It is used in the metabolism of cholesterol and lipogenesis. Acetate is also perceived to play a significant role in human appetite regulation (Frost et al. 2014).

The significance of SCFAs have been demonstrated and documented through various studies. Higher production of SCFAs is associated with lower induced obesity via diet (Lin et al. 2012) and with lower resistance to insulin (Zhao et al. 2018). An *in vivo* study on mice from Lin et al. (2012) demonstrated that both butyrate and propionate were responsible SCFAs (metabolite) for controlling hormones present in the gut along with regulating the appetite and food intake.

2.2 Host-Microbiome Interactions

The enzymes produced by the gut microorganisms contribute to the metabolism of bile acid which then generates secondary and unconjugated bile acids that act as a signalling molecule to initiate significant host pathways (Long et al. 2017). The gut microbiota-host has primarily evolved as a symbiotic relationship due to the presence of commensal microbes, and at the same time, a pathological bond is observed due to pathogenic microbiota co-existing together in the gut. The disintegration of gut microbiota, termed as dysbiosis, is associated with metabolic disorders, including metabolic syndrome, inflammatory bowel disease, obesity, and certain cancers (Sarafian et al. 2017). Other products from the metabolism of gut microbiota have been correlated with human health which includes compounds like trimethylamine and indole propionic acid. Trimethylamine has been found to increase the risk associated with atherosclerosis and cardiovascular diseases (Tang et al. 2013) while on the other hand indolepropionic acid possesses distinct radical scavenging activity *in vitro* reducing the risk for type 2 diabetes (De Mello et al. 2017). In the human body, most bacteria are found within the GIT such that majority of predominantly anaerobic bacteria harbours within the colon region, as shown in Fig. 1.

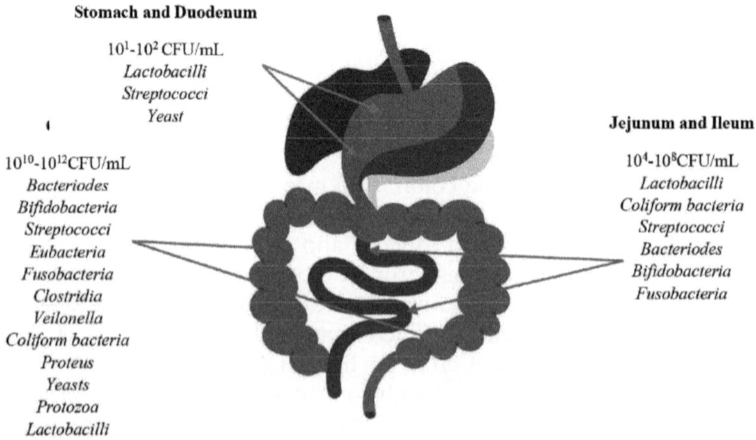

Figure 1. Gut microbiota predominance. Adapted from Cresci and Izzo 2019

2.3 Factors Regulating the Microbiome in the Gut

Composition of the gut microbiome is significantly different among each individual, resting on factors like birth method, medication, geography, age, diet, exercise and stress

2.3.1 Microbiome Inherited during the Birthing Process

During birth, an infant inherits gut microbiome from mother. Notably, the birth method, which includes vaginal microbiota (natural) or cesarean section has a profound significance on the composition of the gut microbiome (Cresci and Izzo 2019). It is during the birthing process; an infant is exposed to the environment full of bacteria and are thus inoculated with microbiota. This brings diversity in gut microbiota among individuals (Cresci and Bawden 2015).

2.3.2 Microbiome Inherited during Breastfeeding

The colonization of gut microbiota continues from infanthood to breast or formula feeding. Breast milk itself consists of more than 600 species of bacteria along with unique and beneficial several species of *Bifidobacteria,* which are probiotic in nature. Out of several beneficial components, breast milk constitutes oligosaccharides which are present as the third-largest components. Oligosaccharides are indigestible polysaccharides that promotethe growth of beneficial *bacteriodetes* as a prebiotic within GIT. Immune support is provided by the proliferation of *Bacteroidetes*. Immunoglobulin A (IgA) is increased by *bacteriodetes* which controls the intestinal immune system and enhances mucosa present in the intestine that can be useful in the prevention of pathogenic bacteria (Cresci and Bawden 2015).

2.3.3 Microbiome Inherited through Diet

Third factor regulating gut microbiota is diet. Diet is very significant to the microbial diversity.It affects the overall composition and presence of microbes in GIT. Ingestion of foods containing macronutrients like carbohydrates, proteins, fat, vitamins and other foods containing bioactive components determine optimum gut health due to an increase in commensal gut microbiota (Cresci and Bawden 2015). A distinct change in gut microbiota can be observed when an infant starts to take solid food which develops its microbiome from infant to grown-up adult-like by three years of age (Herman et al. 2019). Nature of diet and its pattern governs gut microbiota composition and subsequently its functionality. An example of this is *Prevotella* genus which is associated with a plant-based diet while for animal-based diet, *Bactero* are abundant (Xifra et al. 2016).

Along with these factors, a significant increase in the bacteria producing butyrate has been observed with regular exercise and physical activities. This helps in the symbiosis between both pathogenic and commensal bacteria in the gut. Stress, however, promotes dysbiosis, which results in dysregulation, inflammation, impaired intestinal barrier (Gubert et al. 2019). Emergence of antibiotics possess threats to gut microbiota as well. It leads to an imbalance between host-microbiome, disturbing homeostasis, promoting the growth of harmful bacteria (Li et al. 2020). The diversity and the number of gut microbiota are not the same and stable over the lifetime of an individual host body and is most prone to fluctuate with age, change in diet and disease. Moreover, ecological and geographical factors contribute to the change in gut microbiota (Xifra et al. 2016).

3. Mycotoxins

Mycotoxins are the naturally occurring toxic compounds, with low molecular mass (MW ~700 Da) and are produced as secondary metabolites by the different fungal species belonging to the genera mainly *Aspergillus*, *Fusarium*, and *Penicillium* (Ramírez-Guzmán et al. 2018). The term mycotoxin is coined from *myco*-referring mold and *toxin*-referring poison. The syndrome resulting due to mycotoxin poisoning is termed as mycotoxicosis (Haschek et al. 2002). Mycotoxins are responsible for the physiological, pathological and biochemical toxicity to vertebrates and other animals, excluding its threatening effect to bacteria and plant distinguished as antibiotics and phytotoxins, respectively (Peng et al. 2018).

Humans are prone to mycotoxin exposure via ingestion methods, inhalation of mycotoxin itself or contact methods via the skin. The impact of mycotoxin on the health condition of human and animal depends on the level of exposure which might be acute or chronic in nature. There are two possible ways for the mycotoxicosis to occur. Firstly, it can occur directly due to consumption of contaminated food, hence termed as primary mycotoxicosis or indirectly through consumption of animal origin food which were fed with contaminated feeds, termed as secondary mycotoxicosis (Nieto et al. 2018). These fungal species require specific temperature, water activity aeration and substrate condition to produce toxins. The toxin can contaminate the diverse range of agricultural commodity, including cereals, nuts, spices, fruits, oilseeds and others at any stage throughout the food chain (Marín et al. 2018). Globally, 25% of the crops are estimated to be contaminated by fungi, of which aflatoxins, trichothecenes, fumonisin, ochratoxins, zearalenone and sterigmatocystin are of agricultural importance (Stein and Bulboaca 2017). Some of the common fungi and mycotoxins of world-wide importance are listed in Table 1.

Globally, almost 400 mycotoxins have been identified so far with 20 of them found to be present in prominent amounts in food and feeds, thus making it a significant food safety concern. Most of these toxins are produced by fungi of the genera, *Aspergillus*, *Penicillium* and *Fusarium*. In general, aflatoxins are most commonly occurring mycotoxins with its variants being B1, B2, G1, G2 and M1. Similarly, other mycotoxins of significance are ochratoxin A, patulin, citrinin, and the fusarium toxins namely fumonisin (B1, B2 and B3), T-2 and HT-2 toxins, zearalenone, nivalenol, along with deoxynivalenol (DON). Group 1 mycotoxins are classified as human carcinogens with aflatoxins as major mycotoxin. Group 2 mycotoxins are classified as possible human carcinogens with ochratoxins and fumonisin as potent mycotoxins while trichothecenes and zearalenone are not recognized as human carcinogens (Group 3). Besides, some of these mycotoxins suppress the immune system, thereby exposing the consumer to health threats (Ogbuewu 2011). Some of the common fungi and mycotoxins of world-wide importance are listed in Table 1.

Mycotoxins are synthesized as secondary metabolites, which onset when fungi complete initial logarithmic phase as generally associated with the growth curve of microorganisms. The production of mycotoxins is also related and triggered by the onset of conidiation (asexual reproduction of filamentous fungi with spores). Unravelling metabolic pathway leading to the production of mycotoxins is still an area to be explored, most significantly intending to trace out strategies to lessen mycotoxin burden in plants. Some common mycotoxins with their chemical structure and safe limit are shown in Table 2.

3.1 Plant, Feed and Food Chain of Mycotoxin

Mycotoxin contamination of agricultural commodities, feed and other food products are integrated closed-loop cycle as shown in Fig. 2. These commodities are transferred from the farm from plant phase to the collection centre, then from the port of exporting country to port of importing country, and then to distribution centres and finally to the consumer (Chein et al. 2019). During these chains of the cycle from producer plants to feed and finally to humans via food, the commodities are generally stored for various lengths of time. The duration of storage depends on distance involved, speed and mode of transport, meteorological conditions among many others. If the storage conditions are substandard, providing favourable conditions for mold

Table 1. The major fungi and their associated mycotoxins

Fungal species	*Mycotoxins*	*Food*	*References*
Aspergillus flavus *Aspergillus parasiticus*	Aflatoxins B1, B2 Aflatoxins B1, B2, G1, G2	Rice, peanut, beans, herbs and spices, and dried fruits	Ruadrew et al. 2013
Aspergillus ochraceus *Aspergillus carbonarius* *Aspergillus niger* *Penicillium verrucosum*	Ochratoxin A	Cereal grains, dried fruits, wine, coffee and other agricultural products	Bui-Klimke and Wu 2015
Fusarium graminearum	Deoxynivalenol	Oats, barley, corn, wheat, and other grains	Sobrova et al. 2010
Fusarium culmorum, *Fusarium roseum* *Fusarium graminearum*	Zearalenone	Wheat, barley, rice, maize and other crops	Hueza et al. 2014
Fusarium poae	T-2 toxins HT-2 toxin Diacetoxyscirpenol	Grains and feed	Yang et al. 2013
Claviceps purpurea	Ergot alkaloids	Cereals including rye, wheat, barley, oat, millet and others.	Bryła et al. 2019
Fusarium verticilloides *Fusarium proliferatum*	Fumonisins	Maize and maize based foods	Munawar et al. 2019

growth, mycotoxin production in the commodities is likely to occur (Rehmat et al. 2019).

3.1.1 Mycotoxins in Plants

Microbial fungal infections possess threats to both animals and humans. This affects economically with plant diseases, even without the occurrence of mycotoxin contamination in plants. Typical examples of such fungal infestation in plants are wheat and barley head blight (*Fusarium graminearum*), banana wilt (*Fusarium oxysporum*) and potato and tomato blight (*Phytophthora infestans* and *Alternaria solani*). Moreover, there exists a hostile relationship between fungal species and plant pathogens, which is giving birth to the rapid unfolding of novel fungi with acclimatization and adaptational nature and thus in the process becoming more resilient spreading into the plants (Giraud et al. 2010). The effects of globalization due to climate change and human interventions such as nitrogen deposition and rising CO_2 levels between the fungal infestations in plant and environment are also worth to be noted. It is a probability that such interventions are to bring a change in fungus-plant interaction to produce mycotoxins (Mitchell et al. 2003).

3.1.2 Mycotoxins in Feed

Feed contamination of mycotoxin is prevalent worldwide, and thus its residues in animal-derived foods play a vital role in human exposure. European Union

Table 2. Mycotoxins with their chemical structure and safe limit

S. N.	*Mycotoxin name*	*Chemical structure*	*Safe limit*
1	Aflatoxin B1	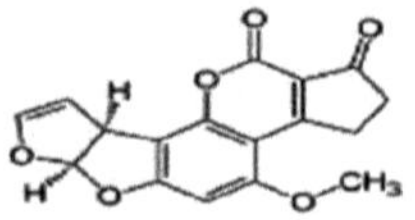	10 μg/kg
2	Aflatoxin M1	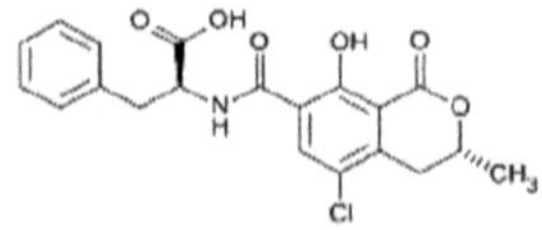	10 μg/kg
3	Aflatoxin G1	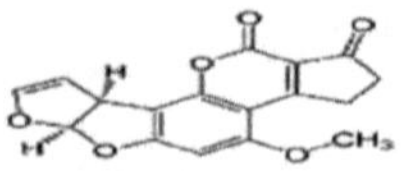	10 μg/kg
4	Deoxynivalenol	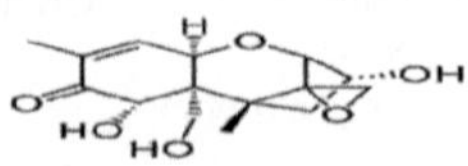	750 μg/kg
5	Zearalenone	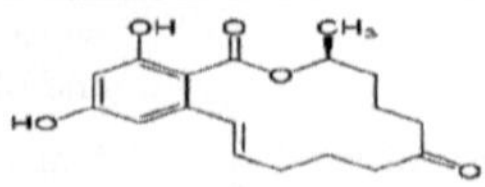	1000 μg/kg
6	Ochratoxin A	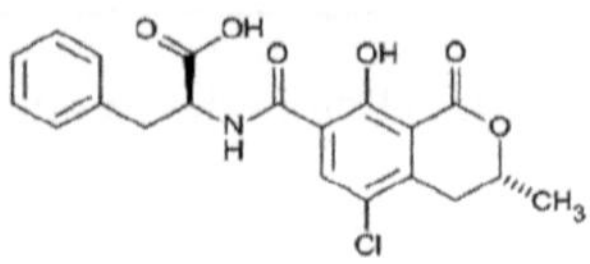	5 μg/kg
7	Fumonisin B1	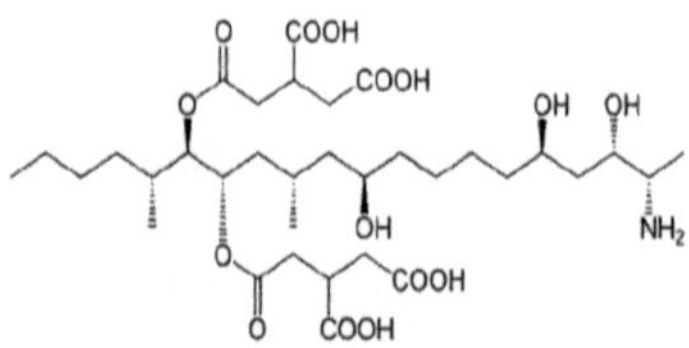	1000 μg/kg
8	Hydrolyzed Fumonisin B1	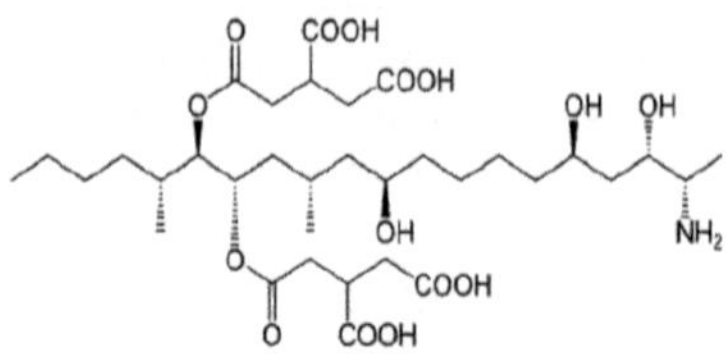	1000 μg/kg

Source: Tola and Kebede 2016

(EU) has harmonized legislation on animal feed which applies to feed for farmed livestock covering farmed fish, pets and horses. The European Food Safety Authority (EFSA) has carried out a risk assessment on a number of

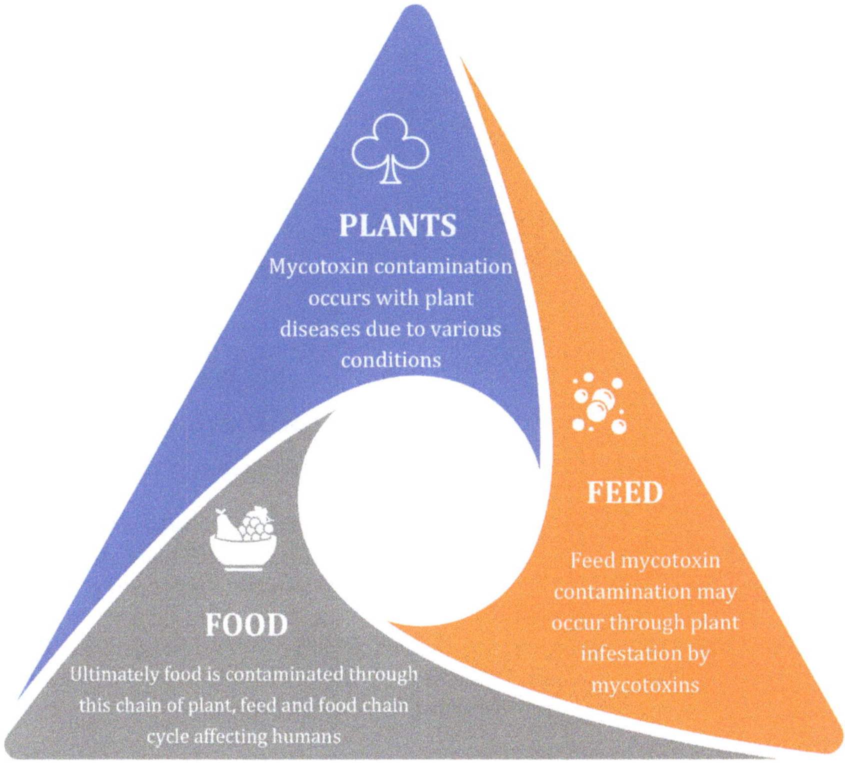

Figure 2. Occurrence and distribution of mycotoxins in plant, feed and food chain.

mycotoxins (AFB1, deoxynivalenol, zearalenone, ochratoxin A, fumonisin, T-2 and HT-2) that are considered to pose health risks to human and animals. Moreover, physical and chemical stability of mycotoxins to food processing techniques also exposes human to mycotoxicosis (Fink-Gremmels and van der Merwe 2019).

3.1.3 Mycotoxins in Food

Mycotoxins prevalent in plants occur mostly in plant-derived foods such as cereals, nuts, oilseeds, (dried) fruits and species and products thereof. Humans get exposed to mycotoxin *via* moldy foods that are spoiled. Visibly fungal contaminated foods are not consumed unless the growth of fungi is preferred for some dairy products like Roquefort or Gorgonzola cheese for the characteristic flavour and part of production or processing (Ladeira 2016). Primary mycotoxicosis is a toxicological condition which onsets through consumption of food where mycotoxins are developed due to spoilage of food. In contrast, secondary mycotoxicosis is a result of bioaccumulation of toxin through animal feed and into humans, as shown in Fig. 3. Dairy products and meats are vulnerable to fungal spoilage when improperly stored due to their

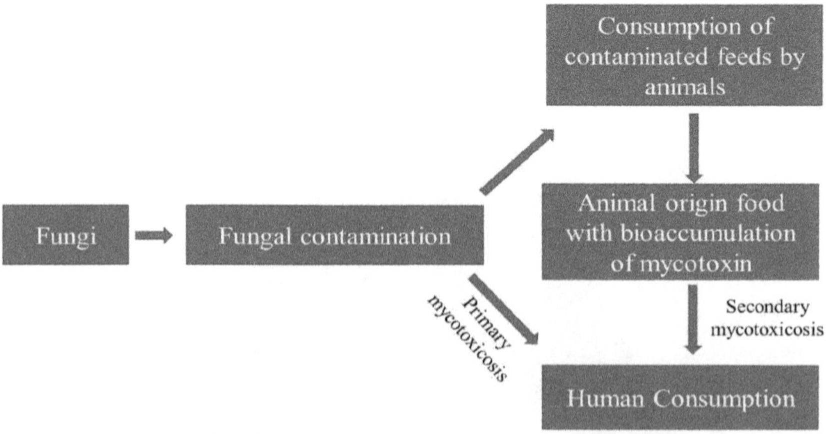

Figure 3. Occurrence of mycotoxins in food and feed in supply chain.

high-water activity and nutrient composition. Although food infested with fungal growth may be visible due to mycotoxins, it is worth to be noted that these metabolites are very small and can migrate within cells of food products into inner layers. Hence, surface removal of apparent mold growth in itself during food processing or common household kitchen food preparation may not eliminate all mycotoxins (Omotayo et al. 2019). This becomes the cause for involuntary exposure to mycotoxins for a prolonged period.

3.2 Toxicological Effects of Mycotoxins in Human Health

Once foods infected with fungus or its mycotoxins are ingested, symptoms are seen either slowly or quickly depending on the dose and toxicity level. This can be categorized into acute and chronic like other toxicological syndromes. Acute toxicity can be observed very quickly, whereas symptoms for chronic toxicity develops over the period (Ladeira 2016).

Multiple aflatoxicosis incidence has occurred in Kenya, India and Malaysia in the past during the late 1990s and early 2000 (Lewis et al. 2005). The symptoms for these chronic toxicity develops with time. Aflatoxicosis caused by aflatoxin are potent hepatotoxin along with carcinogen, immunosuppressant and mutagens in nature. Among the most significant one is AFB_1 that has been associated with liver cancer. Because of this mutagenicity, International Agency for Research on Cancer (IARC) classified it as a Group 1 carcinogen. Humans are exposed to aflatoxin via chronic dietary exposure which have been linked with human hepatocellular carcinomas. Approximately 250,000 deaths are caused by hepatocellular carcinomas in China and Sub-Saharan Africa annually and are attributed to risk factors such as high daily intake (1.4 µg) of AF and high incidence of hepatitis B (Zain 2011). The popularly known aflatoxin is known to cause diseases like *Kwashiorkor Syndrome* and *Reye's syndrome,* which result in immunosuppression in children. Aflatoxins have been found in tissues of children suffering from these diseases and were thought to be a contributing factor to these diseases. Reye's syndrome, which is characterized by encephalopathy and visceral deterioration, results in liver and kidney enlargement and cerebral oedema. Aflatoxins are potent hepatotoxins, immunosuppressant, and mutagens and carcinogens (Hueza et al. 2014).

Fumonisin, on the other hand, has caused acute food-borne disease in India with the onset of symptoms like abdominal pain and diarrhoea through the consumption of maize and sorghum that had fumonisin as a toxin. Fumonisin also has been implicated in oesophageal cancer in China. Fumonisin B_1 is classified as Group 2B carcinogen meaning possible carcinogenic in human. Fumonisin primarily inhibits the folic acid absorption through folic acid receptors, causing early childbirth problems for people living in rural areas and with maize as a staple diet (Zain 2011).

Deoxynivalenol (DON) and Zearalenone (ZEN) toxins from toxic *Fusaria* were associated with grain toxicosis in the developed countries like USA, China, Japan and Australia. The symptoms observed were limited to the feeling of nausea, vomiting, and diarrhoea. Trichothecenes have been suggested as potential biological warfare agents (Liew and Mohd-Redzwan 2018). For example, T-2 toxin was implicated as the chemical agent of 'yellow rain' used against the Lao Peoples Democratic Republic from 1975 through 1981. In an investigation of similar biological warfare agents in Cambodia from 1978 to 1981, T-2 toxin, DON, ZEN, nivalenol, and DAS were isolated from water and leaf samples collected from the affected areas (Brera et al. 2016).

Zearalenone leads to hypoestrogenism with effect on the reproduction system of human while fumonisin is fatal to the liver in animals (Tola and Kebede 2016). In addition to this, age, sex, weight, diet, exposure to infectious agents, the number of toxins exposed, and the presence of other mycotoxins (which activates synergistic effects) are various factors which affect human health due to mycotoxins (Bennett and Klich 2003; Ogbuewu 2011). Various health effects due to mycotoxicosis are shown in Fig. 4.

4. Interactions between Gut Microbiota and Mycotoxins

Several microorganisms including bacteria, viruses, protozoa and fungi constitute gut microbiome that is located in the gastrointestinal (GI) tract. The bacterial flora that are being found are either commensal or pathogenic in nature, competing in the GI tract and are mainly responsible for nutrient absorption, enzyme synthesis, vitamins and amino acids along with the production of short-chain fatty acids (SCFAs). These functionalities ensure sound gut health and healthy microbiome in the gut (Cresci and Bawden 2015).

When it comes to mycotoxins and its effects, the focus is primarily on human health and its nature, whether acute or chronic as discussed above. Since these compounds reach the colon basically in primary structure without any degradation, it is even more important to understand its mechanism

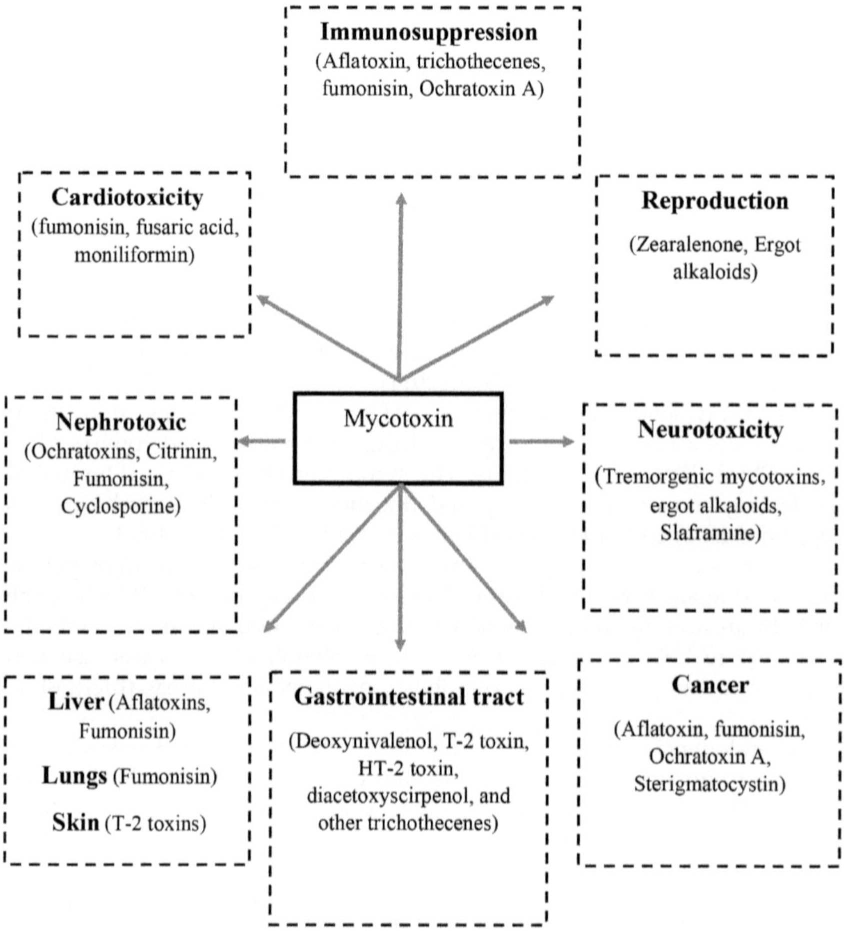

Figure 4. Some of the common effects of mycotoxins (Haschek et al. 2002).

at gastrointestinal tract (GIT) (Grenier and Applegate 2013). As the bioavailability of mycotoxins are different, some are rapidly absorbed while more resilient ones pass along the GI tract. This nature of mycotoxins is very significant. Firstly, cells of colonocytes in the gastrointestinal tract (GIT) are eventually going to be exposed to the mycotoxins ingested in their highest concentrations. As these toxins move along the GIT and are bound to have more opportunities to interact with the gut flora in the intestine. These cells can also be vulnerable to the effects of mycotoxins (Broom 2015), which leads to the dysbiosis in the GI tract. In light to this, recent studies suggest some mechanisms on how microorganisms in the gut help to restore the balance of its components using various mechanisms, as shown in Fig. 5.

Biological methods and interventions have recently been studied with regards to mycotoxin contamination via methods as described above. It is

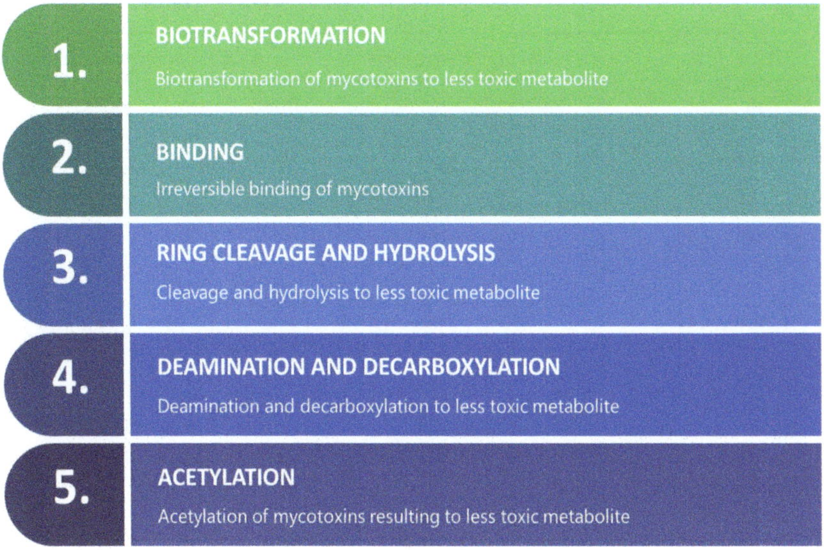

Figure 5. Mechanisms identified from microorganisms in the gut to convert mycotoxins to less toxic compounds.

essential to understand the mechanism of combating the gut microbiota with mycotoxins like transformation of toxins to less toxic metabolite and enzymes from gut microbiota acting specifically on these toxins. The understanding of this biological method of detoxification seems promising as the chemical method leaves residues which do not support the objective of removal of toxins (Ji et al. 2016). Some common detoxifying bacteria or enzymes from bacteria studied so far has been tabulated in Table 3.

Mycotoxins inside the human body are also subjected to exposure with regards to these microbiome interactions and host metabolism. Microbes that are present in the gut, principally aid host in the removal of toxins either by its metabolism by degradation or by binding to the mycotoxins (Liew and Mohd-Redzwan 2018). Studies on microbial species which can biologically transform toxic mycotoxins into lower toxic ones have been gaining momentum, and several works have been documented based on the same. However, this approach is not new as in the past, many studies have revealed on bio-niches to identify the strains for biotransformation of mycotoxins. The partial success of these approaches laid the way for future research with identification of bacterial species anaerobic in nature such as *Genus novus strain BBSH 797* and *Bacillus arbutinivorans strain LS-100.* This strain can convert DON to the deepoxy-deoxynivalenol (DOM-1) metabolite in the animal gut with reduced biological toxicity. The mechanisms involving conversion of mycotoxins to less toxic metabolite apart from the ones discussed above includes degradation or enzymatic transformation of mycotoxins biologically. The microbes identified for biotransformation, metabolism and binding to the mycotoxins act mainly

Table 3. Microorganisms detoxifying the toxicity of mycotoxins

Mycotoxin	*Detoxifying bacteria/enzymes*	*Detoxification mechanism*
Deoxynivalenol	*Bacillus arbutinivorans* strain LS-100	Biotransformation of DON to DOM-1 in an anaerobic atmosphere.
	Genus novus strain BBSH 797	Biotransformation of DON to DOM-1 in an anaerobic atmosphere.
	Devosia mutans strain 17-2-E-8	A soil microorganism capable of epimerising DON to 3-epi-DON and reducing its toxicity at mild temperatures
	An NADH-dependent bacterial cytochrome P450 system (DdnA, Kdx, and KdR)	Transformation of DON into 16-hydroxydeoxynivalenol eliminating its phytotoxicity.
Zearalenone (ZEN)	Zearalenone esterase and laccases	Some copper-containing oxidases (laccases) and a novel esterase have been reported for ZEN detoxification.
Ochratoxin (OTA)	Ochratoxinase, Carboxypeptidase A Carboxypeptidase Y	The ochratoxinase from Aspergillus niger was "reported to be more efficient in ochratoxin A hydrolysis than the other amidases, carboxypeptidase A and Y".
Aflatoxins (AF)	An aflatoxin oxidase (AFO) that was previously known as Aflatoxin-detoxifizyme (ADTZ)	The enzyme targets the 8,9-unsaturated carbon-carbon bond of aflatoxin B1 reducing its overall mutagenic effects.
Fumonins (FUM)	An AP1 amine oxidase	Deamination reaction
	FUMzyme	Originally isolated from *Sphingopyxis* sp. MTA 144 with an esterase activity.

Source: Yousef and Zhou 2016

in the gut (Hathout and Aly 2014). Studies on gut microbes and their role in reducing the toxicity of mycotoxins that have been studied so far, follows.

4.1 Deoxynivalenol

Deoxynivalenol (DON) is the mycotoxins produced by *Fusarium graminearum* and generally found in different food products like oats, barley, corn, wheat, and other grains. Especially, cereal-based crops are more prone to fungal attacks owing to their nutrient and chemical composition. The formation of the molds and toxins occur on a variety of different cereal crops. DON toxins are often found associated with the wheat. Several studies are found on the interaction of DON mycotoxin to the gut microbiota; however, their response against myotoxicity is far less studied. A strain of bacteria *Eubacterium* species *DSM 11798* was studied by Binder et al. (1998) and reported that DON is transformed into DOM-1, the nontoxic, de-epoxide of DON as shown in Fig. 6. *Eubacterium BBSH 797,* a common human gut inhabitant, was in fact, the first microbe to be used as a feed additive for reducing toxicity presented by mycotoxins(Binder et al. 1998). This feed additive was then tested for *in vivo* efficacy by Awad et al. (2006) in chicken broilers with positive results on counteracting the toxic effects of DON at the gut level. The microbial biotransformation of DON was studied by Pierron et al. (2016) against human intestinal cells *in vitro,* and it was observed that de-epoxidation or epimerization of DON altered their interaction with the ribosome, leading to an absence of MAPKinase activation and a reduced toxicity.

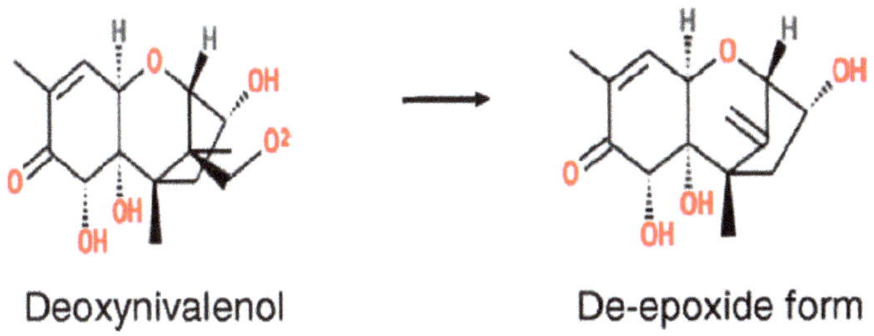

Figure 6. Transformation of DON into de-epoxide form via removal of toxic O_2 chain. (Adapted from Binder et al. 1998)

4.2 Ochratoxin A

Several species of *Aspergillus* and *Penicillium* are found to produce Ochratoxin A (OTA), which is atypical food and feed contaminating mycotoxin. OTA is produced during storage of cereal crops, especially in coffee and dry fruits.A bioreactor-based study was performed to observe the whereabouts of OTA during colon digestion *via* simulation of the human gut. Biodegradation mechanism was analyzed by observing and quantifying OTA concentration over time and screening for intermediary metabolites using LC-MS/MS. Ochratoxin α and ochratoxin β were identified as two metabolites, suggesting that biodegradation by gut microbiota is beneficial for the host, as they are considered less toxic than Ochratoxin A (OTA) (Ouethrani et al. 2013). The researchers also found evidence that descending colon (lower part of GIT) is the predominant site for microbial breakdown of OTA, with an overall mean degradation around 40–50% measured by monitoring of OTA concentrations during the SHIME (Simulator of the Human Intestinal Microbial Ecosystem) kinetic studies.

4.3 Aflatoxin

Aspergillus flavus and *Aspergillus parasiticus* produce aflatoxins which are poisonous and inhabitants in soil, grains and decayed vegetation. Cereal based crops (rice, corn and wheat), oilseeds (cotton seeds, soybean, sunflower and peanut), spices (ginger, chilli peppers, coriander, black pepper, and turmeric) and tree nuts (walnut, almond, pistachio, coconut and Brazil nut) are the crops that are frequently affected by *Aspergillus* spp. The toxins can also be present in the milk of animals that might have been fed with contaminated feed by aflatoxins, and that could transform aflatoxin B in aflatoxin M1 (Brera et al. 2016). Aflatoxicosis occurs due to exposure to large doses of aflatoxins, which can be life-threatening with permanent damage to the liver. Aflatoxins have also been shown to be genotoxic and can damage DNA, causing cancer in animal species. Fochesato et al. (2019) recently reported that *Lactobacillus rhamnosus RC007* (108 CFU/mL) has aflatoxin reducing (AFB_1) activities. The authors studied the adsorption of AFB_1 through Gastro-Intestinal (GI) tract (saliva, stomach and intestine) and found that the adsorption was efficient in gastric and intestinal environment. Another study also has revealed that 2×10^{10} CFU/mL of *Lactobacillus* spp. can reduce the AFB_1 level to 0.1–13% (Apás et al. 2014). It appears that the surface components of probiotic bacteria of the gut are involved in AFB_1-binding (Yan et al. 2018). It is worth to mention that probiotic intervention is potentiated to alleviate AFB_1-induced toxicity as studied by Saladino et al. (2018). They showed that the capacity of probiotic bacteria in the gut was tested *in vitro* in a simulated environment in a food enriched and spiked with probiotic like *Lactobacillus johnsoni CECT 289, Lactobacillus casei CECT 4180, Lactobacillus plantarum CECT 220* and *Lactobacillus reuteri CECT 725*. At the same time, food contaminated with AFB_1 was tested for comparative reduction of mycotoxin. The reduction reached up to 98.66 % showing how gut probiotic microbiome can combat ingested mycotoxins. Similarly, Liew et al. (2019) selected *Lactobacillus casei Shirota (Lcs)* for the AFB_1 removal purpose. During this study, a bilateral interaction were documented. There were significant changes in the gut by $AFB_{1,}$and this was monitored by levels of short-chain fatty acids (SCFAs) in the feces. However, the studied strain was able to restore the AFB_1-induced gut microbiota fluctuations back to normal level. Similar studies of gut microbiota on AFB_2 are far less explored.

4.4 Fumonisins

Fumonisins are the mycotoxins produced by *Fusarium verticilloides* and *Fusarium proliferatum*. They are common contaminants in maize and maize-based products. Niderkorn et al. (2007) studied the potential of bacteria to influence fumonisins, whereas Mokoena et al. (2005) found that lactic acid bacteria (LAB) that are a common inhabitant of the gut were able to decrease fumonisins in maize after the third day of fermentation. Common cell wall material peptidoglycan plays an essential role in binding many mycotoxins, including fumonisins (Mokoena et al. 2005).

4.5 Zearalenone

Zearalenone is produced by fungal species like *Fusarium culmorum, Fusarium roseum* and *Fusarium graminearum* in wheat, barley, rice and maize. Gratz et al. (2017) studied the fate of important masked trichothecenes and various zearalenone compounds like (α and β-zearalenone-glucoside, ZEN14Glc; α-ZEL14Glc, zearalenone-glucoside and β-ZEL14Glc) under gastrointestinal conditions *in vitro*. The study was carried out under simulated gastrointestinal conditions by exposing these mycotoxins to artificial saliva, gastric and intestinal juices and enzymes associated. The results showed that none of the tested masked trichothecenes or zearalenone compounds were hydrolyzed under upper GI tract conditions which showed that mycotoxins are widely exposed to the lower GI tract and are intact. In this study by Gratz et al. (2017), they used a two-compartment Caco-2/TC7 cell model to assess epithelial absorption of modified and free mycotoxins *in vitro*. Upon contact with gut microbiota, ZEN14Glc, α-ZEL14Glc and β-ZEL14Glc were easily hydrolyzed. Colonic bioavailability of masked ZEN compounds was high as they are rapidly hydrolyzed by gut microbiota and released zearalenone compounds are highly reactive with gut microbiota and epithelial colon cells.

4.6 T-2 Toxin

T-2 toxins are very stable mycotoxins produced by *Fusarium poae* in rice and grains.T-2 toxins were incubated *in vitro* with rumen microorganism from a dairy cow at 12 h, 24 h and 48 h. The results reveal the biotransformation of toxins to a variety of de-epoxy and deacylated products like HT-2, T-2 triol and two new metabolites identified as 15-acetoxy-3α,4β-dihydroxy-8α-(3-methylbutyryloxy), trichothec-9,12-diene (de-epoxy HT-2) and 3α,4β,15-trihydroxy-8α-)3-methylbutyryloxy) trichothec-9,12-diene (de-epoxy T-2 triol) (Swanson et al. 1987). These metabolites are broadly classified into T-2 and HT-2 toxins. The toxicity of these compounds has been recommended via a consortium of 13 countries worldwide with maximum legal levels (MLS). In Europe, the European Union (EU) has a commission recommendation (2013/165/EU) as of 27 March 2013 with regards to the presence of T-2 and HT-2 toxin in cereal products. It states tolerable daily intake of 0.06 μg/kg body weight (BW) for the sum of T-2 and HT-2 toxins, which was confirmed in 2002. As new relevant evidence has become available, the Scientific Panel on Contaminants in the Food Chain of the European Food Safety Authority 37 established a full TDI of 100 ng/kg BW for the sum of T-2 and HT-2 toxins (Rodriguez et al. 2014).

5. Conclusion

Mycotoxin contamination of food worldwide has a significant impact on food safety and possesses a potential threat to the economy of the country. Even as pre- and post-harvest technologies are in place to reduce continuous exposure of foods to these toxins, it is insufficient, and contamination is inevitable. Hence, biological transformation, a promising yet challenging process, is emerging as an effective way to combat these mycotoxins or reduce its accumulation. *In vitro* studies have shown that structural integrity of mycotoxins remain intact until it reaches the gut where it exerts its mycotoxicosis. However, the gut microbiome has natural combating properties against the toxins that maintain ecological balance in the small intestine. Several studies have reported to degrade mycotoxins through various mechanisms like binding, acetylation, epimerization and various other structural changes to lower toxic metabolite and thus reducing its toxicity in the lower GI tract. The occurrence of microbiota at the GI tract can act like an intestinal barrier causing different (maximal or limited) bioavailability of these fungal compounds and metabolize them accordingly. Also, these fungal compounds disrupt the gut microbiota balance, impair intestinal functions leading to dysregulation of the host immune system, which may eventually result in chronic mycotoxicosis. This severity can be restored by probiotic administration, especially those which possess mycotoxins reducing ability promoting sound gut health. Despite these, very few enzymes or probiotic gut flora showing such properties from the microbiome in the gut have been identified. In addition to this, further studies are needed in order to elucidate the interaction of gut microbiome towards mycotoxins, consequently preventing mycotoxicosis from mycotoxins thereafter.

References

An, X., Bao, Q., Di, S., Zhao, Y., Zhao, S., Zhang, H., Lian, F. and Tong, X. 2019. The interaction between the gut microbiota and herbal medicines. Biomedicine & Pharmacotherapy 118: 109252.

Apás, A.L., González, S.N. and Arena, M.E. 2014. Potential of goat probiotic to bind mutagens. Anaerobe 28: 8–12.

Awad, W., Böhm, J., Razzazi-Fazeli, E., Ghareeb, K. and Zentek, J. 2006. Effect of addition of a probiotic microorganism to broiler diets contaminated with deoxynivalenol on performance and histological alterations of intestinal villi of broiler chickens. Poultry Science 85: 974–979.

Bennett, J. and Klich, M.C. 2003. Mycotoxins. Clinical Microbiology Reviews 16: 497–516.

Binder, E., Binder, J., Ellend, N., Schaffer, E., Krska, R. and Braun, R. 1998. Microbiological degradation of deoxynivalenol and 3-acetyl-deoxynivalenol. pp. 279–285. *In*: M. Miraglia (ed.). Mycotoxins and Phycotoxins—Developments in Chemistry, Toxicology and Food Safety. Alaken Inc, Fort Collins.

Brera, C., Debegnach, F., Gregori, E., Colicchia, S., Soricelli, S., Miano, B., Magri, M.C. and De Santis, B. 2016. Dietary exposure assessment of European population to mycotoxins: A review. pp. 223–259. *In*: C. Viegas, A.C. Pinheiro, R. Sabino, S. Viegas, J. Brandão and C. Veríssimo (eds.). Environmental Mycology in Public Health. Elsevier, New York.

Broom, L. 2015. Mycotoxins and the intestine. Animal Nutrition 1: 262–265.

Bryła, M., Ksieniewicz-Woźniak, E., Waśkiewicz, A., Podolska, G. and Szymczyk, K. 2019. Stability of ergot alkaloids during the process of baking rye bread. LWT-Food Science and Technology 110: 269–274.

Bui-Klimke, T.R. and Wu, F. 2015. Ochratoxin A and human health risk: A review of the evidence. Critical Reviews in Food Science and Nutrition 55: 1860–1869.

Byndloss, M.X., Olsan, E.E., Rivera-Chávez, F., Tiffany, C.R., Cevallos, S.A., Lokken, K.L., Torres, T.P., Byndloss, A.J., Faber, F. and Gao, Y. 2017. Microbiota-activated PPAR-γ signaling inhibits dysbiotic Enterobacteriaceae expansion. Science 357: 570–575.

Chein, S.H., Sadiq, M.B., Datta, A. and Anal, A.K. 2019. Prevalence and identification of Aspergillus and Penicillium species isolated from peanut kernels in central Myanmar. Journal of Food Safety 39: e12686.

Cresci, G.A. and Bawden, E. 2015. Gut microbiome: What we do and don't know. Nutrition in Clinical Practice 30: 734–746.

Cresci, G.A. and Izzo, K. 2019. Gut microbiome. pp. 45–54. *In*: M. Corrigan, K. Roberts and E. Steige (eds.). Adult Short Bowel Syndrome. Academic Press, New York.

De Mello, V.D., Paananen, J., Lindström, J., Lankinen, M.A., Shi, L., Kuusisto, J., Pihlajamäki, J., Auriola, S., Lehtonen, M. and Rolandsson, O. 2017. Indolepropionic acid and novel lipid metabolites are associated with a lower risk of type 2 diabetes in the Finnish Diabetes Prevention Study. Nature 7: 46337.

De Vadder, F., Kovatcheva-Datchary, P., Goncalves, D., Vinera, J., Zitoun, C., Duchampt, A., Bäckhed, F. and Mithieux, G. 2014. Microbiota-generated metabolites promote metabolic benefits via gut-brain neural circuits. Cell 156: 84–96.

Fink-Gremmels, J. and van der Merwe, D. 2019. Mycotoxins in the food chain: Contamination of foods of animal origin. pp. 1190–1198. *In*: F.J.M. Smulders, I.M.C.M. Rietjens and M. Rose (eds.). Chemical Hazards in Foods of Animal Origin. Wageningen Academic Publishers, Netherlands.

Fochesato, A., Cuello, D., Poloni, V., Galvagno, M., Dogi, C. and Cavaglieri, L. 2019. Aflatoxin B1 adsorption/desorption dynamics in the presence of Lactobacillus rhamnosus RC 007 in a gastrointestinal tract-simulated model. Journal of Applied Microbiology 126: 223–229.

Frost, G., Sleeth, M.L., Sahuri-Arisoylu, M., Lizarbe, B., Cerdan, S., Brody, L., Anastasovska, J., Ghourab, S., Hankir, M. and Zhang, S. 2014. The short-chain fatty acid acetate reduces appetite via a central homeostatic mechanism. Nature Communications 5: 1–11.

García-Cela, E., Ramos, A., Sanchis, V. and Marin, S. 2012. Emerging risk management metrics in food safety: FSO, PO. How do they apply to the mycotoxin hazard? Food Control 25: 797–808.

Giraud, T., Gladieux, P. and Gavrilets, S. 2010. Linking the emergence of fungal plant diseases with ecological speciation. Trends in Ecology & Evolution 25: 387–395.

Gratz, S.W., Dinesh, R., Yoshinari, T., Holtrop, G., Richardson, A.J., Duncan, G., MacDonald, S., Lloyd, A. and Tarbin, J. 2017. Masked trichothecene and zearalenone mycotoxins withstand digestion and absorption in the upper

GI tract but are efficiently hydrolyzed by human gut microbiota in vitro. Molecular Nutrition & Food Research 61: 1600680.

Grenier, B. and Applegate, T.J. 2013. Modulation of intestinal functions following mycotoxin ingestion: Meta-analysis of published experiments in animals. Toxins 5: 396–430.

Gubert, C., Kong, G., Renoir, T. and Hannan, A.J. 2019. Exercise, diet and stress as modulators of gut microbiota: Implications for neurodegenerative diseases. Neurobiology of Disease: 104621.

Haschek, W., Voss, K. and Beasley, V. 2002. Selected mycotoxins affecting animal and human health. pp. 699–721. *In*: W.M. Haschek, C.G. Rousseaux and M. H. Wallig (eds.). Handbook Toxicologic Pathology. Academic Press, New York.

Hathout, A.S. and Aly, S.E. 2014. Biological detoxification of mycotoxins: A review. Annals of Microbiology 64: 905–919.

Herman, D.R., Rhoades, N., Mercado, J., Argueta, P., Lopez, U. and Flores, G.E. 2019. Dietary habits of 2- to 9-year-old American children are associated with gut microbiome composition. Journal of the Academy of Nutrition and Dietetics 120: 517–534.

Hoffmann, V., Moser, C. and Saak, A. 2019. Food safety in low and middle-income countries: The evidence through an economic lens. World Development 123: 104611.

Hueza, I.M., Raspantini, P.C.F., Raspantini, L.E.R., Latorre, A.O. and Górniak, S.L. 2014. Zearalenone, an estrogenic mycotoxin, is an immunotoxic compound. Toxins 6: 1080–1095.

Ji, C., Fan, Y. and Zhao, L. 2016. Review on biological degradation of mycotoxins. Animal Nutrition 2: 127–133.

Konstantinov, S.R. 2017. Diet, microbiome, and colorectal cancer. Best Practice & Research Clinical Gastroenterology 31: 675–681.

Ladeira, C. 2016. Mycotoxins: Genotoxicity studies and methodologies. pp. 343–361. *In*: C. Viegas, A.C. Pinheiro, R. Sabino, S. Viegas, J. Brandão and C. Veríssimo (eds.). Environmental Mycology in Public Health. Elsevier, New York.

Lewis, L., Onsongo, M., Njapau, H., Schurz-Rogers, H., Luber, G., Kieszak, S., Nyamongo, J., Backer, L., Dahiye, A.M. and Misore, A. 2005. Aflatoxin contamination of commercial maize products during an outbreak of acute aflatoxicosis in eastern and central Kenya. Environmental Health Perspectives 113: 1763–1767.

Li, R., Wang, X., Zhou, T., Yang, D., Wang, Q. and Zhou, Y. 2014. Occurrence of four mycotoxins in cereal and oil products in Yangtze Delta region of China and their food safety risks. Food Control 35: 117–122.

Li, X., Liu, L., Cao, Z., Li, W., Li, H., Lu, C., Yang, X. and Liu, Y. 2020. Gut microbiota as an "invisible organ" that modulates the function of drugs. Biomedicine & Pharmacotherapy 121: 109653.

Liew, W.P.P. and Mohd-Redzwan, S. 2018. Mycotoxin: Its impact on gut health and microbiota. Frontiers in Cellular and Infection Microbiology 8: 60.

Liew, W.P.P., Mohd-Redzwan, S. and Than, L.T.L. 2019. Gut microbiota profiling of aflatoxin B1-induced rats treated with Lactobacillus casei Shirota. Toxins 11: 49.

Lin, H.V., Frassetto, A., Kowalik Jr, E.J., Nawrocki, A.R., Lu, M.M., Kosinski, J.R., Hubert, J.A., Szeto, D., Yao, X. and Forrest, G. 2012. Butyrate and propionate protect against diet-induced obesity and regulate gut hormones via free fatty acid receptor 3-independent mechanisms. PloS One 7: e35240.

Lloyd-Price, J., Abu-Ali, G. and Huttenhower, C. 2016. The healthy human microbiome. Genome Medicine 8: 1–11.

Long, S.L., Gahan, C.G. and Joyce, S.A. 2017. Interactions between gut bacteria and bile in health and disease. Molecular Aspects of Medicine 56: 54–65.

Marín, S., Cano-Sancho, G., Sanchis, V. and Ramos, A.J. 2018. The role of mycotoxins in the human exposome: Application of mycotoxin biomarkers in exposome-health studies. Food and Chemical Toxicology 121: 504–518.

Marko, D. 2017. Human microbiota and mycotoxins: Biotransformation, impact on toxicokinetics and relevance for toxicity. Toxicology Letters 280: S26.

Mitchell, C.E., Reich, P.B., Tilman, D. and Groth, J.V. 2003. Effects of elevated CO_2, nitrogen deposition, and decreased species diversity on foliar fungal plant disease. Global Change Biology 9: 438–451.

Mohajeri, M.H., Brummer, R.J., Rastall, R.A., Weersma, R.K., Harmsen, H.J., Faas, M. and Eggersdorfer, M. 2018. The role of the microbiome for human health: From basic science to clinical applications. European Journal of Nutrition 57: 1–14.

Mokoena, M.P., Chelule, P.K. and Gqaleni, N. 2005. Reduction of fumonisin B1 and zearalenone by lactic acid bacteria in fermented maize meal. Journal of Food Protection 68: 2095–2099.

Munawar, H., Safaryan, A.H., De Girolamo, A., Garcia-Cruz, A., Marote, P., Karim, K., Lippolis, V., Pascale, M. and Piletsky, S.A. 2019. Determination of Fumonisin B1 in maize using molecularly imprinted polymer nanoparticles-based assay. Food Chemistry 298: 125044.

Niderkorn, V., Morgavi, D.P., Pujos, E., Tissandier, A. and Boudra, H. 2007. Screening of fermentative bacteria for their ability to bind and biotransform deoxynivalenol, zearalenone and fumonisins in an in vitro simulated corn silage model. Food Additives and Contaminants 24: 406–415.

Nieto, C.H.D., Granero, A.M., Zon, M.A. and Fernández, H. 2018. Sterigmatocystin: A mycotoxin to be seriously considered. Food and Chemical Toxicology 118: 460–470.

Ogbuewu, I. 2011. Effects of mycotoxins in animal nutrition: A review. Asian Journal of Animal Science 5: 1933.

Omotayo, O.P., Omotayo, A.O., Mwanza, M. and Babalola, O.O. 2019. Prevalence of mycotoxins and their consequences on human health. Toxicological Research 35: 1–7.

Ouethrani, M., Van de Wiele, T., Verbeke, E., Bruneau, A., Carvalho, M., Rabot, S. and Camel, V. 2013. Metabolic fate of ochratoxin A as a coffee contaminant in a dynamic simulator of the human colon. Food Chemistry 141: 3291–3300.

Peng, W.X., Marchal, J. and Van der Poel, A. 2018. Strategies to prevent and reduce mycotoxins for compound feed manufacturing. Animal Feed Science and Technology 237: 129–153.

Pierron, A., Mimoun, S., Murate, L.S., Loiseau, N., Lippi, Y., Bracarense, A.P.F., Schatzmayr, G., He, J.W., Zhou, T. and Moll, W.D. 2016. Microbial biotransformation of DON: Molecular basis for reduced toxicity. Scientific Reports 6: 29105.

Ramírez-Guzmán, K.N., Torres-León, C., Martínez-Terrazas, E., De la Cruz-Quiroz, R., Flores-Gallegos, A.C., Rodríguez-Herrera, R. and Aguilar, C.N.

2018. Biocontrol as an efficient tool for food control and biosecurity. pp. 167–193. *In*: A.M. Grumezescu and A.M. Holban (eds.). Food Safety and Preservation. Elsevier, New York.

Rehmat, Z., Mohammed, W.S., Sadiq, M.B., Somarapalli, M. and Anal, A.K. 2019. Ochratoxin A detection in coffee by competitive inhibition assay using chitosan-based surface plasmon resonance compact system. Colloids and Surfaces B: Biointerfaces 174: 569–574.

Rodriguez, A., Rodriguez, M., Herrera, M., Arino, A. and Cordoba, J.J. 2014. T-2, HT-2, and diacetoxyscirpenol toxins from Fusarium. *In*: D. Liu (ed.). Manual of Security Sensitive Microbes and Toxins. CRC Press, New York.

Ruadrew, S., Craft, J. and Aidoo, K. 2013. Occurrence of toxigenic *Aspergillus* spp. and aflatoxins in selected food commodities of Asian origin sourced in the West of Scotland. Food and Chemical Toxicology 55: 653–658.

Saladino, F., Posarelli, E., Luz, C., Luciano, F., Rodriguez-Estrada, M., Mañes, J. and Meca, G. 2018. Influence of probiotic microorganisms on aflatoxins B1 and B2 bioaccessibility evaluated with a simulated gastrointestinal digestion. Journal of Food Composition and Analysis 68: 128–132.

Sánchez-Tapia, M., Tovar, A.R. and Torres, N. 2019. Diet as regulator of gut microbiota and its role in health and disease. Archives of Medical Research 50: 259–268.

Sarafian, M., Ding, N., Holmes, E. and Hart, A. 2017. Effect on the Host Metabolism. pp. 249–253. *In*: M.H. Floch, Y. Ringel and W.A. Walker (eds.). The Microbiota in Gastrointestinal Pathophysiology. Elsevier, New York.

Sender, R., Fuchs, S. and Milo, R. 2016. Revised estimates for the number of human and bacteria cells in the body. PLoS Biology 14: e1002533.

Sobrova, P., Adam, V., Vasatkova, A., Beklova, M., Zeman, L. and Kizek, R. 2010. Deoxynivalenol and its toxicity. Interdisciplinary Toxicology 3: 94–99.

Stein, R. and Bulboaca, A. 2017. Mycotoxins. pp. 407–446. *In*: T. Aldsworth, R. Stein, D. Cliver and H. Riemann (eds.). Foodborne Diseases. Academic Press, New York.

Swanson, S.P., Nicoletti, J., Rood Jr, H.D., Buck, W.B., Cote, L.M. and Yoshizawa, T. 1987. Metabolism of three trichothecene mycotoxins, T-2 toxin, diacetoxyscirpenol and deoxynivalenol, by bovine rumen microorganisms. Journal of Chromatography B: Biomedical Sciences and Applications 414: 335–342.

Tang, W.W., Wang, Z., Levison, B.S., Koeth, R.A., Britt, E.B., Fu, X., Wu, Y. and Hazen, S.L. 2013. Intestinal microbial metabolism of phosphatidylcholine and cardiovascular risk. New England Journal of Medicine 368: 1575–1584.

Tola, M. and Kebede, B. 2016. Occurrence, importance and control of mycotoxins: A review. Cogent Food & Agriculture 2: 1191103.

Wong, J.M., De Souza, R., Kendall, C.W., Emam, A. and Jenkins, D.J. 2006. Colonic health: Fermentation and short chain fatty acids. Journal of Clinical Gastroenterology 40: 235–243.

Xifra, G., Esteve, E., Ricart, W. and Fernández-Real, J.M. 2016. Influence of dietary factors on gut microbiota: The role on insulin resistance and diabetes mellitus. pp. 147–154. *In*: D. Mauricio (ed.). Molecular Nutrition and Diabetes. Elsevier, New York.

Yan, C., Yi, W., Xiong, J. and Ma, J. 2018. IOP conference series: Earth and environmental science. IOP Publishing 128: 012086.

Yang, L., Zhao, Z., Wu, A., Deng, Y., Zhou, Z., Zhang, J. and Hou, J. 2013. Determination of trichothecenes A (T-2 toxin, HT-2 toxin, and diacetoxyscirpenol) in the tissues of broilers using liquid chromatography coupled to tandem mass spectrometry. Journal of Chromatography B942: 88–97.

Yousef, H.I. and Zhou, T. 2016. Mycotoxin: Detection and control. All about feed. 5: 45–51.

Zain, M.E. 2011. Impact of mycotoxins on humans and animals. Journal of Saudi Chemical Society 15: 129–144.

Zhao, L., Zhang, F., Ding, X., Wu, G., Lam, Y.Y., Wang, X., Fu, H., Xue, X., Lu, C. and Ma, J. 2018. Gut bacteria selectively promoted by dietary fibers alleviate type 2 diabetes. Science 359: 1151–1156.

CHAPTER

10

Climatic Change, Toxigenic Fungi and Mycotoxins

Yasmine Hamdouche[1,2*], Riba Amar[2,3] and Didier Montet[4,5]

[1] University of Abdelhamid Ibn Badis Mostaganem, Avenue Hamadou Hossine, 27 000, Mostaganem, Algeria

[2] LBSM Laboratory (Laboratory of Biology and Microbial Systems), ENS (Normal Superior School), Kouba, 16 000, Algiers, Algeria

[3] University of M'Hamed Bougara of Boumerdes, Avenue de l'indépendance, 35000, Boumerdes, Algeria

[4] UMR 95 QualiSud, CIRAD, TA B-95 /16, 73 rue J-F Breton, 34398 Montpellier Cedex 5, France

[5] UMR 95 QualiSud, Univ Montpellier, CIRAD, Montpellier SupAgro, Univ d'Avignon, Univ de La Réunion, Montpellier, France

1. Introduction

Mycotoxins are toxins that contaminate the plants in the field or during storage, in all regions of the world. Climate change seems conducive to the development of these mycotoxins and according to the FAO (Food and Agriculture Organization), nearly 25% of the world's foodstuffs are contaminated by significant amounts of mycotoxins. This concerns cereals, oil seeds, dried fruits, spices and foods intended for human and animal consumption.

The rise in the average surface temperatures of the globe and recorded massive emissions of greenhouse gases are the first expected result of climatic change. Agriculture contributes to this global warming but is equally affected by it. In the recent years, scientists begun to think seriously about the effect of climate change on crop yield, food safety and nutritional foodstuffs quality. The infection of staple commodities by fungal diseases in the pre-harvest and contamination by mycotoxins during the post-harvest processes represents the most important risks in feed and food safety.

This chapter summarizes the principal climate change factors, and how these parameters will influence growth of mycotoxigenic fungi and mycotoxins production. It also shows us the difficulties that scientists will

*Corresponding author: yasminehamdouche@gmail.com

have to face to create analytical methods and to evaluate the toxinogenicity in the near and far future.

2. Environmental Factors Affecting Growth of Toxigenic Fungi and Mycotoxins Production

Fungi could contaminate the entire food chain system from primary production, storage and marketing to consumption. The production of mycotoxins could then be related to environmental factors such as drought, precipitation, temperature and human activities.

Temperature and water activity are the main parameters that affect fungal growth and mycotoxin elaboration. It has been shown that the optimal temperature range of growth of fungi was different to that of mycotoxin production. For example, the optimal temperature range for *Fusarium langsethiae* growth in durum wheat was 20–25°C, while for mycotoxin production it was 15–35°C (Nazari et al. 2014). Mitchell et al. (2004) showed that the optimum conditions for Ochratoxin A (OTA) production by *Aspergillus carbonarius* was very different from those for its growth. The same goes for water activity, the minimum Aw level for growth of *Aspergillus ochraceus* on barley grain was 0.85 and Aw of 0.90 was required for OTA optimal production (Pardo et al. 2004). Generally, mycotoxin production requires lower water activity levels contrary to the mycelium growth, as was the case for aflatoxin B1 production by *Aspergillus flavus* in sorghum grains (Lahouar et al. 2016). The relationship effect between the two parameters, temperature and Aw was studied by Ramirez et al. (2006) who showed that some strains of *Fusarium graminearum* were able to grow at the lowest Aw assayed (0.90), and that the best growth obtained at this minimal Aw was 15–25°C. The study of Oviedo (2011) showed also that the concentration of alternariol (AOH) and alternariol monomethyl ether (AME) produced by *Alternaria alternata* in soya beans varied considerably depending on Aw (range 0.96 to 0.99) and temperature (range 15 to 30°C) interactions. Furthermore, there are also some hygrophilic fungi, which intervene at the field such as *Pyrenophora tritici-repentis*, which produce mycotoxins as emodin, catenarin and islandic (Bouras et al. 2009) and as some *Fusarium* species that produced trichothecene, beauvericin and moniliformin, at Aw between 0.96–0.99 (Kokkonen et al. 2010).

Moreover, the temperature, water activity, and the atmospheric CO_2 were considered as parameters that have a significant effect on the growth of *Aspergillus flavus* and the aflatoxin production in maize grains during storage. In fact, the efficacy of modified atmospheres showed that treatment with 25% CO_2 could be sufficient to reduce *A. flavus* growth but a 50% CO_2 was required to obtain a notable reduction of aflatoxin (Giorni et al. 2008). Also, treatment with a high CO_2 modified atmosphere packaging tested (90% CO_2, 5% O_2 and 5% N_2) prevented mycotoxin production (Riudavets et al. 2018) in maize. Otherwise, Vaughan et al. reported that elevated CO_2 increased *Fusarium verticillioides* proliferation in maize but not fumonisin production (Vaughan et al. 2014). In addition, the combination of the three abiotic factors, Aw ×

temperature × elevated CO_2 may stimulate growth/mycotoxin production by mycotoxigenic species, as was shown on the expression of aflatoxin B1 biosynthetic genes (*aflD*) by varying the three parameters (Medina et al. 2014). At temperature (37°C), Aw (0.95) and CO_2 (650 ppm), a maximum relative expression of *aflD*, was observed, and at the same temperature, Aw (0.92) and CO_2 (1000 ppm) there was a significant *aflR* expression. Furthermore, variation in climatic models in pre-harvest steps was applied to predict the risk of deoxynivalenol (DON) produced by *Fusarium* in wheat and showed that a temperature increase under water stress would have an impact on growth/mycotoxin production (Magan et al. 2011). The above discussion envisages the effect of the abiotic parameters interactions that could help to predict the real impacts of climate change on mycotoxigenic fungi.

The environmental variations affects the distribution of fungi worldwide and some studies showed the distribution of toxinogenic species in cereal grains according to the climate. For example, *Fusarium graminearum* is a predominant species with associated mycotoxins (DON and ZON) in Central and Northern Europe (Krska et al. 1996, Milevoj 1997, Czembor et al. 2015), while *F. verticillioides* preponderates in Southern European countries (Munoz et al. 1990; Aguin et al. 2014). Higher temperatures, relative humidity and rainfall during cultivation could be conducive for one toxin and at the same time may have a decreasing effect on the others. The presence of both deoxynivalenol and zearalenone in cereal grain commodities was related to these climatic factors, but the presence of nivalenol was negatively associated with the same factors (van der Fels-Klerx et al. 2012).

3. Biotic Factors Associated with Environmental Parameters and Mycotoxins Contamination

The effect of the abiotic factors mentioned above could be associated with the effect of other so-called biotic factors such as quality of the foodstuff, insects and mites or other contaminants. The interaction between these factors will influence the safe storage life and the level of contamination with mycotoxins (Magan et al. 2003, 2010). In post-harvest steps, the presence of contaminant ecosystems such as insects and environmental factors will influence fungal growth, mycotoxin production, and strains competitiveness on the occupation of ecological niches. This has been shown in stored grains for both *Fusarium culmorum* and deoxynivalenol production, and *Aspergillus ochraceus*/*Penicillium verruscosum* and ochratoxin production (Magan et al. 2003). The mycotoxins most commonly associated with insect damage are aflatoxin and fumonisin in several commodities such as cottonseed, maize, peanuts, and tree nuts (Dowd, 2003, Danso et al. 2017). A high temperature, drought stress and insect activity increased aflatoxins contamination of corns and peanuts in the pre-harvest steps while warm temperature and high humidity favored the contamination in the post-harvest steps, especially during storage stages (Pessu et al. 2011). Furthermore, Manu et al. (2019) showed that aflatoxins contents were related to the abundance of the insect pests in maize at the pre-harvest stage. Otherwise, work on the same type of cereal showed that plant density was considered as an influencing factor, which increased the occurrence of mycotoxin contamination (Blandino et al. 2008, Krnjaja et al. 2019), because, if the crop was denser, the obtained yield would be lower, and the cereal grain weight would be smaller. Consequently, the plant consumed more water which increased its moisture content and, therefore, increased the mycotoxin contamination. Although, plant density had an impact on the mycotoxigenesis increase, it did not affect mycotoxin distribution, fungal genera and species, which could be related to the climatic conditions (Blandino et al. 2008).

Fontaine et al. (2015) showed that the high variability of roquefortine C and mycophenolic acid contents observed in blue-veined cheese was related to the combination of the effect of biotic factor, which was mycotoxigenic potential of *Penicillium roqueforti* strains and the effect of some abiotic factors (pH, temperature, NaCl and O_2 contents, and C/N ratio).

4. Impact of Pre- and Post-Harvest Conditions on Mycotoxins Distribution in Food Commodities

Environmental factors affect the development of toxinogenic fungi, especially the grain moisture content that differs between field and storage stages. *Alternaria, Fusarium* and *Epicoccum* were the most relevant toxigenic fungi, which contaminates cereal fields, while *Aspergillus* and *Penicillium* were involved during grains storage (Fleurat-Lessard 2017). Sanzani et al. (2016) considered also these fungi as post-harvest rotting fungi of fruits and vegetables in addition to *Alternaria*. Romero et al. (2019) thought that fungi diseases detected at post-harvest steps in fruits had their origin during cropping and had only grown until they found optimal conditions. This was the case of the fungi such as *Colletotrichum* spp., *Fusarium* spp., *Alternaria* spp., *Penicillium* spp. ...etc, which were identified under cultivation and storage. On the other hand, the difference in fungal ecosystems of food during pre- and post-harvest steps influences the mycotoxins levels, as shown by Manu et al. (2019), who found that the aflatoxin levels were low in maize at pre-harvest steps but exceeded safety limits during heaping.

Variation of mycotoxin distribution in transformed products could be related to pre-harvest rainfall. Edwards et al. (2018) studied the impact of different watering regimes on the distribution of mycotoxins produced by *Fusarium* in wheat mill fraction and showed that repeated wetting and drying induced migration of Deoxynivalenol (DON) across the mill fractions. While, high heavy rainfall could cause a decrease of DON in the grain, consequently, an increase within white flour. Xing et al. (2017) showed that cereal grains drying could has an impact on the mycotoxin occurrence, in the pre-drying maize kernels, the level of DON were significantly higher compared with the quantity obtained in the post-drying samples. Drying decreased the moisture content in grains to a safe level as a result, restrained fungal invasion, spore germination and fungal growth.

It is important to know that mycotoxins can persist throughout the production chain despite post-harvest treatment applied; for example, in the field, coffee cherries were highly contaminated with OTA, and despite that the husks were removed during processing; the beans obtained after falling on the soil and after fermentation, were reported as high in OTA content (Paterson et al. 2014).

5. Prediction of Occurrence of Toxinogenic Fungi and Mycotoxins under Climatic Change

During the last few years, researchers were seeking to measure the impact of climate change towards modeling the toxinogenic fungal growth and mycotoxins production in various food matrices. Van de Perre et al. (2015) evaluated the effect of climate change on growth and mycotoxin production on tomatoes according to different temperature models simulating the global warming over the years in the current (1981–2000), near future (2031–2050) and far future (2081–2100) for three European regions. For the current time, data for temperature (°C) were obtained from official weather stations in targeted regions, concerning the periods, near and future, the data were downscaled for four RCPs (Representative Concentration Pathway). They found that the diameter of *Alternaria* spp. in a Petri dish was significantly lower for the far future with the higher temperatures compared with the current time in Spain and Portugal. It was the contrary in Poland where the diameter of the mold was higher in the near and the far future when the temperature becomes optimal for the fungal growth. The results of another work by Van der Fels-Klerx et al. (2013), which estimated impacts of climate change on crop phenology and mycotoxin contamination of wheat and maize cultivated in the Netherlands, showed the same effect using climate change projections data for 2031–2050 of the IPCC A1B (Intergovernmental Panel on Climate Change) emission scenario. The A1B emission scenario represents a mid-range scenario for greenhouse gas emissions. In fact, for this climatic model contamination of wheat and maize with deoxynivalenol increased significantly. Mycotoxins contamination in cereal crops was predicted to become a threat for food safety in Europe, within the next 100 years, especially in the 2°C increase in atmospheric temperature per year scenario (Battilani et al. 2016). In particular, the aflatoxin contamination risk is likely to highly affect maize in South and Central Europe due to favorable climatic conditions for the growth of *A. flavus*. In addition, increasing contamination with *Fusarium* mycotoxins in wheat across Europe during the last few years is a matter for worry (Moretti et al. 2019). Changes in weather conditions have a different influence on the production of mycotoxins. For example, in seasons with extreme drought in Serbia, aflatoxin B1 and ochratoxin A were the most dominant in maize samples, whereas, deoxynivalenol, zearalenone and their derivatives were significantly present during the period with extreme precipitation (Kos et al. 2020). Therefore, the results of the works cited below informs us that contamination by certain mycotoxins would increase in the future.

The review of Medina et al. (2017) has explored many questions, including whether under climatic change scenarios will mycotoxin production patterns change. We quote some: what is going to happen with masked mycotoxins? Are the current control/mitigation strategies going to be effective in the future? These questions need answers in the coming years; indeed, many data are missing and deserve to be replenished by the progress of the researchers in order to control the sanitary quality of foodstuffs, in particular the contamination by mycotoxins.

6. Conclusion

Variations in environmental parameters can influence the growth of toxigenic fungi as well as the production of mycotoxins in different ways. The authors showed that high levels of CO_2 are likely to further contribute to the increased mycotoxins production in infected cultures by some fungi, but could also prevent the production of these metabolites in certain fungi. According to the available data in the literature, mycotoxin contamination of foodstuff will increase and decrease depending on the climate change scenarios. Moreover, mycotoxigenic fungi and crops may change their geographic location due to climate change, and this, by moving towards higher temperatures and drought conditions. Therefore, the production of mycotoxins could vary under other latitudes. For instance, in Europe, DON and aflatoxin B1 contaminations are expected to increase in cereals because of global warming.

Mycotoxin production by fungal strains depends on their environment, biotic factors as well as abiotic factors. To predict mycotoxin contamination under climate change scenarios in different continents and/or countries and/or region, it will be necessary to take into account the effect of other parameters such as the fungal strain resilience and the plant or the host foodstuff behaviors. The combination of different parameters at the same time could make the prediction difficult, it is important to measure the effect of each parameter separately, and then study their interactions. The adaptation of fungi to their changing environment is quite difficult to evaluate and is still a challenge. Obviously, there is an urgent need to find ways to control and prevent mycotoxin contamination to ensure food security in the world. It is also urgent to develop models for predicting mycotoxin contamination and create methods that will permit us to evaluate the toxinogenicity of fungi in the new and future climate.

References

Aguin, O., Cao, A., Pintos, C., Santiago, R., Mansilla, P. and Burtón, A. 2014. Occurrence of Fusarium species in maize kernels grown in northwestern Spain. Plant Pathology 63: 946–951. https://doi.org/10.1111/ppa.12151

Battilani, P., Toscano, P., Van der Fels-Klerx, H.J., Moretti, A., Leggieri, M.C.,

Brera, C., and Robinson, T. 2016. Aflatoxin B1 contamination in maize in Europe increases due to climate change. Scientific Reports 6: 24328. https://doi.org/10.1038/srep24328

Blandino, M., Reyneri, A. and Vanara, F. 2008. Effect of plant density on toxigenic fungal infection and mycotoxin contamination of maize kernels. Field Crops Research 106(3): 234–241. https://doi.org/10.1016/j.fcr.2007.12.004

Bouras, N., Kim, Y.M. and Strelkov, S.E. 2009. Influence of water activity and temperature on growth and mycotoxin production by isolates of *Pyrenophora tritici-repentis* from wheat. International Journal of Food Microbiology 131(2–3): 251–255. https://doi.org/10.1016/j.ijfoodmicro.2009.02.001

Czembor, E., Stępień, Ł. and Waśkiewicz, A. 2015. Effect of environmental factors on *Fusarium* species and associated mycotoxins in maize grain grown in Poland. PloS One 10(7): 0133644. https://doi:10.1371/journal.pone.0133644

Danso, J.K., Osekre, E.A., Manu, N., Opit, G.P., Armstrong, P., Arthur, F.H. and Mbata, G. 2017. Moisture content, insect pests and mycotoxin levels of maize at harvest and post-harvest in the Middle Belt of Ghana. Journal of Stored Products Research 74: 46–55. https://doi.org/10.1016/j.jspr.2019.05.015

Dowd, P.F. 2003. Insect management to facilitate pre-harvest mycotoxin management. Journal of Toxicology. Toxin Reviews 22(2–3): 327–350. https://doi.org/10.1081/TXR-120024097

Edwards, S.G., Kharbikar, L.L., Dickin, E.T., MacDonald, S. and Scudamore, K.A. 2018. Impact of pre-harvest rainfall on the distribution of Fusarium mycotoxins in wheat mill fractions. Food Control 89: 150–156. https://doi.org/10.1016/j.foodcont.2018.02.009

Fleurat-Lessard, F. 2017. Integrated management of the risks of stored grain spoilage by seedborne fungi and contamination by storage mould mycotoxins – An update. Journal of Stored Products Research 71: 22–40. https://doi.org/10.1016/j.jspr.2016.10.002

Fontaine, K., Hymery, N., Lacroix, M.Z., Puel, S., Puel, O., Rigalma, K. and Mounier, J. 2015. Influence of intraspecific variability and abiotic factors on mycotoxin production in *Penicillium roqueforti*. International Journal of Food Microbiology 215: 187–193. https://doi.org/10.1016/j.ijfoodmicro.2015.07.021

Giorni, P., Battilani, P., Pietri, A. and Magan, N. 2008. Effect of Aw and CO_2 level on *Aspergillus flavus* growth and aflatoxin production in high moisture maize post-harvest. International Journal of Food Microbiology 122(1–2): 109–113. https://doi.org/10.1016/j.ijfoodmicro.2007.11.051

IPCC, Intergovernmental Panel on Climate Change.

Kokkonen, M., Ojala, L., Parikka, P. and Jestoi, M. 2010. Mycotoxin production of selected *Fusarium* species at different culture conditions. International Journal of Food Microbiology 143(1–2): 17–25. https://doi.org/10.1016/j.ijfoodmicro.2010.07.015

Kos, J., Hajnal, E.J., Malachová, A., Steiner, D., Stranska, M., Krska, R. and Sulyok, M. 2020. Mycotoxins in maize harvested in Republic of Serbia in the period 2012–2015. Part 1: Regulated mycotoxins and its derivatives. Food Chemistry 312: 126034. https://doi.org/10.1016/j.foodchem.2019.126034

Krnjaja, V., Mandić, V., Stanković, S., Obradović, A., Vasić, T., Lukić, M. and Bijelić, Z. 2019. Influence of plant density on toxigenic fungal and mycotoxin contamination of maize grains. Crop Protection 116: 126–131. https://doi.org/10.1016/j.cropro.2018.10.021

Krska, R., Lemmens, M., Schuhmacher, R., Gresserbauer, M. and Prończuk, M. 1996. Accumulation of the mycotoxin beauvercin in kernels of corn hybrid inoculated with *Fusarium subglutinans*. Journal of Agricultural and Food Chemistry 44: 3665–3667. https://doi.org/10.1021/jf960064m

Lahouar, A., Marin, S., Crespo-Sempere, A., Saïd, S. and Sanchis, V. 2016. Effects of temperature, water activity and incubation time on fungal growth and aflatoxin B1 production by toxinogenic *Aspergillus flavus* isolates on sorghum seeds. Revista Argentina de microbiologia 48(1): 78–85. https://doi.org/10.1016/j.ram.2015.10.001

Magan, N., Hope, R., Cairns, V. and Aldred, D. 2003. Post-harvest fungal ecology: Impact of fungal growth and mycotoxin accumulation in stored grain. Epidemiology of Mycotoxin Producing Fungi 109: 723–730. https://doi.org/10.1007/978-94-017-1452-5_7

Magan, N., Aldred, D., Mylona, K. and Lambert, R.J. 2010. Limiting mycotoxins in stored wheat. Food Additives and Contaminants 27(5): 644–650. https://doi.org/10.1080/19440040903514523

Magan, N., Medina, A. and Aldred, D. 2011. Possible climate-change effects on mycotoxin contamination of food crops pre- and postharvest. Plant Pathology 601: 150–163. https://doi.org/10.1111/j.1365-3059.2010.02412.x

Manu, N., Osekre, E.A., Opit, G.P., Arthur, F.H., Mbata, G., Armstrong, P. and Campbell, J. 2019. Moisture content, insect pests and mycotoxin levels of maize on farms in Tamale environs in the northern region of Ghana. Journal of Stored Products Research 83: 153–160. https://doi.org/10.1016/j.jspr.2019.05.015

Medina, A., Rodriguez, A. and Magan, N. 2014. Effect of climate change on Aspergillus flavus and aflatoxin B1 production. Frontiers in Microbiology 5: 348. https://doi.org/10.3389/fmicb.2014.00348

Medina, A., Akbar, A., Baazeem, A., Rodriguez, A. and Magan, N. 2017. Climate change, food security and mycotoxins: Do we know enough? Fungal Biology Reviews 31(3): 143–154. https://doi.org/10.1016/j.fbr.2017.04.002

Milevoj, I. 1997. Electrophoretic study of proteins in the fungus *Fusarium moniliforme* var. *subglutinans*. Cereal Research Communications 25: 603–606. https://www.jstor.org/stable/23786824

Mitchell, D., Parra, R., Aldred, D. and Magan, N. 2004. Water and temperature relations of growth and ochratoxin A production by *Aspergillus carbonarius* strains from grapes in Europe and Israel. Journal of Applied Microbiology 97(2): 439–445. https://doi.org/10.1111/j.1365-2672.2004.02321.x

Moretti, A., Pascale, M. and Logrieco, A.F. 2019. Mycotoxin risks under a climate change scenario in Europe. Trends in Food Science & Technology 84: 38–40. https://doi.org/10.1016/j.tifs.2018.03.008

Munoz, L., Cardelle, M., Pereiro, M. and Riguera, R. 1990. Occurrence of corn mycotoxins in Galicia (Northwest Spain). Journal of Agricultural and Food Chemistry 38: 1004–1006. https://doi.org/10.1021/jf00094a019

Nazari, L.E.Y.L.A., Pattori, E., Terzi, V., Morcia, C. and Rossi, V. 2014. Influence of temperature on infection, growth, and mycotoxin production by *Fusarium langsethiae* and *F. sporotrichioides* in durum wheat. Food Microbiology 39: 19–26. https://doi.org/10.1016/j.fm.2013.10.009

Oviedo, M.S., Ramirez, M.L., Barros, G.G. and Chulze, S.N. 2011. Influence of water activity and temperature on growth and mycotoxin production by *Alternaria alternata* on irradiated soya beans. International Journal of Food Microbiology 149: 127–132. https://doi.org/10.1016/j.ijfoodmicro.2011.06.007

Pardo, E., Marın, S., Sanchis, V. and Ramos, A.J. 2004. Prediction of fungal growth and ochratoxin A production by *Aspergillus ochraceus* on irradiated barley grain as influenced by temperature and water activity. International Journal of Food Microbiology 95(1): 79–88. https://doi.org/10.1016/j.ijfoodmicro.2004.02.003

Paterson, R.R.M., Lima, N. and Taniwaki, M.H. 2014. Coffee, mycotoxins and climate change. Food Research International 61: 1–15. https://doi.org/10.1016/j.foodres.2014.03.037

Pessu, P.O., Agoda, S., Isong, I.U., Adekalu, O.A., Echendu, M.A. and Falade, T.C. 2011. Fungi and mycotoxins in stored foods. African Journal of Microbiology Research 5(25): 4373–4382. https://doi.org/10.5897/AJMR11.487

Ramirez, M.L., Chulze, S. and Magan, N. 2006. Temperature and water activity effects on growth and temporal deoxynivalenol production by two Argentinean strains of *Fusarium graminearum* on irradiated wheat grain. International Journal of Food Microbiology 106(3): 291–296. https://doi.org/10.1016/j.ijfoodmicro.2005.09.004

Riudavets, J., Pons, M.J., Messeguer, J. and Gabarra, R. 2018. Effect of CO_2 modified atmosphere packaging on aflatoxin production in maize infested with *Sitophilus zeamais*. Journal of Stored Products Research 77: 89–91. https://doi.org/10.1016/j.jspr.2018.03.005

Romero, Y. 2019. Pre- and post-harvest factors that affect the quality and commercialization of the Tahiti lime. Scientia Horticulturae 257: 108737. https://doi.org/10.1016/j.scienta.2019.108737

Sanzani, S.M., Reverberi, M. and Geisen, R. 2016. Mycotoxins in harvested fruits and vegetables: Insights in producing fungi, biological role, conducive conditions, and tools to manage postharvest contamination. Postharvest Biology and Technology 122: 95–105. https://doi.org/10.1016/j.postharvbio.2016.07.003

Van de Perre, E., Jacxsens, L., Liu, C., Devlieghere, F. and De Meulenaer, B. 2015. Climate impact on *Alternaria* moulds and their mycotoxins in fresh produce: The case of the tomato chain. Food Research International 68: 41–46. https://doi.org/10.1016/j.foodres.2014.10.014

Van der Fels-Klerx, H.J., Klemsdal, S., Hietaniemi, V., Lindblad, M., Ioannou-Kakouri, E. and van Asselt, E.D. 2012. Mycotoxin contamination of cereal grain commodities in relation to climate in North West Europe. Food Additives & Contaminants: Part A 29(10): 1581–1592. https://doi.org/10.1080/19440049.2012.689996

Van der Fels-Klerx, H.J., van Asselt, E.D., Madsen, M.S. and Olesen, J.E. 2013. Impact of climate change effects on contamination of cereal grains with deoxynivalenol. PloS One 8(9): 73602. https://doi.org/10.1371/journal.pone.0073602

Vaughan, M.M., Huffaker, A., Schmelz, E.A., Dafoe, N.J., Christensen, S., Sims, J. and Allen, L.H. 2014. Effects of elevated [CO_2] on maize defence against mycotoxigenic *Fusarium verticillioides*. Plant, Cell & Environment 37(12): 2691–2706. https://doi.org/10.1111/pce.12337

Xing, F., Liu, X., Wang, L., Selvaraj, J.N., Jin, N., Wang, Y. and Liu, Y. 2017. Distribution and variation of fungi and major mycotoxins in pre- and post-nature drying maize in North China Plain. Food Control 80: 244–251. https://doi.org/10.1016/j.foodcont.2017.03.055

CHAPTER

11

Are There Advantages of GMO on Mycotoxins Content?

Didier Montet[1*] and Joël Guillemain[2]

[1] UMR 95 QualiSud/CIRAD/Université of Montpellier, TA B-95/16, 73 rue J-F Breton, 34398 Montpellier cedex 5, France

[2] 21, rue de la Vieille Poste; 37520 Limeray, France

1. Introduction

In most developed countries, mycotoxins are reduced by good agricultural practices and storage methods. In developing countries, mycotoxins are still a major health problem for animals and humans. Most people in developed countries have never heard of mycotoxins but they are still outbreaks reported in Europe by the rapid alert system for food and feed (Rasff portal). In these rich countries, mycotoxins are primarily a health concern for livestock. In under-developed countries, the management of mycotoxins is very difficult. Insects damaged seeds and fungi could enter the grain through the insect bites. Most foods in Africa are traded in informal markets. Regulation is usually ineffective and when it exists, it will increase the price of seeds. Even when food is known to be contaminated, it will be eaten by consumers due to food insecurity.

On the 22 of May, 2014, the European Food Safety Authority (EFSA) published an urgent scientific advice for the European Commission (E.U.) concerning mycotoxins in maize. In the EFSA Journal on the 22 May 2014, we could read: "an Evaluation of the increase of risk for public health related to a possible temporary derogation from the maximum level of deoxynivalenol, zearalenone and fumonisins for maize and maize products maximum level for deoxynivalenol (DON), zearalenone (ZON), fumonisins (FUMO) for maize products".

Before 2014, France requested a temporary derogation to the maximum levels of these mycotoxins in maize and maize products for the 2013 harvest. Using these data, EFSA estimated that a temporary increase in the levels of three mycotoxins: deoxynivalenol, zearalenone and fumonisins in maize and maize products could impact significantly on the public's health. They

*Corresponding author: didier.montet@cirad.fr; sesame.jg@wanadoo.fr

estimated that the total exposure from all sources (including other crops) was already close to what was considered as safe level for some consumers.

Binder (2007) from the Biomin Company, a leader in mycotoxins treatment in animal feed, published an important paper on the management of the mycotoxins risk in modern feed production. She proposed many solutions to prevent the presence of mycotoxins in feed. In particular, she proposed logically the application of a Hazard Analysis Critical Control Point (HACCP) system in mycotoxin control. She said that any form of fungi inhibitors or acid mixtures will not affect mycotoxins presence because they are very stable compounds. The only way to really assess the quality of ingredients is the specific testing of mycotoxins or certain groups thereof.

In the case of evident mycotoxin contamination, the most practical approach to date is to redirect the food into feed for less-susceptible animal species or to lower the average contamination by blending of non-contaminated material with material above the limits and thus raise the accepted standards when it is not prohibited by law (as in E.U.).

Various mycotoxin binding agents or adsorbents were tested and some of them are used at industrial levels. These compounds will reduce exposure to mycotoxins by decreasing their bioavailability by inclusion. Various substance groups have been tested and used such as aluminum silicates and in particular clay and zeolitic minerals. The materials were bentonites (sodium and calcium bentonites, organophilic bentonites, acid-treated bentonites, as well as some other special forms), zeolithes, diatomites, and vermiculites (Binder 2007). But binders of mycotoxins are too expensive for African farmers and farming methods to reduce mycotoxins, are not usually adopted. Moreover, even if bentonite itself is probably not more toxic than any other particulate not otherwise regulated and is not classified as a carcinogen by any regulatory or advisory body, the principal exposure pathway of concern is inhalation of breathable dust due to variable amounts of breathable crystalline silica in some bentonite, a recognized human carcinogen (Maxim et al. 2016).

A recent review of Ying et al. (2018) described mycotoxins control by updated techniques. They reviewed the traditional physical methods, the chemical and biological, the biodegradation and transformation of mycotoxins and the detoxification by adsorption. The review focused on the development of innovative strategies on mycotoxin reduction such as natural essential oils, polyphenols and flavonoid inhibitors, magnetic materials and nanoparticles.

If we take a look at the innovation in detoxification, we could read that some natural essential oils (EOs) have advantages as a high efficiency, eco-friendly and low-drug-resistance tool. They are usually active against fungi, for example, turmeric EO against *A. flavus* (Hu et al. 2017), *Menthaspicata* EO against *A. flavus* (Kedia et al. 2016), *Curcuma longa* L. EO against *Fusarium graminearum* (Kumar et al. 2016) and many other EOs.

Ying et al. (2018) suggested the removal of mycotoxins by using magnetic materials bound with other chemicals as citrinin (Magro et al. 2016). AFB1 was removed by magnetic carbon nano-composites prepared from maize wastes (Khan and Zahoor 2014) or by nanocellulose conjugated with retinoic acid. Patulin was decontaminated by chitosan-coated Fe_3O_4 particles (Luo et al. 2016). A chitosan-glutaraldehyde complex was suggested as adsorbent material for reducing multiple mycotoxins (Zhiyong et al. 2015).

Ying et al. (2018) explained in the same review that some volatile bioactive compounds were effective to inhibit fungi such as the bioactive antibiotic compound, allylisothiocyanate produced by oriental and yellow mustard seeds that inhibits the growth of patulin-producing strain *P. expansum* (Saladino et al. 2016). Some aliphatic aldehydes could inhibit fungal growth and patulin production (Taguchi et al. 2013). T-2 toxin and patulin were degraded by glow discharge plasma technology that is high-energy active particles released by glow discharge plasma. The high energy particles could attack cyclic compounds containing epoxy structures (Pu et al. 2017). The authors concluded by saying that in the future mycotoxin control will most likely result not from a single treatment but from a combination of good manufacturing practices, appropriate quality-assurance programmes, and eco-friendly and biosafe postharvest detoxifying procedures throughout the production process.

Another review of Peng et al. (2018) surveys the new strategies to prevent and reduce mycotoxins for feed manufacturing. In this review, the mycotoxin reducing methods are mainly categorized into four methodologies: physical methods, thermal methods, chemical methods, and mycotoxin controlling feed additives. The first three methodologies mainly focus on how to reduce mycotoxins in feed ingredients during processes, while the last one on how to compensate the adverse impacts of mycotoxin-contaminated diets in animal bodies.

The results showed that most of the methods reviewed show evidence of mycotoxin reducing effects, but of different consistencies. On the other hand, many practical factors that can affect the feasibility of each method in practical manufacturing are also discussed in this review. In conclusion, mycotoxin prevention management and the processing stage of cleaning and sorting are still the most efficient strategies to control mycotoxin hazards in current feed manufacturing.

Assaf et al. (2019) in their review, proposed a strategy to reduce AFM1 contamination in milk by microbial adsorbents. Bacteria and yeasts were used to complex AFM1 in milk and they commented the binding stability of the AFM1-microbial complex. Corassin et al. (2013) showed that the use of heat-killed cells is actually more favorable for milk decontamination than viable cells. They explained in particular that heat treatment affects the hydrophobic nature of cell wall components by denaturation of proteins (Haskard et al. 2001).

Phytopathogenic and toxinogenic *Fusarium* spp. may infect maize throughout its cultivation. In parallel with these infections, mycotoxins are often produced. They could accumulate in affected tissues and thus pose a significant risk for animals and humans. European maize is usually more infected by fungal species belonging to the *Fusarium* sections Discolour and Liseola (Oldenburg et al. 2017).

A very large intraspecific polymorphism was found between species in the Discolor section including *Fusarium crookwellense, F. graminearum* and *F. culmorum* and thus of the section Liseola including *F. verticillioides, F. proliferatum* and *F. subglutinans* found through the analysis of intergenic spacer (IGS) sequences of *Fusarium* strains infested with wheat (Mycsa collection 2019).

The Discolor section is more prevalent in cooler and humid climate regions and the Liseola section predominates in warmer and dryer areas. Several pathogenic *Fusarium* spp. usually coexist in growing maize under field conditions. It may lead to multi-contamination with mycotoxins like trichothecens, zearalenone and fumonisins (Oldenburg et al. 2017).

Blandino et al. (2015) described the importance of the European corn borer (*Ostrinianubilalis*) on the contamination of maize. The maize samples were analyzed by HPLC-MS/MS which led to the detection of 13 *Fusarium* mycotoxins: fumonisins, fusaproliferin, moniliformin, bikaverin, beauvericin, fusaric acid, equisetin, deoxynivalenol, deoxynivalenol-3-glucoside, zearalenone, culmorin, aurofusarin and butenolide.

Bt grains were created to avoid the attack of some insects. Bt maize is a corn variety to which the *cry1Ab* gene from the bacterium *Bacillus thuringiensis* (abbreviated as Bt) has been added. This gene allows the plant to produce an insecticidal protein. This insecticide is used to kill Corn Borer (*Ostrinia nubilalis*) and Corn Moth (*Sesamia nonagrioides*). In 1996, the company Monsanto put on the market a corn variety resistant to glyphosate herbicide in addition to insect resistance. The benefit alleged by GMP (Good Manufacturing Practices) in the corn grain had not been strategically anticipated. It remains an unintentional additional effect which should be measured using rigorous evaluation methods to assess the real impact and the benefits to animal and human health. This argument could be opportunely used by professionals in the GMP sector in a strategy to develop their activities. It could help to restore the image of GMP-derived products and help to promote the acceptability to society (Afssa 2004).

2. Does Bt Corn Reduce or not Insect Damage that could in turn Reduce Mycotoxins?

This section is an overview of the advantages and disadvantages of genetically modified seeds on mycotoxins content in food.

Bt maize has revolutionized pest control and many farmers have benefited, but some people remain skeptical of this new technology. Hellmich and Hellmich (2012) from USDA-ARS, Corn Insects and Crop Genetics Research Unit, Dept of Entomology USA published a paper on the use and impact of Bt Maize. These genetically modified (GM) plants produce crystal (Cry) proteins or toxins derived from the soil bacterium, *Bacillus thuringiensis* (Bt). Bt maize is now used for pest control in many countries. These authors focused on the opportunities and challenges of Bt maize and showed that there are still questions about its use and impact.

In the case of mycotoxins, they reported the work of Munkvold et al. (1997) and Dowd (2000) who showed a reduction of the occurrence of ear molds due to a reduction of insect damage that provides a site for infection by fungi especially fumonisin and deoxynivalenol producers.

In corn grain collected from Bt hybrids grown in 107 locations across the United States between 2000 and 2002, Hammond et al. (2004) concluded that fumonisins levels were frequently lower in Bt hybrids compared to their near-isogenic controls. For the authors, these results were consistent with those previously reported with Bt hybrids grown in the United States (Munkvold et al. 1997, Dowd 2000), France and Spain (Bakan et al. 2002), Italy (Pietriand Piva 2000), Turkey and Argentina (Hammond et al. 2004).

Wu et al. (2004) suggested that USA farmers save 23 million dollars annually through mycotoxin reduction. Wu (2006) showed a significant health benefit in the world where Bt maize is used in the diet. A group of experts of the French Food Safety Authority (Afssa 2004) evaluated the bibliography linked to the incidence of the Bt gene on mycotoxin contamination of corn. At this time, the results of the available studies were indisputable and highly significant for fumonisins produced by fungi very common in the climates of Europe. They really highlighted a reduction in mycotoxin contents in Bt corn. The results relative to trichothecenes and zearalenone are still provisional pending a definitive conclusion.

It is in fact more difficult to measure the impact on the health of farm animals and on the quality of the animal products consumed by humans. Piva et al. (2001a, b) observed better growth in pigs and chickens fed with Bt corn, which was less contaminated with fumonisin B1 than conventional isogenic corn. This fact could encourage the creation of new varieties resistant to the development of fungi by conventional plant breeding method (Munkvold and Desjardins 1997, Moreno and Kang 1999) or by the introduction of gene encoding proteins with an antifungal activity (Jach et al. 1995) or involved in the biosynthesis of antifungal compounds (Hain et al. 1993).

Ostry et al. (2010) reviewed 23 studies of mycotoxins in corn. Animal feed with maize are more severely contaminated with Fusarium mycotoxins, e.g. fumonisins (FUM), zearalenone (ZEA) and deoxynivalenol (DON). A positive side-effect in reducing mycotoxin levels was observed when Bt maize was used. In fact, Bt maize is highly resistant to European corn borer larval feeding due to Bt toxin (δ toxin) production. They found that 19 out of 23 studies on Bt maize were less contaminated with mycotoxins (FUM, DON, ZEA) than the conventional control variety in each case. They concluded that Bt maize was an important potential tool for insect pest protection.

Bakan et al. (2002) compared the fungal growth and fusarium mycotoxin content in isogenic traditional maize and GM maize grown in France and Spain. Maize grains from Bt hybrids and near-isogenic traditional hybrids were collected in France and Spain in 1999. They grow under natural conditions. The fungal biomass formed on Bt maize grain measured by ergosterol level, was 4–18 times lower than that on isogenic maize. Fumonisin B concentrations ranged from 0.05 to 0.3 ppm for Bt maize and from 0.4 to

9 ppm for isogenic maize. There was no difference in trichothecenes and zearalenone concentrations measured on transgenic and non-transgenic maize. They concluded that Btmaize plants are a good solution to protect the plant against insect damage (European corn borer and pink stem borer).

In the rat feeding study performed during the GMO90+ project, Chereau et al. (2018) analyzed a comprehensive set of contaminants in MON810 and NK603 genetically modified (GM) maize, and their non-GM counterparts. Both the maize grains and the manufactured pellets were characterized. Among mycotoxins tested, fumonisin and deoxynivalenol were the compounds present in the highest amounts, but at concentrations that were largely below acceptance reference values. Moreover, data reporting at slightly lower levels of fumonisin in MON810 compared to its non-GM counterpart corroborate the lower susceptibility of insect resistant Bt maize to fumonisin-producing fungi.

To assess the safety of French maize food, investigations were carried out by the National Biological Risk Monitoring network from Bt maize MON 810 and its isogenic counterpart growing in South Western France. Fumonisins B_1 and B_2, deoxynivalenol and zearalenone were analyzed. Bt maize decreased concentrations of fumonisins by 90% and zearalenone by 50%, whereas deoxynivalenol was slightly increased (Folcher et al. 2010).

The French experts Delos et al. (2014) commented on the management of mycotoxins in cereal crops in 2013 and what it was possible to expect from biotechnology against these contamination. They considered the year 2012 as a break year for the cereal contamination in Europe due certainly to a better climate for fungal growth. They proposed in order to reduce the risk of contamination different strategies such as crop rotation, choice of sowing and harvest dates for corn, balanced fertilization and the use of irrigation, cleaning of harvesting or transporting machines, biological control tools, systems that can hardly do without fungicides, insecticides and herbicides and thought what could be the consequences on the interest of using genetically modified plants? They concluded that the situation of the United States and that of France are therefore absolutely not comparable. In both cases, however, biotechnologies, thanks to the molecular techniques used in varietal selection, accelerate the development of varieties which are generally more tolerant of *Fusarium* but without recourse to transgenesis. On the other hand, the situation of Serbia is intermediate between that of France and US with recourse to irrigation and exposure to the corn beetle, the question as it was asked on the interaction between Aflatoxin contamination and damage of corn beetles to the roots was justified in this country.

Recently, Barroso et al. (2017) studied the growth of *F. verticillioides* and its fumonisin production in Bt and non-Bt maize cultivated in Brazil. They found that some GM hybrids presented lower *F. verticillioides* frequency than other GM hybrids samples. However, they did not find a statistical difference between fumonisin contamination when Bt and non-Bt samples were compared. They concluded logically that other environmental parameters could possibly trigger fumonisin production during plant development in the field.

3. Possible Metabolism or Actions of Fungal Infection

Trujillo et al. (2001) identified three levels of plant protection:

- Expression in the plant of an insecticide (protein) to control insect damage and fungal spore invasion,
- Control fungal growth after plant invasion, and
- Disruption of the fungal biosynthetic pathways leading to aflatoxin formation.

The main activity that makes the Bt plant different from a conventional plant is its resistance to insects. The plant is less pricked by insects and therefore fungi cannot access the interior of the seed that provides the nutrients necessary for its growth. Certain environmental factors can also promote the presence of molds such as high temperatures or the presence of fungi on the plants *via* clonal infection of seeds and plant debris. Maize natural resistance to contamination is also an important factor and the interaction between both factors.

3.1 Environmental Factors

Logically, fungi growth is influenced by different biotic and abiotic factors. But what is about GM plants?

Miller (2001) described the factors that affect the occurrence of fumonisin produced by *Fusarium graminearum* and *F. verticillioides* and some allied species growing under different environmental conditions. Ear rot and fumonisin accumulation are associated with drought and insect stress. In addition *F. graminearum* grows well only between 26 and 28 °C and requires rain for disease progression. *F. verticillioides* grows well at higher temperatures ($\leq$ 28 °C), while *F. graminearum* is favored when temperature is cooler.

They think that the best available strategy for reducing the risk of fumonisin in maize is to find hybrids that are adapted to the environment and to limit drought stress and insect herbivory. They also proposed some ideas in research consisting of the production of hybrids that contain enzymes capable to degrade fumonisin when it is produced.

Schaafsma and Hooker (2007) proposed the development of predictive tools that they called climatic models to predict occurrence of *Fusarium* toxins in wheat and maize. Several disease incidence models have been commercialized for wheat, but only one toxin prediction model from Ontario, Canada, "DONcast", has been validated extensively and commercialized to date for wheat. DONcast® is actually commercialized by Bayer Crop Science. The DONcast® software was developed by Weather INnovations Incorporated and uses climate data, forecast and observed weather data, as well as plot agronomic data. DONcast® can be used to evaluate the need for a fungicide anti-fusarium treatment at flowering in and to evaluate the accumulation of DON in wheat grains.

Considering that the variation in toxin levels is usually associated with year and agronomic factors, they estimated that environmental effects accounted for 48% of the variation in deoxynivalenol (DON) in wheat. In maize, hybrid accounted for 25% of the variation of either DON or fumonisin, followed by environment (12%). They developed the DON forecast model that accounted for up to 80% of the variation in DON.

Oldenburg et al. (2017) studied the *Fusarium* diseases of maize associated with mycotoxin contamination used for feed and food and described how the fungi gain access to the target organs of the plant. They explained that they are in relation to specific symptoms of typical rot diseases regarding ears, rudimentary ears, kernels, roots, stem, leaves, and seeds.

Forage maize is often affected by both *Gibberella* and *Fusarium* and ear rots are of major importance in affecting the safety of grain in animal and human nutrition. Thus, rudimentary ears contain high amounts of fusarium toxins. In forage maize, it is only the grains that are eaten representing a low portion of the whole plant. The impact of foliar diseases on grain contamination is thus low, as *Fusarium* infections are restricted to some parts on the leaf husks and sheaths.

When *Fusarium* toxins are produced in the rotten lower part of the stem, they usually remain in the stubble after harvest. To avoid the progression of the disease, they proposed to harvest grains at the appropriate maturity stage to keep contamination as low as possible (Oldenburg et al. 2017).

Blandino et al. (2015) proposed to find a strategy based on the reduction of ECB (European Corn Borer) damage to avoid contamination by *Fusarium* spp. mycotoxins. They compared the insect attacks in anormal and in a protected area. They showed that the ears collected in the protected areas were free of ECB attack, while those subject to natural insect attacks had an important damage varying from 10% to 25%.

In conclusion, the environmental factors that affect mycotoxin production are the same as those that influence fungi growth. The most important thing is to limit attacks of insects favoring the penetration of fungal pathogens. New sequencing techniques could help in following changes in the spectrum of dominating *Fusarium* pathogens involved in mycotoxin contamination of maize to ensure safety in the food and feed chain.

3.2 Clonal Infection of Seeds and Plant Debris

The ECB plays an important role in promoting *F. verticillioides* infections and in the consequent fumonisin contamination in maize grain in temperate areas (Blandino et al. 2015). The objective of this study was to evaluate whether the ECB feeding activity could also affect the occurrence of emerging mycotoxins in maize kernels. During the 2008–2010 period, natural infestation of the insect was compared, in field research, with the protection of infestation, which was obtained by using an entomological net. The ears collected in the protected plots were free from ECB attack, while those subject to natural insect attacks showed a damage severity that varied from 10% to 25%. The maize samples were analyzed by means of an LC-MS/MS-based multi-mycotoxin method, which led to the detection of various metabolites: fumonisins (FUMs), fusaproliferin (FUS), moniliformin (MON), bikaverin (BIK), beauvericin (BEA), fusaric acid (FA), equisetin (EQU), deoxynivalenol (DON), deoxynivalenol-3-glucoside (DON-3-G), zearalenone (ZEA), culmorin (CULM), aurofusarin (AUR) and butenolide (BUT). The occurrence of mycotoxins produced by *Fusarium* spp. of Liseola section was affected significantly by the ECB feeding activity. The presence of ECB injuries increased the FUMs from 995 to 4694 $\mu g.kg^{-1}$, FUS from 17 to 1089 $\mu g.kg^{-1}$, MON from 22 to 673 $\mu g.kg^{-1}$, BIK from 58 to 377 $\mu g.kg^{-1}$, BEA from 6 to 177 $\mu g.kg^{-1}$, and FA from 21 to 379 $\mu g.kg^{-1}$. EQU, produced by *F. equiseti* section Gibbosum, was also increased by the ECB activity, by 1-30 $\mu g.kg^{-1}$ on average. Instead, the content of mycotoxins produced by *Fusarium* spp. of Discolor and Roseum sections was not significantly affected by ECB activity. As for FUMs, the application of a strategy that can reduce ECB damage could also be the most effective solution to minimize the other mycotoxins produced by *Fusarium* spp. of Liseola section.

Schaafsma and Hooker (2007) used climatic models to predict the occurrence of *Fusarium* toxins in maize and wheat. They showed that forecasting *Fusarium* infections is more useful to help reducing their entry into the food chain. These predictive tools are permitted to estimate the variation in toxin levels associated with year and agronomic effects. For wheat, environment effects accounted for 48% of the variation in deoxynivalenol (DON) across all fields, followed by variety (27%), and previous crop (14 to 28%). For maize, hybrid accounted for 25% of the variation of either DON or fumonisin, followed by environment (12%), and when combined 42% of the variability. The forecast of DON and fumonisins was more difficult in maize because of its greater exposure to infection, the role of wounding in infection, the important role of hybrid susceptibility, and the number of uncharacterized hybrids.

Bacon et al. (2001) described that during the biotrophic endophytic maize contamination, *F. moniliforme* produces fumonisins. This fungus is transmitted vertically and horizontally to the next generation of plants *via* clonal infection of seeds and plant debris. They explained that horizontal infection is the manner by which this fungus is spread contagiously. This infection could be reduced by application of certain fungicides. They focused on an important point by explaining that the endophytic phase of *F. moniliforme* is vertically transmitted. This type of infection is not controlled by applications of fungicides, and it remains the reservoir from which infection and toxin biosynthesis takes place in each generation of plants. They developed a biological control system using an endophytic bacterium, *Bacillus subtilis* and showed a high reduction of mycotoxin accumulation during the endophytic (vertical transmission) growth phase.

3.3 Genetic Factors (Maize Resistance)

Different research teams have tried to find natural resistance to pathogenic insects. Duvick (2001) proposed to reduce fumonisin contamination of maize through genetic modification. They considered developing a *Fusarium* ear mold-resistant maize germplasm without creating undesirable agronomic traits.

One of their three strategies was the use of transgenic maize expressing *B. thuringiensis* (Bt) toxin, targeted to the European corn borer. Their second strategy was based on the overexpression of specific antifungal proteins and metabolites to enhance the plant's own defense systems by controlling the ability of *Fusarium* to infect and colonize the ear. Their third strategy was a transgene strategy that aimed to prevent mycotoxin biosynthesis or to detoxify mycotoxins in plants. They said that some enzymes that degrade Fumonisins have been identified in a filamentous saprophytic fungus isolated from maize. Corresponding genes have been cloned and tested in transgenic maize.

Munkvold (2003) considered that cultural practices, including crop rotation, tillage, planting date, and management of irrigation and fertilization, had limited effects on infection and mycotoxin accumulation. The author reviewed the efforts of researchers to control infection or mycotoxin development through conventional breeding and genetic engineering and described the role of transgenic insect control in the prevention of mycotoxins in maize. It was concluded that cultural and genetic approaches for the management of mycotoxins are now available and the greatest tool to manage mycotoxin problem is through genetic resistance.

Mycotoxigenic fungi resistance exists naturally in maize but has not been sufficiently exploited because of its usual polygenic nature and the poor agronomic performance of resistance sources. Available commercial hybrids generally lack adequate resistance levels. Transgenic resistance strategies are being pursued intensively, but the fruits of this research will not be available for several years or more. Munkvold (2003) also concluded that the Bt approach is not sufficiently robust to constitute a long-term solution for fumonisins, because *Fusarium* spp. can enter kernels unassisted by insects.

More recently, Santiago et al. (2015) reviewed research on the influence of environmental variables on fumonisin accumulation, the genetics of maize resistance to fumonisin accumulation, and the search for the biochemical and/or structural mechanisms of the maize plant that could be involved in resistance to fumonisin contamination. They explore also the outcomes of breeding programs and risk monitoring of undertaken projects.

Samayoa et al. (2019) proposed conventional breeding by a complex genetic architecture of resistance to fumonisin accumulation and marker-assisted selection as an efficient alternative. In their study, Genome-Wide Association Study (GWAS) has been performed for the first time for detecting high-resolution Quantitative Trait Loci (QTL) for resistance to fumonisin accumulation in maize kernels complementing published GWAS results for *Fusarium* ear rot.

They found 39 SNPs significantly associated with resistance to fumonisin accumulation in maize kernels and clustered into 17 QTL. They highlighted the genes probably implicated in resistance to pathogens.

Poland is the fifth largest European country in terms of maize production. Ear rots caused by *Fusarium* spp. are significant diseases affecting yield and causing grain mycotoxin contamination. Inbred lines, which are commonly used in Polish breeding programs, belong mostly to two distinct genetic categories: flint and dent. However, historically used lines belonging to the heterotic Lancaster, Iodent Reid (IDT) and Stiff Stalk Synthetic (SSS) (expansion) groups were also present in previous Polish breeding programs. In the current study, 98 inbred lines were evaluated across a 2-year-long experiment, after inoculation with *F. verticillioides* and under natural infection conditions (Czembor et al. 2019). Lancaster, IDT, SSS and SSS/IDT groups were characterized as the most susceptible ones and flint as the most resistant. Based on the results obtained, the moderately resistant and most susceptible genotypes were defined to determine the content of fumonisins (FBs) in kernel and cob fractions using the HPLC method. Fumonisin's content was higher in the grain samples collected from inoculated plants than in cobs. The association of visible *Fusarium* symptoms with fumonisin concentration in grain samples was significant. Conversely, the cobs contained more FB_1 under natural infection, which may be related to a pathogen's type of growth, infection time or presence of competitive species. By using ddRADseq genome sampling method, it was possible to distinguish a basal relationship between moderately resistant and susceptible genotypes. Genetic distance between maize genotypes was high. Moderately resistant inbred lines, which belong to IDT and IDT/SSS belong to one haplotype. Genotypes which belong to the flint, dent or Lancaster group, characterized as moderately resistant were classified separately as the same susceptible one. This research has demonstrated that currently grown Polish inbred lines as well the ones used in the past are a valid source of resistance to *Fusarium* ear rot. A strong association was observed between visible *Fusarium* symptoms with fumonisin concentration in grain samples, suggesting that selection in maize for reduced visible molds should reduce the risk of mycotoxin contamination. NGS (expansion) techniques provide new tools for overcoming the long selection process and increase the breeding efficiency.

The antifungal mechanisms of natural EOs are thought to be related to the disruption of membrane and fungal cell organization (Ying et al. 2018). Alternatively, EOs might inhibit some key enzymes related to carbohydrate catabolism and mycotoxin production (Tian et al. 2011). The inhibitory activity of polyphenol and flavonoid compounds has been investigated on the occurrence of mycotoxins. The results confirmed that chlorogenic and gallic acids have important protective effects on the occurrence of AFB1 (Telles et al. 2017). Quercetin and umbelliferone were confirmed to be effective in preventing patulin accumulation at the transcript level. The expression of five genes involved in the patulin biosynthetic pathway (*msas, IDH, p450-1, p450-2* and *peab1*) was down-regulated in the presence of the two phenolic compounds (quercetine and umbelliferone), singly and in combination (Sanzani et al. 2009). Flavanones such as neo-hesperidin, hesperidin, naringin and hesperetinglucoside are usually extracted from by-products of the citrus industry. All flavanones tested inhibited patulin production and reduced patulin accumulation at least 95% compared to the control (Salas et al. 2012).

4. Conclusion

The genetically modified plants therefore have many advantages in the fight against mycotoxins. The fight against insects that attack the outer membranes of the plant or seeds is essential. Some authors emphasize the importance of environmental factors but also of the plant's ability to naturally resist attack by insects. The risk of fungi contamination could be reduced by different strategies such as crop rotation, choice of sowing and harvest dates, balanced fertilization and the use of irrigation, cleaning of harvesting or transporting machines, biological control tools, systems that can hardly do without fungicides, insecticides and herbicides. GMOs therefore have their role to play in a control policy which would like to reduce the impact of chemical pesticide molecules on the environment.

References

Assaf, J.C., Nahle, S., Chokr, A., Louka, N., Atoui, A. and El Khoury, A. 2019. Assorted methods for decontamination of aflatoxin M1 in milk using microbial adsorbents. Toxins 29: 11(6). DOI: 10.3390/toxins11060304

Aumaitre, L.A., Besançon, P., Branchard, M., Branlard, G., Corthier, G., Guillemain, J., Houdebine, L.M., Joudrier, P., Lindley, N., Moulin, G., Paris, A., Pineau, T., Raynal, A., Zalta, J.P., Barthomeuf, C., Casse, F., Montet, D., Spinnler, H.E., Urdaci, M., Schwartz, M. and Gallotti, S. 2004. GMOs and food: Is it possible to identify and assess health benefits? Afssa Working Group. https://www.vie-publique.fr/sites/default/files/rapport/pdf/054000074.pdf

Bacon, C.W., Yates, I.E., Hinton, D.M. and Meredith, F. 2001. Biological control of *Fusarium moniliforme* in maize. Environmental Health Perspectives 109(Suppl 2): 325–332. DOI: 10.1289/ehp.01109s2325

Bakan, B., Melcion, D., Richard-Molard, D. and Cahagnier, B. 2002. Fungal growth and Fusarium mycotoxin content in isogenic traditional maize and genetically modified maize grown in France and Spain. Journal of Agricultural Food Chemistry 13: 50(4): 728–731. DOI: 10.1021/jf0108258

Barroso, V.M., Rocha, L.O., Reis, T.A, Reis, G.M., Duarte, A.P., Michelotto, M.D. and Correa, B. 2017. *Fusarium verticillioides* and fumonisin contamination in Bt and non-Bt maize cultivated in Brazilian Mycotoxin Research 33(2): 121–127. DOI: 10.1007/s12550-017-0271-4.

Binder, E.M. 2007. Managing the risk of mycotoxins in modern feed production. Animal Feed Science and Technology 133(1–2): 149–166. DOI: 10.1016_j.anifeedsci.2006.08.008

Blandino, M., Scarpino, V., Vanara, F., Sulyok, M., Krska, R. and Reyneri, A. 2015. Role of the European corn borer (*Ostrinia nubilalis*) on contamination of maize with 13 Fusarium mycotoxins. Food Additives & Contaminants: Part A 32(4): 533–543. DOI: 10.1080/19440049.2014.966158

Chereau, S., Rogowsky, P., Laporte, B., Coumoul, X., Moing, A., Priymenko, N., Steinberg, P., Wilhelm, R., Schiemann, J., Salles, B. and Richard-Forget, F. 2018. Rat feeding trials: A comprehensive assessment of contaminants in both genetically modified maize and resulting pellets. Food Chemical Toxicology 121: 573–582. DOI: 10.1016/j.fct.2018.09.049

Corassin, C.H., Bovo, F., Rosim, R.E. and Oliveira, C.A.F. 2013. Efficiency of *Saccharomyces cerevisiae* and lactic acid bacteria strains to bind aflatoxin M1 in UHT skim milk C.H. Food Control 31: 80–83. https://doi.org/10.1016/j.foodcont.2012.09.033

Czembor, E., Waśkiewicz, A., Piechota, U., Puchta, M., Czembor, J.H. and Stępień, Ł. 2019. Differences in ear rot resistance and *Fusarium verticillioides*-produced Fumonisin contamination between Polish currently and historically used maize inbred lines. Frontiers in Microbiology 18(10): 449. DOI: 10.3389/fmicb.2019.00449

Delos, M., Regnault, R.C. and Joudrier, P. 2014. Les mycotoxines dans les récoltes de céréales. Quelle gestion en 2013? Qu'attendre des biotechnologies contre ces fléaux? Académie d'Agriculture de France, Groupe de Réflexion "Plantes Génétiquement Modifiées" 12 janvier 2014.

Dowd, P.F. 2000. Indirect reduction of ear molds and associated mycotoxins in *Bacillus thuringiensis* corn under controlled and open field conditions: Utility and limitations. Journal of Economic Entomology 93: 1669–1679.

Duvick, J. 2001. Prospects for reducing fumonisin contamination of maize through genetic modification. Environmental Health Perspectives 109(Suppl. 2): 337–342. DOI: 10.1289/ehp.01109s2337

EFSA urgent advice. 2014. https://www.efsa.europa.eu/en/press/news/140522-1.

EFSA Journal. 22 May 2014. Evaluation of the increase of risk for public health related to a possible temporary derogation from the maximum level of deoxynivalenol, zearalenone and fumonisins for maize and maize products maximum level (ML), deoxynivalenol (DON), zearalenone (ZON), fumonisins (FUMO), maize products, occurrence, exposure. https://www.efsa.europa.eu/en/efsajournal/pub/3699

Folcher, L., Delos, M., Marengue, E., Jarry, M., Weissenberger, A., Eychenne, N. and Regnault-Roger, C. 2010. Lower mycotoxin levels in Bt maize grain. Agronomy for Sustainable Development 30: 711–719. DOI: https://doi.org/10.1051/agro/2010005

Hain, R., Reif, H.J., Krause, E., Langebartels, R., Kindl, H., Vornam, B., Wiese, W., Schmelzer, E., Schreier, P.H. and Stocker, R.H. 1993. Disease resistance results from foreign phytoalexin expression in a novel plant. Nature 361: 153–156. DOI: 10.1038/361153a0

Hammond, B.G., Campbell, K.W., Pilcher, C.D., Degooyer, T.A., Robinson, A.E., McMillen, B.L., Spangler, S.M., Riordan, S.G., Rice, L.G. and Richard, J.L. 2004. Lower fumonisin mycotoxin levels in the grain of Bt corn grown in the United States in 2000-2002. Journal of Agriculture Food Chemistry Mar 10: 52(5): 1390–1397. DOI: 10.1021/jf030441c

Haskard, C.A., El-nezami, H.S., Kankaanpa, P.E., Salminen, S. and Ahokas, J.T. 2001. Surface binding of aflatoxin B1 by lactic acid bacteria. Applied and Environmental Microbiology 67: 3086–3091. DOI: 10.1128/AEM.67.7.3086–3091.2001

Hellmich, R.L. and Hellmich, K.A. 2012. Use and impact of Bt maize. Nature Education Knowledge 3(10): 4.

Hu, Y., Zhang, J., Kong, W., Zhao, G. and Yang, M. 2017. Mechanisms of antifungal and anti-aflatoxigenic properties of essential oil derived from turmeric (*Curcuma longa* L.) on *Aspergillus flavus*. Food Chemistry 220: 1–8. https://doi.org/10.1016/j.foodchem.2016.09.17

Jach, G., Gornhardt, B., Mundy, J., Logemann, J., Pinsdorf, E., Leah, R., Schell, J. and Maas, C. 1995. Enhanced quantitative resistance against fungal disease by combinatorial expression of different barley antifungal proteins in transgenic tobacco. The Plant Journal 8: 97–109. DOI: 10.1046/j.1365–313x.1995.08010097.x

Kedia, A., Dwivedy, A.K., Jha, D.K. and Dubey, N.K. 2016. Efficacy of *Mentha spicata* essential oil in suppression of *Aspergillus flavus* and aflatoxin contamination in chickpea with particular emphasis to mode of antifungal action. Protoplasma 253(3): 647–653. DOI: 10.1007/s00709-015-0871-9

Khan, F. and Zahoor, M. 2014. In vivo detoxification of aflatoxin B1 by magnetic carbon nanostructures prepared from bagasse. BMC Veterinary Research 10(1): 255. DOI: 10.1186/s12917-014-0255-y

Kumar, M., Kothari, N. and Gupta, B.D. 2016. Autoimmune haemolytic anaemia in a child due to chickenpox. Indian Journal of Hematology and Blood Transfusion 32: 522–524. DOI: https://doi.org/10.1007/s12288-016-0693-8

Luo, Y., Zhou, Z. and Yue, T. 2016. Synthesis and characterization of nontoxic chitosan-coated Fe_3O_4 particles for patulin adsorption in a juice-pH simulation aqueous. Food Chemistry 221. DOI: 10.1016/j.foodchem.2016.09.008

Magro, M., Esteves Moritz, D., Bonaiuto, E., Baratella, D., Terzo, M., Jakubec, P., Malina, O., Čépe, K., Falcao de Aragao, G.M., Zboril, R. and Vianello, F. 2016. Citrinin mycotoxin recognition and removal by naked magnetic nanoparticles. Food Chemistry 203: 505–512. https://doi.org/10.1016/j.foodchem.2016.01.147

Maxim, L.D., Niebo, R. and McConnell, E.E. 2016. Bentonite toxicology and epidemiology – A review. Inhalation Toxicology 28(13): 591–617. DOI: 10.1080/08958378.2016.1240727

Miller, J.D. 2001. Factors that affect the occurrence of fumonisin. Environmental Health Perspectives 109(Suppl. 2): 321–324. DOI: 10.1289/ehp.01109s2321

Moreno, O.J. and Kang, M.S. 1999. Aflatoxins in maize: The problem and genetic solutions. Plant Breed 118: 1–16.

Munkvold, G.P. and Desjardins, A.E. (1997) Fumonisins in maize: Can we reduce their occurrence? Plant Disease 81: 556–565. DOI: 10.1094/PDIS.1997.81.6.556

Munkvold, G.P. 2003. Cultural and genetic approaches to managing mycotoxins in maize. Annual Review of Phytopathology 41: 99–116. https://doi.org/10.1146/annurev.phyto.41.052002.095510

Mycsa collection. 2019. https://www6.bordeaux-aquitaine.inra.fr/mycsa/Des-ressources-biologiques-uniques/La-collection-de-Fusarium-toxinogenes

Oldenburg, E., Höppner, F., Ellner, F. and Weinert, J. 2017. Fusarium diseases of maize associated with mycotoxin contamination of agricultural products intended to be used for food and feed. Mycotoxin Research 33(3): 167–182. DOI: 10.1007/s12550-017-0277-y.

Ostry, V., Ovesna, J., Skarkova, J., Pouchova, V. and Ruprich, J. 2010. A review on comparative data concerning Fusarium mycotoxins in Bt maize and non-Bt isogenic maize. Mycotoxin Research 26(3): 141–145. DOI: 10.1007/s12550-010-0056-5

Peng, W.X., Marchal, J.L.M. and Van der Poel, A.F.B. 2018. Strategies to prevent and reduce mycotoxins for compound feed manufacturing. Animal Feed Science and Technology 237: 129–153. https://doi.org/10.1016/j.anifeedsci.2018.01.017

Pietri, A. and Piva, G. 2000. Occurrence and control of mycotoxins in maize grown in Italy. Proceeding of the 6th International Feed Products Conference. Piacenza, Italy 226–236.

Piva, G., Morlacchini, M., Pietri, A., Rossi, F. and Prandini, A. 2001a. Growth performance of broilers fed insect protected (MON810) or near isogenic control corn. Journal of Animal Science 79(Suppl. 1): 320 (Abst. 1324).

Piva, G., Morlacchini, M., Pietri, A., Piva, A. and Casadei, G. 2001b. Performance of weaned piglets fed insect-protected (MON810) or near isogenic corn. Journal of Animal Science 79(Suppl. 1): 106 (Abst. 441).

Pu, L., Bi, Y., Long, H., Xue, H., Lu, J., Zong, Y. and Kankam, F. 2017. Glow discharge plasma efficiently degrades T-2 toxin in aqueous solution and patulin in apple juice. Advanced Techniques in Biology & Medicine 5: 2. DOI: 10.4172/2379-1764.1000221

Rasff portal, https://webgate.ec.europa.eu/rasff-window/portal/?event=SearchForm&cleanSearch=1

Saladino, F., Manyes, L., Luciano, F.B., Mañes, J., Fernandez-Franzon, M. and Meca, G. 2016. Bioactive compounds from mustard flours for the control of patulin production in wheat tortillas. LWT-Food Science and Technology 66: 101–107. https://doi.org/10.1016/j.lwt.2015.10.011

Salas, M.P., Reynoso, C.M., Céliz, G., Daz, M. and Resnik, S.L. 2012. Efficacy of flavanones obtained from citrus residues to prevent patulin contamination. Food Research International 48(2): 930–934. https://doi.org/10.1016/j.foodres.2012.02.003

Samayoa, L.F., Cao, A., Santiago, R., Malvar, R.A. and Butrón, A. 2019. Genome-wide association analysis for fumonisin content in maize kernels. BMC Plant Biology 27: 19(1): 166. DOI: 10.1186/s12870-019-1759-1.

Santiago, R., Cao, A. and Butrón, A. 2015. Genetic factors involved in fumonisin accumulation in maize kernels and their implications in maize agronomic management and breeding. Toxins 20: 7(8): 3267–3296. DOI: 10.3390/toxins7083267

Sanzani, S.M., De Girolamo, A., Schena, L., Solfrizzo, M., Ippolito, A. and Visconti, A. 2009. Control of Penicillium expansum and patulin accumulation on apples by quercetin and umbelliferone. European Food Research and Technology 228(3): 381–389. DOI:10.1007/s00217-008-0944-5

Schaafsma, A.W. and Hooker, D.C. 2007. Climatic models to predict occurrence of Fusarium toxins in wheat and maize. Int J. Food Microbiology 20: 119(1-2): 116–125. DOI: 10.1016/j.ijfoodmicro.2007.08.006

Telles, A.C., Kupski, L. and Furlong, E.B. 2017. Phenolic compound in beans as protection against mycotoxins. Food Chemistry 1: 214: 293–299. DOI: 10.1016/j.foodchem.2016.07.079

Taguchi, T., Kozutsumi, D., Nakamura, R., Sato, Y., Ishihara, A. and Nakajima, H. 2013. Effects of aliphatic aldehydes on the growth and patulin production of *Penicillium expansum* in apple juice. Bioscience, Biotechnology, and Biochemistry 77(1): 138–144. DOI: 10.1271/bbb.120629

Trujillo, S., Njapau, H., Park, D.L., Price, W.D., Pohland, A.E., Fremy, J.M. and Dragacci, S. 2001. In GMO and food. Proceedings of the International Conference organized by Afssa, Paris 17–18 December 2001.

Wu, F., Miller J.D. and Casman, E.A. 2004. The economic impact of Bt corn resulting from mycotoxin reduction. Journal of Toxicology: Toxin Reviews 23(2–3): 397–424. DOI: 10.1081/TXR-200027872

Wu, F. 2006. Mycotoxin reduction in Bt corn: Potential economic, health, and regulatory impacts. Transgenic Research 15: 277–289. https://doi.org/10.1007/s11248-005-5237-1

Ying, L., Xiaojiao, L. and Jianke, L. 2018. Updating techniques on controlling mycotoxins – A review. Food Control 89: 123–132. DOI: 10.1016/j.foodcont.2018.01.016

Zhiyong, Z., Na, L., Lingchen, Y., Jianhua, W., Suquan, S., Dongxia, N., Xianli, Y., Jiafa, H. and Aibo, W. 2015. Cross-linked chitosan polymers as generic adsorbents for simultaneous adsorption of multiple mycotoxins. Food Control 57: 362–369. https://doi.org/10.1016/j.foodcont.2015.05.014

Index

For Product Safety Concerns and Information please contact our EU
representative GPSR@taylorandfrancis.com
Taylor & Francis Verlag GmbH, Kaufingerstraße 24, 80331 München, Germany

www.ingramcontent.com/pod-product-compliance
Ingram Content Group UK Ltd.
Pitfield, Milton Keynes, MK11 3LW, UK
UKHW022210190926
475702UK00003B/286
* 9 7 8 1 0 3 2 0 0 8 3 7 0 *